Copernicus Books

Sparking Curiosity and Explaining the World

Drawing inspiration from their Renaissance namesake, Copernicus books revolve around scientific curiosity and discovery. Authored by experts from around the world, our books strive to break down barriers and make scientific knowledge more accessible to the public, tackling modern concepts and technologies in a nontechnical and engaging way. Copernicus books are always written with the lay reader in mind, offering introductory forays into different fields to show how the world of science is transforming our daily lives. From astronomy to medicine, business to biology, you will find herein an enriching collection of literature that answers your questions and inspires you to ask even more.

Kersten Hall • Ralf Dahm

The Dawn Fisherman

Friedrich Miescher and the Discovery of DNA

Kersten Hall
School of Philosophy, Religion and History of Science
University of Leeds
Leeds, UK

Ralf Dahm
Institute of Molecular Biology
Mainz, Rheinland-Pfalz, Germany

ISSN 2731-8982 ISSN 2731-8990 (electronic)
Copernicus Books
ISBN 978-3-032-14218-4 ISBN 978-3-032-14219-1 (eBook)
https://doi.org/10.1007/978-3-032-14219-1

01-XX, 01-01, 01-02

This Springer imprint is published by the registered company Springer Nature Switzerland AG
The registered company address is: Gewerbestrasse 11, 6330 Cham, Switzerland

To my children—Amelie, Lena, Zoe, and Yorik—whose curiosity about the world keeps reminding me of the pursuit of knowledge that has driven science forward. As you grow and explore, I hope you will always find excitement in the unknown and enjoy your very own journeys of discovery. R.D.

To Michelle, Matthew and Ed for patiently listening over the past few years whilst I went on about old bandages, pus, and the testes of the Rhine salmon. And to the memory of Bill Astbury (1954–2024). Thanks, Bill, for all the tireless support and interest you showed when I was researching the life and work of your grandfather, and the memories of him that you kindly shared with me. You were a gentle soul in a world that is often way too harsh. You'll be missed. K.T.H.

Acknowledgements

Writing this book has required us to pester a lot of people on such matters as the location of archival documents and old photographs and we'd like to say a heartfelt thanks to them all for their patience and hard work. In particular to Rebecca Turpin and all the staff at the Brotherton Library, University of Leeds, for their perseverance in chasing up my frequent requests for ever-more-obscure, nineteenth-century scientific journals; University of Basel Library (Stefan Krauss); Medizinmuseum Davos (Dr. Peter Flury); WaldHotel, Davos (Marietta Zürcher); Davos Library (Astrid Schneider, Annick Ryf); The Museum of the University of Tübingen (Lucas Rau and Ernst Seidl); University of Tübingen Archives (Susanne Riess Stumm); Tübingen City Archives (Antje Zacharias); Tübingen Stadtmuseum (Dr. Bruno Wiedermann); Heidelberg University Archives (Karin Zimmermann); Basel Historical Museum (Daniel Suter); Basel City Archive (Sabine Strebel and Patricia Eckert); Sigma Xi, The Scientific Research Honor Society (Mia Evans); University of Leipzig Archive (Sandy Muhl Stockmann); Max Planck Institute for Biology, Tübingen (Beatriz Lucas); Friedrich Miescher Institute for Biomedical Research, Basel (Katrin Markopoulos); Rockefeller Archive Center (Bob Clark); The American Philosophical Society (Michael P. Miller); Cold Spring Harbor Laboratory Archives (Stephanie Satalino); Churchill Archives Centre, Cambridge, UK (Katherine Thompson); Special Collections and Archives Research Center, Oregon State University (Rachel Lilley); and Dublin Institute for Advanced Studies (Michelle Tobin, Senior Executive, and George Rogers, Librarian in the School of Theoretical Physics). All the team at the Press Office of Greene King Limited for kindly providing much higher quality photographs than our own of the plaques adorning the walls of 'The Eagle' pub in Cambridge. And thanks to Dr. Alison Li, biographer of Canadian scientist James Collip, for taking the time and trouble to dig out Collip's wonderful quotation about the nature of scientific discovery. Also, to Dr. Pete McHugh, University of Oxford for his lucid explanation of the difference between siRNA and miRNA, and to my colleagues (KH) in the Centre for History and Philosophy at the University of Leeds, especially Professor Jon Topham for his welcome warnings about the pitfalls of heroic narratives when writing the history of science.

To Professor David Harris who, as my undergraduate tutor at St. Anne's College, Oxford, first lit my fascination for biochemistry—I hope (KH) that with this book I can redeem myself for any below par academic performances on my part in tutorials all those years ago!

A very special debt of thanks on two counts must also go to Dr. Neeraja Sankaran of Ahmedabad University. Firstly, it was Neeraja who first persuaded one of us (KH) to collaborate with her on producing the first-ever complete translation into English of Miescher's 1871 paper in which he described his discovery of DNA. This translation (along with an accompanying commentary) was published in 2021 in *The British Journal for the History of Science* and we'd like to thank Cambridge University Press for kindly granting us permission to include it as Appendix II in this book.

But secondly, without Neeraja this book might never even have been written in the first place. For it was thanks to her that Ralf and I were first introduced to each other when she invited us as guest contributors to a scientific podcast series that she was putting together with Babak Ashrafi of the Consortium for the History of Science, Technology and Medicine. Called 'The DNA Papers', this series (now available to listen at https://www.chstm.org/video/144) contains 15 episodes in which a panel of scientists and historians discuss landmark scientific papers in the story of DNA. These discussions have served as inspiration and a source of invaluable knowledge throughout the writing of this book, and we'd like to thank Neeraja for her impressive energy in putting it together, Babak for all his technical skill in making it a reality, and all the guests for their contributions.

It was through recording the very first episode of this series which focussed on Miescher that Ralf and I began to collaborate on writing this book. An undertaking of this scale however usually turns out to be something of an odyssey and this was no exception, so thanks must also go to our editor Frida Trotter at Springer Nature not only for her ongoing support and encouragement but also for showing admirable patience and willingness to accommodate our endless revisions.

Very special thanks must also go to Dr. Syabira Yusoff for both her impressive baking skills and generosity. In 2022, Syabira's semi-final entry of a DNA double-helix baked from Krokan biscuits helped her to win the popular UK TV show 'The Great British Bake-Off' (or 'The Great British Baking Show' as it is known in Canada and the USA where it has also proved popular). In an interview,[1] Syabira explained that her inspiration for this edible masterpiece came from her work as a molecular geneticist at King's College, London, but for us it was one of the most powerful examples of how DNA has become an icon in popular culture that we have ever seen. So, a heartfelt thanks, to you, Syabira, and Love Productions—the photographs that you have kindly allowed us to use were very much the icing on our own cake (if you'll pardon the dreadful—but nevertheless irresistible—pun).

To Crum, Liz, and Simon not only for listening politely as I harped on about the discovery of DNA, but also with whom the weekly games of 'Horrified', 'Pandemic' and 'Dark Castle' offered welcome respite from the ups and downs of writing a book about it. And finally, a very warm thanks to all members of our respective families who, whilst we were immersed in writing this book, showed support, patience, and interest, despite having to hear time and time again from us about why the story of a long dead scientist who washed pus from discarded surgical bandages really does matter.

[1] (https://geneticsunzipped.com/transcripts/2023/04/20/syabira-yussoff-bake-off)

Competing Interests The authors have no competing interests to declare that are relevant to the content of this manuscript.

Contents

1

Introduction

Contents

Abstract From toothpaste tubes to sports trophies, the double-helical shape of DNA has become an icon in popular culture. Sports coaches talk about the DNA of their team; politicians reassure voters that certain policies are in the ideological DNA of their party; and one contestant in the semi-final of a popular British TV baking show even baked a double-helix in pastry as a tribute to Rosalind Franklin. Yet if asked to name the discoverer of DNA, few people would be able to give an answer. Even scientists are no exception to this. For we both admit that despite our training in molecular biology we had worked with DNA on a daily basis for many years without having a clue as to who had discovered it. This is far from unusual and we reflect on how Miescher has come to be eclipsed by the achievement of James Watson & Francis Crick, how we both came to know about Miescher and why we believe his story matters.

Keywords DNA • Miescher • Discovery • Science • History • Watson Crick • Structure • Nucleic acids • Biology • Genetics • Popular culture • Icon • Symbolism • Misconception • Oversimplification • Complexity •

K. Hall, R. Dahm, *The Dawn Fisherman*, Copernicus Books,
https://doi.org/10.1007/978-3-032-14219-1_1

System • Destiny • Determinism • Free will • Ethics • Society • Politics • Legacy • Heritage • Innovation • Progress • Inspiration • Sisyphus

As someone who was unashamedly ill at ease with contemporary culture not to mention modernity in general, it's unlikely that the biochemist Erwin Chargaff (1905–2002) paid much attention to Hollywood blockbusters. But had he seen the 1993 film adaptation of Michael Crichton's novel 'Jurassic Park' he might well have given it a quiet nod of approval. For the sight of CGI dinosaurs cloned from prehistoric DNA and now rampaging across a theme park were not only a vivid expression of Chargaff's long held fears about our potential to misuse science, but they also confirmed his sense of himself as a minor prophet.[1] Writing about DNA nearly 20 years earlier, Chargaff had observed that the genetic molecule was undergoing a profound change in its status. 'The double helix,' he had written, '…quite apart from its many undeniable scientific merits, has become a mighty symbol.'[2] Two decades later, with DNA becoming a household name thanks to a story of dinosaurs brought back from extinction, he was shown to be spectacularly correct.

Chargaff had recognised that DNA—a name once unknown outside science laboratories—was now taking on the status of an icon in popular culture. Nearly half a century after he wrote those words, they ring even more true today. DNA and the winding coils of its double-helical structure have become an icon that resonates in the popular consciousness. It can be invoked to confer a sense of scientific authority on a product, such as when adorning the side of the tube from which I (KH) squeezed toothpaste this morning. In the world of business, meanwhile, the double-helical structure of DNA has been used as an organisational model[3] or to convey a core sense of identity, such as when a 2008 article in the Harvard Business Review asked 'What's Your Company's DNA?'[4] More recently, when UK luxury carmaker Jaguar announced a plan to replace their long standing famous leaping cat logo, critics accused the company of having abandoned its 'brand DNA.'[5] Sports coaches have similarly invoked the image of DNA to design training regimes

[1] In his autobiography, 'Heraclitean Fire', Chargaff did actually describe himself as 'a minor apocryphal prophet' (Chargaff, 1978); p. 176.

[2] (Chargaff, 1976); p. 290.

[3] (https://www.mckinsey.com/capabilities/people-and-organizational-performance/our-insights/the-organization-blog/become-flexible-and-speed-up-with-a-helix-model-part-one)

[4] (Dobni 2008); https://hbr.org/2008/02/whats-your-companys-dna

[5] Jaguar's boss denies going 'woke' after online critics pounce on bold rebrand.' Kana Inagaki & Daniel Thomas, The Financial Times, 23rd November 2024.

based on the premise that their team is defined by a particular style of play,[6] while politicians have seized on the concept of ideological DNA to both bolster their own policies—and denigrate those of their opponents.[7]

But it is through cinema that DNA has permeated most effectively into the public consciousness. After all, what would 'Jurassic Park', or 'Jurassic World' be without ancient DNA from which to resurrect dinosaurs? (Probably a much more tranquil and pleasant holiday destination). And although, in a reflection of Cold War atomic age anxieties, Peter Parker gained his superpowers in the original Marvel comics thanks to the bite of a radioactive spider, more recent cinematic versions have reimagined his transformation into Spiderman as having occurred due to the hybridisation of his own DNA with that of a stray laboratory arachnid.

Alongside the cinema screen, TV has also played its part in making DNA a household name. For two consecutive years, global fans of horse racing watching the world's most famous steeplechase, the UK Grand National—will have seen the winning jockeys receive a trophy featuring a sculpture of the double-helical structure of DNA.[8] And in 2022, research scientist Dr. Syabira Yusoff achieved fame in the UK by winning a popular TV baking competition thanks to her semi-final entry of an edible double-helical DNA molecule baked in Krokan (Fig. 1.1).[9]

Yet despite its iconic status in popular culture,[10] if asked to name the person who discovered DNA, very few people -including most who work in the life sciences—would be able to do so. They might, at best, say it was Cambridge scientists, James Watson and Francis Crick (Fig. 1.2). We too were once among their number. Despite both of us having spent much of our careers working with DNA, neither of us can recall exactly when we first heard of Friedrich Miescher (Fig. 1.3), but we agree on one thing: that we must very soon afterwards have forgotten all about him.

---

[6] 'Ashworth believes 'England DNA' will breed era of success' Sam Wallace, The Independent newspaper (London) 5th December 2014.

[7] A recent example of this was when, during a TV debate in the run up to the 2024 UK general election, Prime Minister Rishi Sunak warned a studio audience that raising taxes was ingrained in the DNA of the opposition Labour Party, whose leader Sir Keir Starmer then responded a few minutes later by reassuring potential voters that support for the country's National Health Service ran through his party's DNA.

[8] The idea behind the double-helix sculpture was in recognition of sponsorship by a healthcare diagnostics company.

[9] Krokan is a traditional Swedish dessert made from multiple layers of almond flour. (https://geneticsunzipped.com/transcripts/2023/04/20/syabira-yussoff-bake-off)

[10] Readers who would like even more examples of how DNA has become an icon in popular culture as well as an exploration of this phenomenon are referred to 'The DNA Mystique: The Gene as a Cultural Icon' (Nelkin and Lindee 1995).

Fig. 1.1 An example of how DNA has become an icon in popular culture. (**a**) An edible double helix which impressed the judges (from left to right, Paul Hollywood, Prue Leith, Noel Fielding and Matt Lucas) in episode 9 of the popular UK TV show "The Great British Bake Off (series 13)." Made from Krokan biscuits it was baked by Dr. Syabira Yusoff (**b**) who then went on to win the final. (Reproduced with kind permission of Love Productions Limited. © Love Productions Limited)

Fig. 1.2 James Watson (1928-) and Francis Crick (1916–2004) taking a stroll along 'The Backs' in Cambridge in 1953. Nine years later, along with Maurice Wilkins they shared the Nobel Prize in Physiology or Medicine for the discovery of the structure of DNA. (Reproduced with kind permission of Cold Spring Harbor Laboratory Archives)

Like many biologists of our generation, Miescher's name, unlike those of Watson and Crick for example, didn't mean much to either of us when we were young scientists. If encountered at all, Miescher was most likely found only as a brief footnote in some textbook on molecular genetics during our

Fig. 1.3 Portrait of Friedrich Miescher (1844–1895) taken circa 1878 when he was in his mid-30 s. Photographer: Jakob Höflinger (1819–1898). (Image reproduced with kind permission of the University of Basel Library)

undergraduate years. And afterwards, as we both embarked on post-doctoral research, the relentless grind of having to churn out papers and win grants left little time or energy for us to reflect on the achievements of obscure, long dead scientists.

So how then, did we end up writing a book about him? And perhaps more importantly, why?

Ralf: For me, the pivotal year was 2003. This was the 50th anniversary of Watson and Crick's discovery of the structure of DNA–a scientific milestone that was celebrated all over the world. At the time, I was working at the Max Planck Institute for Developmental Biology[11] in Tübingen, Germany—a town, which, as I would soon find out, had an intimate, important, and much overlooked connection with Watson and Crick's achievement.

[11] Now called the Max Planck Institute for Biology.

Watson and Crick's discovery of the double-helical structure of DNA explained how the molecule was able to pass on genetic information from one generation to the next. This was a turning point in the development of molecular biology and certainly worthy of being celebrated 50 years later. But, justified as they were, something struck me as rather odd about these celebrations. Because amidst the countless articles, editorials, and opinion pieces that appeared in scientific journals and popular media, I found no mention of who had discovered DNA in the first place.

A casual reader flicking through these articles could easily be forgiven for thinking that the history of research into DNA had begun only 9 years before Watson and Crick's discovery with the demonstration by Oswald T. Avery in 1944 (Fig. 1.4) that DNA could pass on hereditary traits in bacteria. Thankfully, Christiane Nüsslein-Volhard who was my post-doctoral advisor at the time, knew better. From her, I heard the occasional odd snippet about this enigmatic figure, Friedrich Miescher who, some 135 years before my arrival in Tübingen, had investigated the molecular composition of cells in the bowels

Fig. 1.4 Microbiologist Oswald Avery (1877–1955) whose demonstration in 1944 that nucleic acids alone can confer the property of virulence in bacteria was crucial to our understanding of DNA as the genetic material. (Image reproduced with kind permission of the Rockefeller Archive Center)

of the castle there and, doing so, found DNA. Being a curious person by nature, I was naturally intrigued to find out more.

So, I began to dig deeper into the history of the man who had discovered DNA and whom history seemed to have largely forgotten. Searching a local library, I found a biography of Miescher by his uncle, Wilhelm His (Fig. 1.5) which also contained Miescher's collected publications. The arcane and sometimes convoluted German in which these articles were written hardly made for typical bedtime reading. But I was nevertheless hooked. Having begun to read them, I found myself unable to stop. The more I read, the more I was astounded by the depths of Miescher's finding and insights. Far from having happened to stumble across a new molecule and not understanding what he had found, Miescher knew what he was doing and that he had discovered something of great importance. He had moved to Tübingen only shortly before with no less of an aim than to uncover the molecular basis of life. And,

Fig. 1.5 Photograph of Miescher's Uncle, Wilhelm His (1831–1904) taken some time between 1863 and 1870 when he would have been advising Miescher during the early years of his career—something which he continued to do for the rest of his life. (Image reproduced with kind permission of the University of Basel Library)

when he had found DNA, he knew that it was an important molecule with key functions inside the cell.

His working conditions seemed an unlikely place in which to make a discovery of such importance. I still vividly recall for the first time seeing historic photographs of the laboratory Miescher worked in at Tübingen castle (Fig. 1.6) and visiting the place myself in 2003 (when it was little more than a storage facility of Tübingen University's archaeological department): the place looked to be a far cry from the high-tech molecular biology laboratories in which I had worked for the previous 10 years. Instead, it conjured up images of knights sleeping off stinking hangovers on the hard stone floor having stumbled from an evening's overindulgence in the nearby banqueting hall. And the tools available to Miescher at the time would have looked more at home in the dank basement of a Prohibition era moonshiner than in a modern biochemistry laboratory.

Few people, even in Tübingen, appeared to be aware of Miescher or the importance of what he had done in their town all those years ago. I decided that I wanted to change this. After having researched the history of the discovery and early characterisation of DNA, I wrote a piece on Miescher and his

Fig. 1.6 The laboratory in the former kitchen of the castle in Tübingen as it was in 1879. It was in this room that Miescher had discovered DNA 10 years earlier. The equipment and fixtures available to Miescher at the time would have been very similar, with a large distillation apparatus in the far corner of the room to produce distilled water and several smaller utensils, such as glass alembics and a glass distillation column on the side-board. (Photograph by Paul Sinner, 1879, Tübingen)

seminal work for the local Tübingen newspaper, the *Schwäbisches Tagblatt*, in 2003. It was to be the first of many, including one for a magazine published by the Max Planck Society—the *Max Planck Research*. The latter was noticed by an editor of the scientific journal *Developmental Biology* who happened to be visiting our institute at the time, and I ended up writing a more scholarly article on the topic for this journal. Over the years, I published a fair number of articles, both for the general public and for the scientific community, highlighting various aspects of Miescher's life, work and reception.

In 2015, Tübingen finally recognised Miescher's achievements more broadly and established a museum situated in what had once been Miescher's former laboratory in Tübingen castle (Fig. 1.7). This museum is now open to the general public, thanks to conversion and renovation of the old castle kitchen being made possible through financial support from Tübingen-based biotechnology company CureVac. Four years later a symposium commemorating the 150th anniversary of Miescher's discovery was held in the former banqueting hall of the castle, next to the new museum at which I was fortunate enough to be invited to give a talk on Miescher's life and achievements.

But despite this success, there was still more to be done. After nearly two decades of writing articles in the hope of restoring Miescher to his rightful place in the history of biology, I felt that the best way to achieve this would be to bring all this disparate information together in a single and, most importantly, an accessible publication.

In taking on this task, I was lucky on two counts. Firstly, I mentioned my interest in Miescher to Stefan Müller-Stach, a mathematician and Vice-President at Johannes Gutenberg University of Mainz, Germany, who suggested that it would make a great subject for a popular science book. Secondly, in 2022 I was invited by Dr. Neeraja Sankaran to take part in a podcast about Miescher for the Consortium for the History of Science, Technology and Medicine. Kersten was a fellow guest on the podcast and although this was the first time we had met, I knew that in seeking to bring the story of Friedrich Miescher to the wider world, he should be a part of telling it. So, I am very grateful to both, because without either, this book on Miescher's seminal discovery and how it transformed the way we understand life in all its wonderful forms, would never have come about.

Kersten: I suspect that most middle-aged men sometimes find themselves shouting at the television, and I am no exception. My outbursts however tend not to be provoked by perceived injustices on the sports field or pronouncements from politicians during the evening news, but instead during the British TV quiz show, 'University Challenge.'

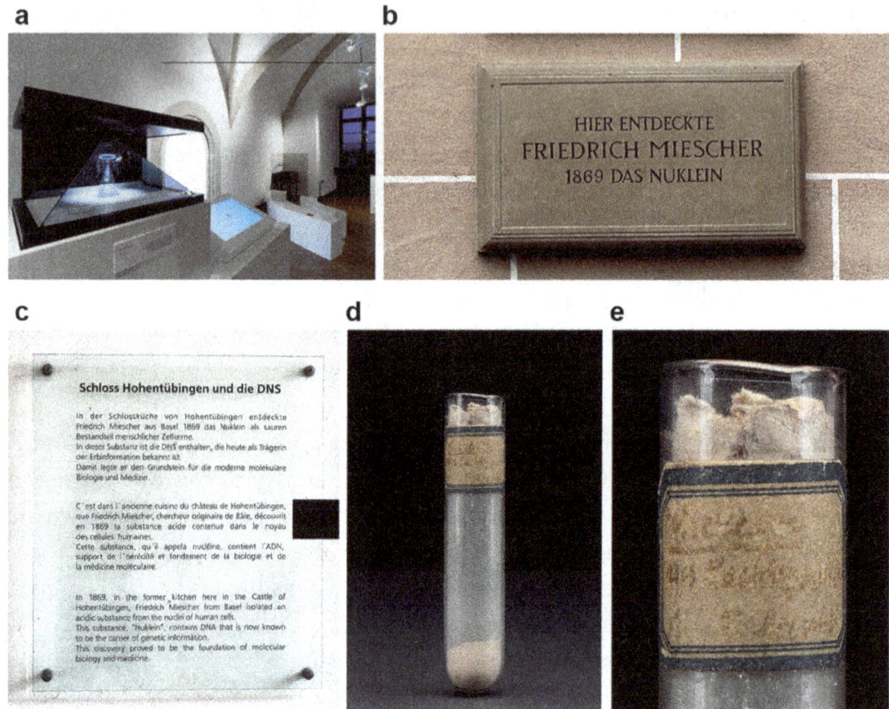

Fig. 1.7 (**a**) The room which once housed Miescher's lab—as it appears today where it is part of the Museum of the University of Tübingen hosting displays about the discovery of DNA. (Photograph by V. Marquant MUT and reproduced with kind permission of the Museum of the University of Tübingen (MUT)). (**b** and **c**) Commemorative plaques on the wall outside Miescher's former laboratory at what is now the Museum of the University of Tübingen. The inscription on the older of the two plaques (**b**) reads 'Here, in 1869, Friedrich Miescher discovered nuclein' whilst on the more recent plaque (**c**) states that nuclein 'contains DNA that is now known to be the carrier of genetic information.' (Photograph of plaque (**b**) taken by Fabian Kurze © MUT I F. Kurze). (**d**) Glass vial containing nuclein isolated from salmon sperm by Friedrich Miescher while working at the University of Basel. (**e**) Close-up of the faded label which reads 'Nuclein aus Lachssperma, F. Miescher' (*Nuclein from salmon sperm, F. Miescher*). Possession of the Interfakultäres Institut für Biochemie (Interfacultary Institute for Biochemistry), University of Tübingen, Germany; photograph by Alfons Renz, University of Tübingen, Germany. (Photographs reproduced with kind permission of the Museum of the University of Tübingen (MUT))

My ears pricked up during an episode last year when contestants were asked a question about the scientist Rosalind Franklin whose X-ray studies of DNA were crucial in solving its famous double-helical shape. As we'll see later in this book, Franklin's work was without doubt vital in unravelling the structure of the DNA molecule, but on hearing the question describe her as having done 'important work in the discovery of the DNA molecule' I shouted out loud, 'But what about Friedrich Miescher?'

It was in 1869, whilst working in a freezing cold laboratory that had once been the kitchen of a medieval castle, that Miescher made his landmark discovery. Having isolated white blood cells from pus washed out of discarded surgical bandages, he identified within them a completely novel and previously unknown class of molecule that he called 'Nuclein' but which is today known as 'nucleic acid.'

Thanks to the hair product ads, and cinema blockbusters featuring superheroes and cloned dinosaurs, the most famous member of this group was, for a long time, deoxyribonucleic acid, or DNA. In fact, anyone who did not work in a molecular biology lab could easily have been forgiven for thinking that DNA was the only member of this group. But during the global Covid-19 pandemic of 2020–2022 this changed. With the production of vaccines against SARS-CoV2 using another nucleic acid, RNA, the general public became slowly aware that there was more to the story of nucleic acids than just DNA.

Having studied biochemistry at university, I'd known about both DNA and RNA for a long time and how nucleic acids were one of the four classes of long chain molecules, alongside proteins, lipids and carbohydrates, that were essential to life. During my PhD and subsequent work as a molecular biologist I had spent nearly every day of my working life immersed in the task of isolating, purifying, sequencing and analysing both these types of nucleic acid. But all this was done without the faintest idea of who had discovered them in the first place. And even had I known, I suspect that to my shame I might well have dismissed such knowledge as irrelevant. What, after all would have been the point of being able to recall the name of some long dead scientist when professional survival in the hyper-competitive world of research depended upon ever more papers being published and grant applications written?

It was only when I finally hung up my lab coat and turned to writing about science that I came across Miescher whilst researching a book that I was working on at the time about how wool had played a little known, but important role in unravelling the structure of DNA. But -and it pains me to write this—I gave him nothing more than a minor walk-on part as a member of the supporting cast.

A few years later however, I was sitting in a quiet corner of 'The Eagle' pub in Cambridge when my thoughts turned to the little that I knew about Miescher. This might have been because the beer I was enjoying was called 'DNA' but it was more likely thanks to a group of tourists who had clustered around the wall opposite to point their phones at two commemorative plaques and read the inscriptions upon them with the hushed awe and reverence of medieval pilgrims visiting a holy shrine (Fig. 1.8).

Reading the inscription on the first of these plaques, it took an immense effort of self-restraint not to lean across and gently explain to them that what it said was…well, wrong:

> *Discovery of DNA: On this spot on 28th February 1953 Francis Crick and James Watson made the first public announcement of the discovery of DNA with the words 'We have discovered the secret of life'. Throughout their early partnership, Watson and Crick dined in this room on six days every week.*

The story of how Francis Crick and James Watson, euphoric at having solved the structure of DNA, went bounding through the streets of Cambridge and burst into 'The Eagle' to announce that they had just discovered the secret of life has been told many times and become something of a legendary scene in popular science. Back when I was 17 and studying A level biology, our teacher

Fig. 1.8 Plaques in 'The Eagle' pub, Cambridge commemorating the discovery of the structure of DNA by James Watson and Francis Crick in 1953. (Images provided and reproduced with kind permission of Greene King Ltd)

encouraged us to watch a BBC drama called 'Life Story' which told the story of Watson and Crick's discovery.[12] At the climax (spoiler alert), Watson, played by a young Jeff Goldblum[13] and Crick, played by British actor Tim Piggott-Smith dash out of their laboratory and race through the streets of Cambridge in a dramatic slow-motion sequence to burst through the doors of 'The Eagle' and proclaim that 'We've found the secret of life.' In response to this announcement, the entire pub, which is packed solid with drinkers as if it were New Years Eve (and not just an ordinary weekday lunchtime), erupts in jubilation.

It's great drama, but poor history. In a 2009 biography written by the historian of science Robert Olby, Crick said that he had no recollection of ever having burst into 'The Eagle' pub to make this announcement and ascribed this entire episode to the poetic license and imagination of James Watson—who himself has since confessed to having made the whole thing up[14].

Despite this episode being a figment of Watson's imagination, the idea that DNA somehow equates to being 'the secret of life' has taken hold in the popular consciousness. In TV debates preceding the 2024 UK general election, then Prime Minister Rishi Sunak warned a studio audience—'Labour will put up your taxes – it's in their DNA' whilst only a few minutes later, when the debate had shifted onto funding of the UK National Health Service (NHS), his opponent Sir Keir Starmer reassured voters that the 'NHS is in our DNA.' Along with newspaper headlines proclaiming that scientists have supposedly found 'a gene for' vegetarianism or a predisposition to playing violent video games, there is a widely held perception that DNA is the essence of life; that DNA is destiny.[15]

There is no denying that Watson and Crick's discovery was, without doubt, a landmark. The double-helical structure of DNA that they proposed explains how the molecule can copy itself to pass on the genetic information required to stock fictional cinematic theme parks with cloned dinosaurs or allow young

[12] Ever since first seeing this drama in 1986, I had been interested to watch it again but had never been able to find it anywhere. So a big thanks to Professor Matthew Cobb of the University of Manchester, UK for bringing to my attention that it can be watched in two parts at the following links: https://www.dailymotion.com/video/xitlyu and https://www.dailymotion.com/video/xitmu3

[13] Jeff Goldblum's early acting career seems to have been intertwined with DNA. In the same year as 'Life Story' was aired, he played scientist Seth Brundle whose pioneering teleportation experiment resulted in disaster when his own genetic material becomes merged with that of an insect in the remake of 1950s classic horror film, 'The Fly.' Seven years later, as mathematician Ian Malcolm, he was warning about the perils of resurrecting dinosaurs from prehistoric DNA samples in 'Jurassic Park.'

[14] See Crick cited in Olby, R. Francis Crick: Hunter of Life's Secrets. Cold Spring Harbor Press, 2009; Footnote 26 to Chapter 9; p.467; 'Happy 100th birthday, Francis Crick (1916–2004)' by Prof. Matthew Cobb, 8th June 2016 - https://whyevolutionistrue.com/2016/06/08/happy-100th-birthday-francis-crick-1916-2004/).'

[15] (Daily Mail, 14th Feb 2014; Daily Mail, 21st Feb 2014).

Peter Parker to swing across the Manhattan skyline with threads of spider's silk fired from his wrists. But if, in some parallel universe, they really did burst into 'The Eagle' pub, what they should really have said was not that 'We've found the secret of life', but rather 'We've found one of life's biggest secrets.' Admittedly, this is not quite as catchy a line in a TV drama, but it is nevertheless far more accurate. Importantly, it is less misleading, for it is a gross oversimplification to think that the living world in all its wonderful complexity can be reduced simply to DNA. Despite its importance as the bearer of hereditary information, DNA can't do much on its own. It functions not in isolation, but as part of a complex system and as we hope will become clear later in the book, reductive oversimplifications such as describing it as 'the secret of life'—whether in science or history, can be not just misleading, but dangerous.

So, Watson and Crick didn't go bursting into 'The Eagle' to announce they had discovered the secret of life. But crucially, nor did they discover DNA, as the plaque claims. This accolade goes to Friedrich Miescher who, almost a century before Watson and Crick revealed its molecular structure, first isolated the substance itself.

Yet despite the totemic status that DNA, occupies in popular culture, the name of its discoverer remains largely unfamiliar not only to the wider public, but also to most in the life sciences who work day-in-day out with the material that he first isolated.

Does any of this really matter? It probably comes as no surprise that as a science historian I would say that it most definitely does. And this is not simply for the satisfaction of academic pedantry. Because, as we hope this book will show, it matters that we get our history of science right. When we get it wrong, the consequences can reverberate far beyond the academic lecture theatre or seminar room and, as we discuss in the closing chapters, can even be tragic.

Thankfully, my discretion prevailed on that afternoon in 'The Eagle', and the tourists were spared a lecture about the misleading wording of the plaque and Miescher's discovery of nucleic acids. But as they moved away towards the bar, I wondered what Miescher might have felt had he known that 150 years after his death visitors would be flocking to see a plaque that ascribed his discovery to someone else. Or even that the very beer I was sipping was named in honour of Watson and Crick legendary entrance into 'The Eagle'—and not Miescher's efforts in Tübingen.

I was still pondering this question when, a few years later, Ralf invited me to work with him on this book. By this time, having been invited by Dr. Neeraja Sankaran (now at Ahmedabad University, India) to collaborate with her on the first complete translation from German into English of Miescher's landmark 1871 paper, I was well acquainted with Miescher. And thanks to Ralf's kind invitation, I now have an answer to my question at last. I'm pretty sure that if Friedrich Miescher could have joined me in 'The Eagle' that afternoon, the 'DNA' beer might well have left him with a bitter taste in his mouth, but seeing the plaque and the tourists clustered around it would not. Rather, it would have brought a smile of relief, satisfaction, and quiet contentment.

By telling his story, we hope you'll see why—and more importantly why it matters that we get our history of science right. And if, as a result, a future series of 'University Challenge' includes a question on Friedrich Miescher, then we'll know that we've been successful. In any case, we very much hope that you will enjoy following us on our little trip back through time to the beginnings of how nucleic acids were first discovered, how this changed our understanding of the living world, and the promise it holds for the future.

1.1 Note About the Structure of This Book

We've presented Miescher's story as a drama in three acts. Act 1 (Chaps. 2–7) describes his discovery of DNA and its mixed reception at the time by the scientific community. This act then closes on a note of tragedy—not simply because Miescher died in 1895 at the relatively young age of 51, but rather that he did so, burdened by a gnawing sense of failure and missed opportunity. As the curtain rises on Act 2 (Chaps. 8–11) we explore the scientific developments in the century that followed Miescher's death and how they showed such feelings to have been grossly misplaced. Finally in Act 3 (Chaps. 12 and 13) we explore the ongoing legacy of Miescher's discovery together with some of the important social, ethical and political challenges it may well raise in the future.

Finally, in case you've had a quick look / flick through the book and were wondering why each chapter closes with a cartoon of what looks to be a crash-test dummy pushing a boulder up the spiralling coils of the DNA double helix, let us briefly explain: Miescher once compared his struggles with those of Sisyphus, the character from Greek mythology who was punished by the gods of Olympus to spend eternity rolling a boulder up a mountain, only to have it roll back down again when he had nearly reached the summit. It is a feeling no doubt familiar to many a PhD student or post-doc doing lab research. It also stands in sharp contrast to the popular belief that science progresses thanks to lone geniuses crying 'Eureka' in sudden moments of inspiration or discovery.

In writing this book, we hope to show that science typically advances incrementally and through struggles like that of Sisyphus rather than from 'Eureka' moments. Importantly, scientific discovery is mostly not the result of lone geniuses but much more often results from the contributions of large numbers of people, many of whom remain anonymous.

It was to convey this sense of all those who toil at the lab bench to roll the boulder slowly upwards and yet remain anonymous, that Sisyphus was transformed from a rugged figure from Greek mythology into a featureless crash-test dummy that stands in for all those men and women we have since forgotten.

We hope that you enjoy the book.

References

Chargaff, E. 1976. The path to the double helix by Robert Olby (Review). *Perspectives in Biology and Medicine* 19:289–290.

Chargaff, E. 1978. *Heraclitean Fire: Sketches from a Life Before Nature*. New York, NY: The Rockefeller University Press.

Dobni, C. B. 2008. The DNA of innovation. *Journal of Business Strategy* 29 (2): 43–50. https://doi.org/10.1108/02756660810858143.

Nelkin, D., and M. S. Lindee. 1995. *The DNA mystique: The gene as a cultural icon*. New York: W. H. Freeman.

Olby, R. 2009. Francis Crick: Hunter of Life's Secrets. Cold Spring Harbor Press.

2

'The Quiet in the Land'

Contents

Abstract **The Quiet in the Land** was how the biochemist Erwin Chargaff once described the general lack of recognition amongst scientists of Friedrich Miescher and his discovery of DNA. This chapter traces how, had it not been for an illness contracted during his youth and the influence of his uncle, the eminent physiologist Wilhelm His, Miescher may never have made this discovery in the first place. Left with impaired hearing by his illness, Miescher is forced to reconsider his plans to follow in his father's footsteps and train as a doctor. Acting on the advice of his uncle he goes to Tübingen where Professor Felix-Hoppe Seyler is pioneering the new science of physiological chemistry (or biochemistry as it is known today). Unlike traditional methods of studying cells that relied on microscopy, physiological chemistry, Miescher sought to understand the cell in terms of its chemical constituents and so, in the freezing cold of the former kitchen of Tübingen's medieval castle, Miescher sets out on this ambitious undertaking.

K. Hall, R. Dahm, *The Dawn Fisherman*, Copernicus Books,
https://doi.org/10.1007/978-3-032-14219-1_2

Keywords DNA • Friedrich Miescher • Discovery • History of Science • Biology • Life Sciences • Chemistry • Physiological Chemistry • Molecular Genetics • Molecular Biology • Cell theory • Scientific revolution • 19th Century Science • Biochemistry • Genetic material • Nucleic acids

A setback due to illness offers a surprising new career opportunity for Friedrich Miescher

Long before sunrise, in the dark of a winter morning, Friedrich Miescher slid from the warmth of his bed and slipped silently out of the house on a mission. With his breath casting clouds of mist before him, he made his way through the deserted streets of Basel down to the banks of the river Rhine.

Beneath its dark waters, Miescher hoped to find treasure. According to an old Germanic legend, a vast hoard of gold lay hidden deep beneath the waters of the river Rhine. The legend told how it had been mined by the Nibelungs, a race of ancient magical dwarves, and that it promised both wealth and power to those who possessed it. But the treasure that Miescher sought beneath the cold waters of the Rhine on those dark winter mornings was not a glittering hoard of mythical gold. It was the salmon that had swum 500 miles here to breed. Or more accurately, it was their swollen gonads. For within these tissues there lay a biochemical treasure—a previously unknown cellular compound that Miescher had discovered only a few years earlier whilst working in a freezing cold laboratory that had once been the kitchen of the medieval castle in Tübingen, Germany. Observing that this substance was localised in the nuclei of the cells, he had christened it 'Nuclein' and every instinct told him that he was on the cusp of a momentous discovery—one which promised to unlock the mystery of how living systems function. In the coming century, his hunch would be proven to be spectacularly correct. Nuclein—albeit under the new name of nucleic acid of which DNA and RNA are both examples—would not only transform our understanding of biology but now promises seismic changes in medicine.

As we hope will become clear in this book, the discovery of DNA has since brought opportunities for both power and great wealth. According to the old legend, so too did the Rhinegold of the Nibelungs. But with its wealth and power came a curse: that all those who possessed the treasure of the Rhine must forever renounce any hope of feeling or giving love, and it would eventually bring doom upon them.

It would probably be too strong to say that the salmon which Miescher caught on his dawn fishing expeditions[1] brought down a similar curse upon him, but his discovery of DNA nevertheless came at a personal cost. For despite being utterly convinced that Nuclein played a central role in biology, its precise function eluded him. And this took its toll. In 1890 he contracted tuberculosis and became a patient at a sanatorium in the Swiss resort of Davos. But sublime as they were, the stunning alpine views of snow-capped peaks or thickly forested slopes brought little consolation.

Because long before the tuberculosis bacterium began to eat at his lung tissue, a sickness of a very different sort had gnawed at his psyche. As the years had passed, Miescher felt himself ever more plagued by an insidious sense of missed opportunities and having failed to live up to his potential. In a letter to his old friend the German pharmacologist Rudolf Boehm, he despaired that 'I will never know the happiness that belongs to the man who has lived up to their station in a harmonious way to the satisfaction of themselves and others, so my basic mood is the uncomfortable feeling of one who has lost a button from their braces.'[2] Elsewhere he describing himself as going to bed each evening, 'feeling like a schoolboy who has failed to complete his tasks.'[3]

Like Tantalus in Classical mythology, grasping for fruit that will remain forever out of his grasp, Miescher was tortured by the knowledge that although he had discovered Nuclein, its function remained a mystery to him. But in a letter written in 1888, just 7 years before his death, it was to another character from mythology that he turned when trying to express this sense of despair:

> *Since the middle of September, I have worked without a single day's interruption, he wrote, 'from 6 in the morning until late into the night, in the hope of rolling the Stone of Sisyphus back up the mountain'.*[4]

In Greek mythology, Sisyphus was a mortal who, as a punishment by the gods of Olympus, had been doomed to spend eternity rolling a boulder up a mountain only to reach the summit and have it roll all the way back down again. It's a poignant, powerful image for what Miescher felt to have been a life of

[1] See, for example Meuron-Landolt, M. de. 1969. 'Johannes Friedrich Miescher: Sa Personalite et l'importance de Son Oeuvre'. Bulletin Der Schweizerischen Akademie Der Medizinischen Wissenschaften (Bulletin of the Swiss Academy of Medical Sciences) 25: 9–24; p.16. But while Meuron-Landolt even speculates as to the type of net Miescher would have used, we ourselves have a strong suspicion that Miescher must have realised pretty soon that the local fishing industry offered a far easier source of salmon than having to rise early on a winter morning to catch the fish himself!

[2] (His 1897); p. 31.

[3] Ibid.

[4] Miescher, 23rd November 1888, Letter LXIII, in (His 1897); p. 106.

relentless, unforgiving and, increasingly he feared, pointless toil. Little could he have known, how utterly wrong he would be.

But at least the name of Sisyphus is reasonably well-known. Writing about Miescher many years later, the biochemist Erwin Chargaff described him as being 'one of the quiet in the land.'[5] It was intended as a compliment, a reference to Miescher's modest and unassuming manner. But it also hints at the way in which Miescher's name had become eclipsed by the monumental legacy of his discovery. Miescher's name, said Chargaff, was one 'that our time—long in bibliography but short in memory—likes to forget.'[6] And had it not been for two crucial influences that shaped the course of his life, Miescher might have remained not just quiet, but utterly silent.

2.1 The Louse That Changed the Course of Science[7]…?

One of these was Miescher's uncle, Wilhelm His, (1831–1904) who held a Professorial Chair in Anatomy and Physiology at the University of Basel and was a regular visitor to his nephew's home where a group of scientists and intellectuals were often to be found engaging in lively discussions. Listening in on these conversations stirred the flames of intellectual curiosity in the young Miescher and, after showing an early interest in theology, he opted to study the physical world, rather than the metaphysical one. So, at the age of only 17, he embarked on a course in medicine at the University of Basel in the hope that he might follow in the footsteps of both his father and uncle.

When Miescher was born on 13th Aug 1844, as the first of five children, it was clear from the start that his father Johann Friedrich (1811–1887) (Fig. 2.1) had high expectations of him. After all, why else would Johann Friedrich Miescher the elder name his new son after himself? And the young Miescher would have much to live up to. His father was a Professor of Anatomy at the University of Basel, who had studied in Berlin under the eminent physiologist Johannes Müller (1801–1858), before writing a dissertation on the structure of bones and their disease states, which had set him on a

[5] (Chargaff 1963); p. 164.

[6] (Chargaff 1971); p. 637.

[7] A special thanks must go to Professor Gareth Williams for coining this memorable moniker on p. 10 of his 2019 book 'Unravelling the Double Helix: The Lost Heroes of DNA'—'Miescher's childhood was steeped in music, literature and intellectual discussion and was spoiled only by meeting a louse that, in its modest way, changed the course of science.'

Fig. 2.1 Portrait of Miescher's father, Friedrich Miescher-His (1811–1887) taken some time between 1867–1887 by Jakob Höflinger. (Image reproduced with kind permission of the University of Basel Library)

course to become a renowned physician. In addition to his academic prestige, Miescher's father was also musically gifted and sometimes sang in concerts—a talent which he is said to have passed on to his eldest son. But although Miescher shared his father's choral talents, he was not destined to emulate his success as a physician.

This was due to a hearing impairment, the exact cause of which is not entirely clear. Miescher's uncle Wilhelm His writes that his nephew had developed the condition during his youth giving the impression that he was already suffering from it before he embarked on his studies in medicine. But he then goes on to mention that sometime between 1865-1866, Miescher's medical studies were interrupted by a bout of typhus. This disease can, in some circumstances, result in impaired hearing and, as it is often spread by lice, has led to speculation that Miescher may have been hard of hearing after having been bitten by a louse. If so, then as Professor Gareth Williams has said, it was a louse bite which 'in its modest way, changed the course of science.'[8] But whatever the origin of his condition, it brought an end to Miescher's hopes of being a doctor. Concerned that his poor hearing might impair his bedside manner, Miescher abandoned his plan to be a doctor and began looking at alternative directions for his career.

[8] (Williams 2019); p. 10.

2.2 The Discovery of the Cell

But what to do? His father was adamant that he ought to put his intellect and skills to practical use, but it was Miescher's uncle Wilhelm who saw how his nephew might best achieve this, by steering him in the direction of a new area of research which had already preoccupied his own interests—the study of the cell. It has been estimated that an adult human body contains roughly 37 trillion cells, all of which have arisen from the successive division of a single fertilised egg cell. Over the course of the development of the embryo, these cells differentiate into over 200 types, each of which is specialised to perform a specific function in the body. Muscle cells, for example, contract, while white blood cells patrol the body scanning for invading pathogens, and nerve cells conduct information to and from the brain in the form of electrical impulses.

Wilhelm His was interested in how this specialisation of cells from the fertilised egg occurs, particularly that of the nervous system and he was sure that the answers must lie deep within the structures found inside cells. But nearly nothing was known about the molecules that make up cells at the time. Unravelling the mysteries of the innermost workings of the cell therefore seemed to him the ideal research subject for his nephew to tackle.

The term 'cell' had first been coined in the seventeenth century by the English natural philosopher Robert Hooke who used it to describe the box-like structures he had observed whilst gazing through his microscope at slices of cork. But the modern cell theory, or concept that the cell is the fundamental unit from which all living tissues are built, is usually credited to the German scientists Matthias Schleiden (1804–1881) and Theodor Schwann (1810–1882). In 1838, Schleiden had proposed that all plant tissue is composed of cells while, a year later, Schwann published a monograph in which he proposed that cells are the elementary units of tissues, and that their formation is a universal principle of the development for organisms.[9]

The names of Schleiden and Schwann have since been hailed as being as important in the formation of cell theory as those of James Watson and Francis Crick for the discovery of the double-helical structure of DNA.[10] But their claim to this title has not gone unchallenged. Other contenders include the French aristocrat Henri du Trochet, (who, to avoid the guillotine during the French Revolution, hastily changed his surname to the less grandiose sounding 'Dutrochet'), and Jan Purkinje (1787–1869), a Professor at the University of Breslau, in what is now Wroclaw in Poland.

---

[9] (Schwann 1911); p. 41.

[10] (Mazzarello 1999); p. E14; (Wolpert 1996); p. 226.

The question of who deserved the credit for first proposing the cell theory may have been open for debate, but one thing was clear. If cells were found in the tissues of both plants and animals, then they must share some common features which allowed them to be identified as a single distinct entity. Or put another way, what did cells contain?

Crucial advances in the field of microscopy such as those made by the Austrian pioneer of optical instruments Simon Plössl (1794–1868) shed important new light on this question.[11] Improvements in microscopy revealed that plant and animal cells contained a common structure which, believing it to be the source of new cells, Schleiden called the 'cytoblast.'[12] In a paper given in 1833 to the Linnaean Society on microscopical study of plant tissue, meanwhile, the Scottish botanist Robert Brown (1773–1858) suggested two alternative names for this structure. Of these, Brown's term 'the areolus' has, like Schleiden's 'cytoblast', slipped into obscurity, but the other has endured and is the name by which this organelle is known today: the nucleus.[13] But if, as appeared to be the case, the nucleus was a universal feature of both plant and animal cells, then it must surely play some crucial role in the life of the cell. Just what this function might be became a major question of the time for cytologists.

For Schleiden, the answer was clear: he proposed that new cells grew endogenously from the nuclei of pre-existing ones.[14] If this were correct, it would solve one of the biggest questions puzzling cell biologists at the time which was how do cells reproduce themselves? But there were other possibilities. His colleague Schwann for example, argued that, in animal tissue at least, new cells arose exogenously through the steady growth of nuclei in an extracellular material that he termed the cytoblastem.

There was also a third possibility. In 1744, the Swiss naturalist, Abraham Trembley (1710–1784) wrote to the President of the Royal Society of London, describing how he had observed cells of the freshwater polyp, Hydra in the process of dividing in two, to give rise to new cells.[15] Other evidence that cells multiplied by a process of binary fission came from the study of the cleavage of eggs in animals such as toads and frogs as well as the observation made in 1841 by the German-Polish biologist Robert Remak (1815–1865) that red blood cells in the chick embryo took on the appearance of pears, joined by the

---

[11] (Drews 1999).

[12] Such as mammalian red blood cells, which lose their nucleus as they differentiate into mature erythrocytes.

[13] (Mazzarello 1999); E14.

[14] (Harris 1999); p. 98.

[15] Cited in (Harris 1999); p. 56.

stalks, which led him to conclude that this was because they were in the process of division.[16]

As with the cell theory, the origins of the idea that cells divide by binary fission are still disputed, but it is the German physician Rudolf Virchow (1821–1902) who is usually accorded the accolade of being the first to propose that cells multiply by binary fission.[17] Sometimes hailed as 'the father of modern pathology', and even 'the Pope of Medicine', in 1858 he made what has since been described as 'one of the grandest inductions in biology' - 'Omnis cellula e cellula'. (Every cell is derived from another cell.)[18]

Strictly speaking, Virchow's pronouncement does not actually rule out either Schwann or Schleiden's ideas about how new cells arose. Nor was it entirely original. In 1825, the French scientist Francois-Vincent Raspail (1794–1878), had already used this phrase to describe the endogenous generation of new cells from pre-existing ones, as Schleiden had envisaged.[19] But thanks to Virchow, the pronunciation 'Omnis cellula e cellula' came to be associated with the idea that cells multiply by a process of binary fission.

One of those who had contributed to the growing body of evidence that cells divide by this process was the botanist Hugo von Mohl (1805–1872). In his inaugural lecture, given at the University of Tübingen in 1835, von Mohl described how he had observed cells of the filamentous green algae *Conferva* dividing by this process. When the lecture appeared in print 2 years later under the title 'On the multiplication of plant cells by division' 2 years later, von Mohl drew the modest conclusion that 'The observations cited above will suffice to prove that the increase of cells by division is not an altogether rare phenomenon among the Confervae.'[20] History has been less modest in its assessment of his work, hailing his observations as a 'turning-point in the history of the study of cell-multiplication.'[21] And thanks to Miescher's arrival there just over 30 years after von Mohl gave his lecture, Tübingen would become the scene of a second—and seismic—turning point in biology.

[16] Ibid., p. 433.

[17] (Silver 1987); p. 83.

[18] (Baker 1953); p. 435.

[19] (Harris 1999); p. 135.

[20] von Mohl cited in (Baker 1953); p. 426.

[21] (Baker 1953); p. 425.

2.3 The Road to Tübingen

What had brought Miescher from Basel to Tübingen? Until now, microscopy had been the one and only means of studying the structure of the cell. Botanists and naturalists peered through the lenses of their instruments using chemical stains to characterise the cell in terms of the morphology and structure of its various organelles. But an alternative approach was emerging - one which Miescher's uncle Wilhelm felt offered a great opportunity for his nephew to make a name for himself:

> *I pointed him in the direction of histochemistry as, from my own histological work, I had become ever more convinced that the final questions about the development of tissues were to be answered on chemical grounds. The differences between the nucleus and the cell, from cells and the intercellular substances and so many other subsequent problems could hardly be realised until we had made the material relations between these different things from one another comprehensible.*[22]

The innovative approach of using chemicals to tease out the secrets of the cell is usually credited to the French scientist Francois Vincent Raspail (1794–1878), who had enjoyed plenty of time to puzzle over such matters whilst languishing within a cell of a very different kind. Being active in politics as well as science, Raspail's opposition to the July Monarchy that had taken power after the revolution of 1831 earned him a 27-month prison sentence on a charge of subversion. He did however put his incarceration to good use and wrote two books, 'A New System of Organic Chemistry' and 'A New System of Plant Physiology and Botany'. The former, published in 1833 was almost immediately translated into German, English—and Arabic.

Another 'accolade' for 'A New System of Organic Chemistry' was that it appeared on the Vatican's Index of Prohibited Books. This was most likely because its materialist conviction that the cell could be envisaged as a kind of laboratory in miniature, and that the living world could be best understood in terms of chemistry had caused offence in certain theological circles.

Raspail's inclination for stirring revolution extended beyond his politics and into his science. In the eighteenth century, chemistry had not existed as an independent subject and was only taught in European universities as part of the medical curriculum. Physicians themselves felt little need to know any practical chemistry as this was done largely by pharmacists—or those rich enough to afford private laboratories in which they could dabble it as a hobby.

[22] (His 1897); p. 7.

But by the 1840s, things had begun to change. Chemistry was no longer merely the handmaiden of medicine and was starting to gain recognition as an independent discipline taught in public institutions. And by leading in the vanguard of this seismic shift, Raspail laid the foundations for Miescher's work in two crucial ways. Firstly, his innovation of using iodine to show the distribution of starch in plant tissue, gave birth to the field of histochemistry. This approach, which used specific stains and dyes to study the localisation of specific components and structures within cells offered Miescher a starting point for his studies of the nucleus. But secondly, Raspail had pioneered an approach which sought to understand the cell in terms of its chemical constituents. Following up on his success with iodine, he had gone on to discover reagents that could identify the presence of sugar, silica, mucin, resins, chloride and iron.[23] But he also went further. By incinerating cellular material on a platinum spoon in the presence of specific reagents, he was able to quantify the different chemical elements within the cell.

Raspail's method would become an essential part of Miescher's own toolkit. With the development of this method, Raspail had not only helped to give chemistry a new status, but had brought it together in a powerful new union with microscopy, or as he put it himself far more poetically:

> *Seeing the micrographer content to draw and dissect organs, and the chemist alter, mix, or destroy them in order to have the pleasure of rediscovering and recomposing them, I seemed to see two men walking side by side without knowing it, on two roads that would never meet. And I decided to follow them no longer, but to merge their work. I would no longer be sometimes a chemist, physicist, or botanist, or physiologist, but all of these at once, all the time. I therefore needed to abandon known methods and create new ones, establish new rules, for I was to work in a new kind of laboratory.*[24]

Raspail had long dreamt of earning recognition from the scientific community. Now, thanks to having pioneered this novel approach to the study of the cell he had gained just that. But the journey had not been easy. Having presented his first paper to the Academy of Sciences in 1824, he had come away with a crushing sense of disappointment when his work was met not with the 'fatherly solicitude' for which he had hoped, but instead with 'general indifference, noise and chatter.'[25] This set-back might well have brought his scientific career to a premature end, had it not been for the intervention of the

[23] (Harris 1999); p. 34.
[24] (Weiner 1968); p. 91.
[25] Ibid., p. 85.

distinguished biologist, Geoffroy Saint-Hilaire (1772–1844) who took the despondent Raspail to one side as he left the meeting and offered the young man some much needed words of encouragement. Saint-Hilaire's intervention paid off and almost 10 years later, by which time Saint-Hilaire had become President of the Academy of Sciences, he wrote to Raspail, praising him for his pioneering achievements in the study of the cell:

> *Your microscopic research has revealed...certain molecular properties and you have brought new materials within the reach of science and thus amassed treasures of enormous potentialities... ideas that are both original and suggestive of further research.*[26]

Saint-Hilaire's words proved to be prophetic. Others did indeed beat a path further down the avenue of research that Raspail's chemical approach to biology had opened. And of all who blazed this trail, one of the most eminent was Miescher's mentor in Tübingen, Felix Hoppe-Seyler (1825–1895) (Fig. 2.2).

2.4 All in the Blood: The Rise of Physiological Chemistry

Hoppe-Seyler's determination to distinguish himself may well have been forged by the painful experience of finding himself an orphan by the age of 11. After being brought up in an institution in Halle, Germany, he began studying natural sciences at the university there, before moving to Leipzig and finally receiving the degree of Doctor of Medicine in Berlin in 1850.

His doctoral dissertation on the chemical and histological aspects of cartilage had lit a spark that would later blaze into a passion for understanding living tissue through its chemical composition.[27] But for now, unlocking the mysteries of life at the chemical level would have to wait. Having completed his studies, Hoppe-Seyler began working as a practicing clinician, and served as the medical officer in a Berlin workhouse. But throughout this time, research remained his true passion and when time permitted, he continued his studies on the structural and chemical properties of bone, tooth and connective tissue.

His dedication was rewarded when he took up a post in 1856 as a research assistant to the eminent pathologist Rudolf Virchow under whom his work

[26] Cited in Ibid., p. 126.

[27] (Gamgee 1895); 10th Oct Nature, p. 575.

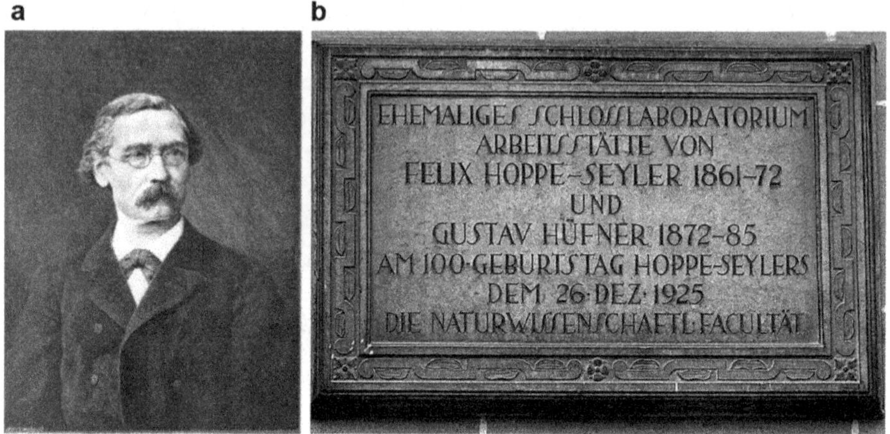

Fig. 2.2 (**a**) Portrait of Felix Hoppe-Seyler (1825–1895) taken circa 1872, a couple of years after Miescher had worked in his laboratory at the University of Tübingen. By that time, Hoppe-Seyler had moved to the University of Strasburg. (**b**) Memorial plaque for Hoppe-Seyler at the entrance to the castle laboratory (Photo: Hedwig Storch, CC BY-SA). The inscription on the plaque reads: 'Former castle laboratory, workplace of Felix Hoppe-Seyler between 1861-1872 and Gustav Hüfner between 1872-85. [Unveiled on] the 100th anniversary of Hoppe-Seyler's birth on 26 Dec 1925. Faculty of Natural Sciences.' (Photographs reproduced with kind permission of the Museum of the University of Tübingen)

took an important change of direction. Until now Hoppe-Seyler's interest had focussed on bone, tooth and connective tissue but in 1857, he published a paper on a new subject—one that would become a milestone in his career.

The paper described the observation that carbon dioxide prevents the blood pigment, haemoglobin from carrying oxygen and, although it was only short, it marked a major change in the direction of Hoppe-Seyler's research.[28] Three years later, when Hoppe-Seyler took up a Professorial Chair at the University of Tübingen, haemoglobin had come to be the focus of his work.

One of his major observations was that when sunlight is focussed and filtered through a prism before being passed through a cuvette containing a dilute sample of blood, a spectrum is produced in which there are two distinct dark bands—one in the yellow, and the other in the green region. It was already known at that time that passing light of certain 'refractivities' (the term 'wavelength' was still unknown at the time) through certain pigments such as indigo or chlorophyll[29] in solution gave rise to a spectrum in which certain bands of colour were missing. This indicated that these wavelengths of

[28] (Hoppe-Seyler 1857).

[29] The material in plant cells that gives them their green colour.

light were being absorbed by specific chemical groups within the indigo and chlorophyll pigments. Now Hoppe-Seyler had found that this was also the case for the blood protein haemoglobin.

The same effect was observed with red blood cells and aqueous solutions of blood taken from a dog, an ox, a sheep, a pig, a pigeon or a fish. But crucially, Hoppe-Seyler observed that treatment with acetic or tartaric acid or strong alkali made these dark bands disappear. His conclusion was that red blood cells contain a pigment that can be split by weak acid or bases into two components—a colourless protein and an iron containing group called haematin, which was responsible for the absorption and gave haemoglobin its distinctive red appearance.

With his analysis of haemoglobin, Hoppe-Seyler was in the vanguard of an emerging new discipline known as 'physiological chemistry' or 'animal chemistry.' This sought to characterise the cell not through merely observing morphological features under the microscope, but rather by analysing the chemical composition of its constituents. It had grown out of the work of Justus von Liebig (1803–1873) at the University of Giessen, Germany, whose studies of the chemical processes involved in plant and animal metabolism left him convinced that chemistry was essential to understanding living systems. In the preface to his 1842 book 'Animal chemistry or, Organic chemistry in its applications to physiology and pathology', he declared that the aim of his approach was:

> *to direct attention to the points of intersection of chemistry with physiology, and to point out those parts in which the sciences become, as it were, mixed up together... These questions and problems will be resolved and we cannot doubt that we shall have in that case a new physiology...In the hands of the physiologist, organic chemistry must become an intellectual instrument, by means of which he will be enabled to trace the causes of phenomena invisible to the bodily sight... The path which has led to it will open up other paths; and this I consider as the most important object to be gained.*[30]

Until this time, chemistry in Germany as in France, had merely been subservient to other fields of enquiry such as pharmacy, metallurgy and mining. But by the mid-nineteenth century, chemistry was gaining institutional status as an independent discipline and by 1850 Germany had a thriving chemical industry.

The Eberhard-Karls University in Tübingen, founded in 1477, was quick to recognise the important intellectual changes that were taking

[30] (von Liebig 1842); pp. xiii–xix.

place. In 1863 it became the home of Germany's first independent Faculty of Natural Sciences, comprising Chairs in Mathematics, Physics, Astronomy, Mineralogy, Pharmacology, Botany, Zoology—and, of course, Chemistry.

2.5 Miescher Arrives in Tübingen

On his arrival in Tübingen therefore, Miescher's first port of call was to spend a semester working in the field of chemistry, with one of its most eminent figures at that time. This was Adolph Strecker (1822–1871) who had been a student of von Liebig and had shown for the first time how amino acids, which are the chemical building blocks of proteins, could be synthesized in the laboratory from simple compounds such as urea and hydrogen cyanide.

As a student, Miescher had already gained some familiarity with organic chemistry, having spent some time in Göttingen, Germany, working with Friedrich Wöhler (1800–1882) who had shown that the compound urea could be synthesized from simple inorganic substances. But although Germany was by now renowned for its expertise in organic chemistry, the success and prestige of which were perceived by many to outshine that of physiological chemistry, Miescher had no real interest in pursuing a career assembling distillation apparatus and reflux condensers. Instead, as he explained to his father in a letter written in 1868, his sights were set on much grander ambitions.

> *It was only with the physiological lectures that the whole glory of research on the organic dawned on me. The facts and questions about the processes in living organisms, about the interlocking of all parts and the ever more transparent centralization in the whole machine, about what is common to these processes in whole organic fields, and again about the cause of differences, about the roots of all this becoming and happening in the becoming and happening of organic nature—all this filled me with a kind of religious reverence.*
>
> *It seemed to me that this was actually the most direct task of natural research, which in my view was being worked on—to free man from all the countless threads with which, unconsciously to him, his thinking, feeling and acting everywhere and again is rooted in the material substrates of his existence. Precisely because these dependen-*

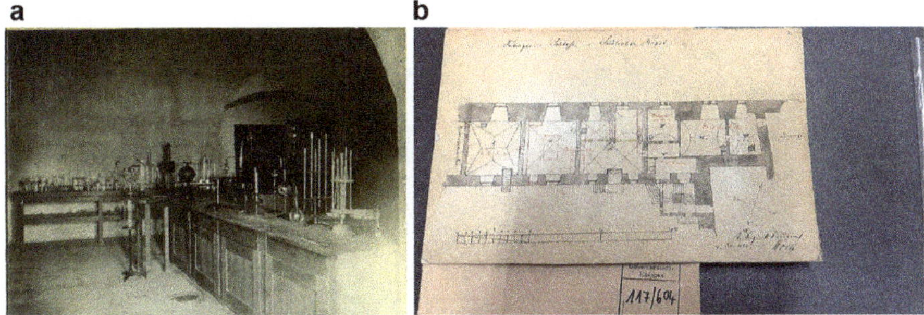

Fig. 2.3 (**a**) Hoppe-Seyler's laboratory around 1879. Prior to becoming the chemical laboratory of Tübingen University in 1823, this room was the laundry of the castle. (Photograph reproduced with kind permission of the Museum of the University of Tübingen). (**b**) Floor plan dated 1885 showing the area of Tübingen castle that would have housed the laboratories of Felix Hoppe-Seyler (IV, V, & VI) and Friedrich Miescher (III) 15 years earlier. These rooms, which had previously been the laundry and kitchen of the castle, had become laboratories of Tübingen University in 1823 and 1818, respectively. (Images reproduced with kind permission of the Tübingen University Archive)

> *cies are not yet transparent enough for him, he is unable to detach and raise his pure self from this network.*[31]

The location in which Miescher set about this task was no less grand than his youthful ambition. In 1845, the University of Tübingen had established the very first chair in physiological chemistry—a post now occupied by Hoppe-Seyler.

Whilst Hoppe-Seyler carried out his experiments in what had once been the laundry (Fig. 2.3) of Tübingen's impressive medieval castle that overlooks the river Neckar (Fig. 2.4), Miescher was given a laboratory in what had been the kitchen of the castle, which he likened to the laboratory of a medieval alchemist. It was a description which, metaphorically at least, would prove to be rather apt. Just as in the popular imagination, lead had been the starting material from which the alchemists sought to make gold, so too did Miescher's work begin with a substance that was rather unremarkable and ordinary. But while the efforts of the alchemists proved to be in vain, Miescher's work did indeed result in a dramatic transformation, and one that would ultimately revolutionise our understanding of life.

[31] Miescher (1868) cited in (Signer 1969); p. 29.

Fig. 2.4 (**a**) Tübingen Castle from the Upper Neckar (Color Lithograph by Adam Gatternicht, around 1855) and (**b**) as it looks today where it now houses the Museum of the University of Tübingen. (Images reproduced with kind permission of the Museum of the University of Tübingen)

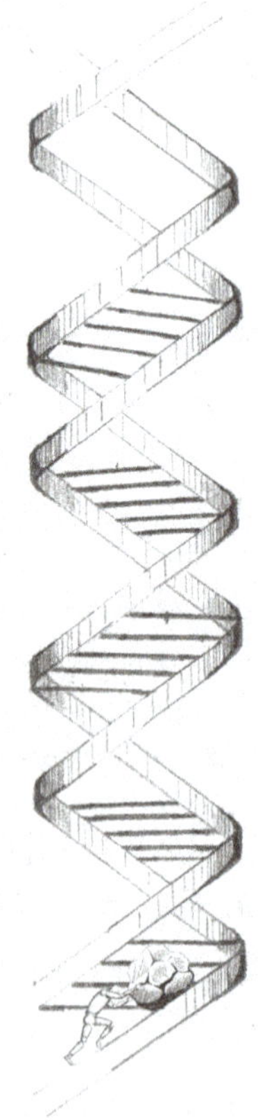

Drawing by Kersten Hall

References

Baker, J. 1953. The cell theory: A restatement, history and critique: Part IV. The multiplication of cells. *Quarterly Journal of Microscopical Science* 94:407–440.

Chargaff, E. 1963. *Essays on Nucleic Acids*. Amsterdam: Elsevier.

Chargaff, E. 1971. Preface to a grammar of biology. *Science* 172:637–642.

Drews, G. 1999. Ferdinand Cohn: A founder of modern Biology. *ASM News* 65:547–553.

Gamgee, A. 1895. The late professor Hoppe-Seyler I. *Nature* 52:575–576.

Harris, H. 1999. *The birth of the cell*. New Haven, CT: Yale University Press.

His, W. 1897. *Die Histochemischen Und Physiologischen Arbeiten von Friedrich Miescher*. Leipzig: F. C. W. Vogel.

Hoppe-Seyler, F. 1857. Ueber Die Einwirkung Des Kohlenoxydgases Auf Das Hämatoglobulin. *Archiv für Pathologische Anatomie und Physiologie und für Klinische Medicin* 11:288–289.

Mazzarello, P. 1999. A unifying concept: The history of Cell theory. *Nature Cell Biology* 1:E13–E15.

Schwann, T. 1911. Mikroskopische Untersuchungen Über Die Uebereinstimmung in Der Struktur Und Dem Wachstum Der Thiere Und Pflanzen. *Nature* 86:41.

Signer, R. 1969. Die Entwicklung Der Erforschung Der Nukleinsäuren. *Bulletin Der Schweizerischen Akademie Der Medizinischen Wissenschaften*25–31.

Silver, G. A. 1987. Virchow, the heroic model in medicine: Health policy by accolade. *American Journal of Public Health* 77:82–88.

von Liebig, J. F. 1842. *Animal chemistry: Or, organic chemistry in its applications to physiology and pathology*. Luton: Taylor & Walton.

Weiner, D. B. 1968. *Francois-Vincent Raspail, 1794–1878. Scientist and Reformer*. New York: Columbia University Press.

Williams, G. 2019. *Unravelling the double helix: The lost heroes of DNA*. London: Weidenfeld and Nicholson.

Wolpert, L. 1996. The evolution of "the Cell theory". *Current Biology* 6:225–228.

3

Lifting the Veil

Contents

Abstract Working in what he once compared to the laboratory of a medieval alchemist in Tübingen Castle, Miescher isolates leukocytes from pus washed out of discarded surgical bandages. By meticulous trial and error, he develops a protocol by which he isolates a number of distinct proteins but is disappointed to obtain only already known ones. He is, however, intrigued when he finds another cellular material which, unlike proteins, cannot be digested with enzymes that break down proteins, is soluble in alkali and precipitates out of solution when acid is added, and has a high content of phosphorus. He further shows that it is not a carbohydrate or a lipid—leading him to conclude that he has discovered an entirely new type of biomolecule. As this substance is localised in the nuclei of the cells, he gives it the name 'nuclein'—a name still preserved in its modern name 'desoxyribonucleic acid' or DNA. Miescher is convinced that he has found something of great importance that will 'lift the veil' on living processes and is eager to make his discovery public. But his mentor Hoppe-Seyler urges caution – and for good reason.

Keywords Friedrich Miescher • DNA • Discovery • Nuclein, • White blood cells • Leukocytes • Physiological Chemistry • Laboratory • Felix

K. Hall, R. Dahm, *The Dawn Fisherman*, Copernicus Books,
https://doi.org/10.1007/978-3-032-14219-1_3

Hoppe-Seyler • University of Tübingen • Phosphorus • Cell growth • Microscopy • Biochemistry • Molecular biology • Genetics • Scientific research • Experimentation • Cellular structures • Organelles • Biological molecules • History of science

Initial disappointment turns to success when Miescher discovers a mysterious new substance in the cell

The substance with which Miescher began his work was not just an everyday material—it was also an utterly revolting one. But despite this, he nevertheless sang its praises as being ideal for his task of determining the chemical make-up of cells:

> *[I was] faced with the task of determining, as completely as possible, the chemical building blocks whose diversity and arrangement determines the structure of the cell. For this purpose pus is one of the best materials. Hardly with anything else would it be possible to obtain such histological purity....*[1]

Pus is a thick yellow liquid that is produced as a by-product of infection, a cellular junkyard containing the debris of pathogens that have been destroyed by the body's white blood cells. Although it is usually far more likely to evoke disgust than delight, Miescher described pus as being 'one of the most beautiful materials' with which to work.[2] The beauty of pus was that, unlike solid tissues, which contained a mixture of different cells, it was rich in a type of cell known as a leukocyte. Because these are a type of white blood cell, they float freely in suspension and were therefore far easier for Miescher to isolate and purify than cells from solid whole tissues. With such a relatively pure source of a single cell type to hand, his task of determining the chemical composition of these cells would be much more straightforward.

Leukocytes had not however been Miescher's first choice of material to study. Hoppe-Seyler had originally suggested that he use lymphocytes, a type of white blood cell found in the lymph glands for his work:

> *In full agreement with Hoppe-Seyler, I had set myself the task of elucidating the constitution of lymphoid cells. I was captivated by the thought of tracking down the basic prerequisites of cellular life on this simplest and most independent form of animal cell.*[3]

---

[1] Miescher 26th Feb 1869 in (His 1897); p. 34.

[2] Ibid.

[3] Ibid., p. 33.

When these cells proved to be too difficult to isolate from lymph glands, Miescher turned his attention to leukocytes, the white blood cells found in pus. Before he could begin his analysis of their cellular constitution, however, he faced a much more immediate practical challenge—how to procure a supply of pus from which to extract leukocytes in the quantities he needed for his experiments?

The solution came thanks to another pioneer who was working in Tübingen at around the same time as Miescher. As Professor of Surgery at the University of Tübingen, Viktor von Bruns (1812–1883) was wrestling with a serious problem faced by clinicians at the time. Patients undergoing surgery in the mid-nineteenth century were as likely to die due to post-operative blood poisoning and gangrene from wounds that had become infected as from the injuries themselves. Reflecting on this dismal situation, the renowned Scottish surgeon Sir James Young Simpson (1811–1870) made the grim observation that a patient 'laid on an operating-table in one of our surgical hospitals is exposed to more chance of death than the English soldier on the field of Waterloo.'[4]

In 1867, Joseph Lister (1827–1912), Regius Professor of Surgery at the University of Glasgow described how this problem might be treated, by means of a novel method that he called 'the antiseptic treatment.'[5] Being familiar with the work of Louis Pasteur (1822–1895), Lister recognised that conditions such as gangrene were caused not by contact with air as such, but rather with microbes carried in the air. If wounds could be treated in such a way as to kill these microbes, then the risk of infection for the patient would be greatly reduced.

Such a means was readily to hand thanks to carbolic acid, or phenol. First isolated from coal tar by the German industrial chemist Friedlieb Runge (1797–1867), this substance was already being used in the treatment of sewage and had been shown to be effective against typhoid. Reasoning that it might kill disease-causing bacteria in wounds, Lister treated several of his patients using a combination of spraying the operating theatre with carbolic acid whilst dressing their wounds with lint soaked in carbolic acid and linseed oil around which was packed absorbent wool.

Lister is often lauded in medical textbooks as a pioneer who revolutionised medicine with this antiseptic treatment, but even at the time, he faced claims that his approach was not entirely original. His fellow surgeon Sir James Young-Simpson claimed that Lister had 'been long forestalled by the

[4] Cited in (Waller 2002); p. 163.

[5] (Lister 1867); p. 356.

experiences of our continental neighbours' such as the French physician Jules Lemaire (1814–1873) who had not only theorised about the anti-microbial properties of carbolic acid, but had been using it to prevent infections in wounds several years before the publication of Lister's first reports.[6,7] Moreover, Young-Simpson pointed out that surgeons in France, Spain and Germany had already been using carbolic acid in the dressing of wounds well before Lister had done so.

One of whom was Viktor von Bruns, who recognised that the lint, (or charpie as it was sometimes known) used for these dressings was not produced under particularly hygienic conditions. Often it was made of fibres plucked by factory workers or even patients themselves, from softened sheets of old canvas and sometimes it was even re-used.[8] So the search was on to find a better material from which wound dressings could be made.

One potential candidate was cotton, but this suffered from one major drawback—in its raw condition, it was not very absorbent and would therefore be unable to soak up antiseptic substances such as carbolic acid. Drawing on his experiences with other materials, however, Viktor von Bruns realised that this was due to fatty substances present in the raw cotton fibres, which formed a water-repellent coating on their surface.[9] Yet by treating the cotton with ether and a 4% solution of caustic soda, before then drying it out, von Bruns found that he could remove this fatty coating. In so doing, he developed a simple and cheap means of producing highly absorbent cotton that could now be easily rendered sterile simply by soaking in carbolic acid.

The first attempts by von Bruns to make this new material are thought to have been carried out around 1863–1864 and they quickly proved their worth. The outbreak of the Franco-Prussian War in 1870 led to a massive demand for wound dressings and the important contribution to medicine made by von Bruns is commemorated today by a modest plaque that stands on the Neue Strasse in Tübingen (Fig. 3.1).[10]

But in addition to saving lives both on and off the battlefield, the new wound dressings may also have played an important role in Miescher's

[6] Ibid.

[7] (Young-Simpson 1867).

[8] (Klose 2007).

[9] (Saltzwedel 1977); p. 130.

[10] (Klose 2007); In his book 'Victor von Bruns (1812-1883): Leben und Werke', Gerhard Saltzwedel points out that the date of 1871 is wrong (Saltzwedel 1977); p. 132) The inscription on the plaque reads: In dieser Apotheke wurden um 1871 unter Prof. Viktor von Bruns die ersten Versuche gemacht, aus Baumwolle Watte herzustellen. (In this pharmacy around 1871 the first attempts were made under Prof. Viktor von Bruns to produce absorbent cotton.)

Fig. 3.1 The pharmacy in Tübingen centre which today bears a plaque commemorating the development by Viktor von Bruns of the absorbent cotton bandages that would prove to be so invaluable for Miescher's work. (Photograph reproduced with kind permission of Stadtarchiv Tübingen, Fotosammlung)

discovery.[11] This was because raw, untreated cotton fibres did not soak up pus very well. The water-repellent properties of the fatty substances that coated their surface meant that instead of being soaked up, pus from wounds tended to just run down the wounded limb.[12] But when the cotton fibres were treated with von Bruns' method to remove these fatty substances, the bandages made from them were able to soak up and retain large quantities of pus.

In the opening of his first paper in which he described the isolation of nuclein, Miescher thanked Dr. Bevers and Dr. Koch of the surgical clinic in Tübingen for providing him with a supply of discarded bandages. Since von Bruns was the director of this institution it seems highly likely that the bandages Miescher received would have been ones prepared by the von Bruns method.[13] But even though these provided a regular supply of pus, Miescher still had reason to feel frustrated:

> *Luck has hardly favoured me during the course of my investigations. At no time did I have at my disposal a large amount of pus derived from abscesses, which would*

---

[11] Miescher mentions von Bruns in Letter II ('On the Behaviour of Pus in Common Salt Solutions') (His 1897); p. 38.

[12] Ibid, p. 132.

[13] Statistik sämtlicher in der chirurgischen Klinik in Tübingen von 1843 bis 1863 vorgenommenen (Verlag von Ebner & Seubert, Stuttgart, 1863). Foreword by Professor Viktor von Bruns.

> *have been essential for a detailed qualitative investigation. The quantities made available to me were highly variable and rarely amounted to even a couple of ounces*[14]

Getting a regular supply of pus-soaked bandages was a good start—but it was not enough. Miescher had to find some way of concentrating the leukocytes obtained from them. To do this, he began by subjecting his starting material to a somewhat unsophisticated form of analysis.[15] By sniffing the bandages he was able to determine which were fresh and therefore suitable for use, and which should be discarded. His next challenge was to isolate the precious leukocytes from contaminating materials such as oil, carbolic acid, and traces of cotton wool left over from the bandages. The conventional means of doing this was to sediment the cells with saline solution, as was done when isolating red blood corpuscles. But when Miescher attempted to apply this same technique to the separation of leukocytes, the result was what he simply described as 'a slimy mess'.[16]

Undeterred, Miescher ploughed on. Testing a range of salts of alkaline and alkaline earth metals, he found through trial, error, and an awful lot of patience, that by using a solution of sodium sulphate diluted with nine parts water, the leukocytes could be washed clean out of the bandages, without becoming a lump of slime. After passing the solution through a filter to remove any residual fibres of cotton wool from the bandages, Miescher then had to think about how to collect the cells he had isolated and concentrate them into a smaller volume of liquid.

Had he done this work only a few years later, this task would have been so much easier. In 1875, Alexander Prandtl (1840–1896), a Professor at the Agricultural College Weihenstephan in Freising, developed an idea that had first been suggested by his brother Antonin 11 years earlier, to build a machine that used a centrifugal field to separate butterfat from milk by spinning them at high speeds.[17]

Within only a matter of a few short years, Miescher would become actively involved in adapting the Prandtl brothers' invention from its use in the dairy industry to become the bench centrifuge - a key scientific tool with which any undergraduate student today can quickly isolate cells from suspension in a

[14] Miescher, F. 'Ueber die chemische zusammensetzung der Eiterzellen'(1871) Medizinisch-Chemische Untersuchungen 4: pp. 441–460 ; p. 442. Translation by N. Sankaran and K. Hall, 'DNA translated: Friedrich Miescher's discovery of nuclein in its original context ' (2021) British Journal for the History of Science, 54: 99–107.

[15] Ibid.

[16] (Miescher 1871); p. 442.

[17] (Vogel-Prandtl 2004); p. 10.

matter of minutes.[18] But when Miescher first began his work in the kitchens of Tübingen castle, no such device was yet available to him. Instead, he had simply to rely upon his reserves of patience, by leaving his solutions to stand on the lab bench and watch the hours tick away as the leukocytes sank slowly to the bottom of the test tubes.

Today this might well be considered an ideal means for cultivating mindfulness but before Miescher could proceed any further, he needed to know that the long hours he had spent watching leukocytes sink to the bottom of a tube had not been a complete waste of time. Examining his harvest of precious white blood cells under the microscope, he confirmed that they had not been damaged by the isolation procedure, before moving on to the next stage of his work which was to begin his analysis of their chemical composition.

In undertaking this ambitious task, he had a powerful champion. After his vigorous public defence of Charles Darwin's book 'The Origin of Species', the eminent English biologist and populariser of science, Thomas Henry Huxley (1825–1895) had already earned himself the nickname of 'Darwin's Bulldog', but the names of Miescher and Hoppe-Seyler might equally be inserted into this memorable epithet. Speaking at a lecture given in Edinburgh in 1868, Huxley issued a rallying cry that the mystery of living systems was to be solved through chemistry.

Huxley was sceptical about the cell theory as proposed by the likes of Schleiden and Schwann. Not that he denied the existence of cells—but rather, he doubted whether they were the fundamental unit of living systems. For him, the truly fundamental substance of life was the protoplasm, which had first been proposed by the French biologist Félix Dujardin (1801–1860) as being responsible for carrying out all the different chemical processes of the living cell, and was described by the German biologist Ernst Haeckel (1834–1919) as a 'single, homogenous matter… the active substrate of all vital motions and of all vital activities: nutrition, growth, motion, and irritability.'[19]

Huxley shared this view, believing protoplasm to be 'the formal basis of all life. It is the clay of the potter: which, bake it and paint it as he will, remains clay, separated by artifice, and not by nature from the commonest brick or sun-dried clod.'[20] More importantly, he believed that an understanding of protoplasm in terms of its chemical composition would refute once and for all

[18] (Miescher and Schmiedeberg 1896); p. 104.

[19] (Haeckel 1868); p. 108.

[20] Huxley T.H. cited in (Welch 1995); p. 482.

the notion of vitalism—that living systems were distinguished from inanimate matter by some inherent mystical animating life force.

And of all the various chemical components that might make up the protoplasm, there was for Huxley, one of particular importance:

> *The researches of the chemist have revealed... a striking uniformity of material composition in living matter... All protoplasm is proteinaceous.*[21]

If Miescher was to succeed in his ambition of describing the cell in chemical terms, he needed to isolate and identify the many different proteins present. And this would prove to be far from easy. Proteins had first been identified as major constituents of food thanks to their key role in familiar and highly visible changes such as the curdling of milk due to acidification, or the coagulation of egg-whites upon heating. It was in fact due to this latter effect that they acquired their original name of 'Eiweisskörper', or 'egg-white bodies' in German. But it was the Swedish chemist Jöns Jacob Berzelius (1779–1848) who, in recognition of their fundamental role in living systems, first suggested to his former student, the Dutch chemist Gerrit Jan Mulder (1802–1880) the name by which such substances are known today:

> *The word protein that I propose to you for the organic oxide of fibrin and albumin, I would wish to derive from proteios, because it appears to be the primitive or principal substance of animal nutrition that plants prepare for herbivores, and which the latter furnish to the carnivores.*[22]

It is certainly true that proteins are the principal substance of a tasty meal, but they are also much more. They are Nature's nanomachines—able carry out a vast range of essential functions within living systems. Some, such as collagen or keratin in hair and nails are purely structural. Others, such as haemoglobin which carries oxygen in the blood, fulfil the role of transporting other substances. Protein hormones such as insulin, which is essential for maintaining balanced levels of blood sugar, act as chemical messengers, while enzymes such as pepsin catalyse chemical reactions.

Even as Mulder popularised the term 'protein', it was becoming clear thanks to albumin in eggs, casein in milk, and fibrin in blood, that this was a highly diverse family of substances. And Mulder believed that the cause of this diversity was rooted in differences in the chemical composition of different proteins:

---

[21] Huxley, T.H., 'The Physical Basis of Life' Fortnightly Review NS5: 129-145; cited in (Fruton 1999); p. 167.

[22] Cited in (Fruton 1999); p. 167.

In plants as well as in animals there is present a substance which is produced in the former, constitutes the part of the food of the latter, and plays and important role in both. It is one of the very complex compounds, which very easily alter their composition under various circumstances...it is without doubt the most important of all the known substances of the organic kingdom, and without it, life on our planet would probably not exist...It combines with sulphur or phosphorus, or both, and thereby exhibits differences in its appearance and physical properties. The substance has been named protein, because it is the origin of very different substances and therefore may be regarded as a primary compound.[23]

The very first studies of protein chemistry were not made, however, with the aim of probing deep into the mysteries of life, but rather, in response to an urgent and pressing practical need. Following the French Revolution of 1789, a shortage of food led to riots in France while in England the enclosure of farmland during the eighteenth century had resulted in a steep rise in the price of flour. As a result, many English towns were now suffering a shortage of bread at the same time as they were experiencing a rapid rise in population due to the growth of factories.

In response to this drastic situation, research was directed towards new means of food production—such as extracting sugar from beet, and a better understanding of the chemical composition of dietary components. From this grew a drive towards the analysis of the complex organic molecules found in living systems. The French chemist Michel Eugène Chevreul (1786–1889) made the first successful analysis of such a compound when he found that treating sheep fat with alkali, gave rise to two components—glycerol and a range of fatty acids. But Chevreul wanted to go much further, insisting that the ultimate task was to determine the amounts of certain elements such as carbon, hydrogen, nitrogen and oxygen that were present in complex molecules.

Mulder attempted to do this for proteins and from his analyses, proposed that although all proteins shared the same basic chemical formula of carbon, nitrogen, oxygen and hydrogen, what made one type of protein such as casein distinct from another, such as albumin, was the number of sulphur or phosphorus atoms present in their structure.

But the methods of analysis available to Mulder were limited, and by the 1850s, it was becoming quite apparent that proteins were far more complex than his hypothesis had suggested. By treating proteins with strong sulphuric acid, for example, the French chemist Henri Braconnot (1780–1855) isolated

[23] Cited in (Fruton 1999); p. 171.

crystals of leucine and glycine, both of which belong to the family of chemicals known as amino acids and are the basic molecular building blocks of proteins. There was also a growing recognition that different proteins had different physico-chemical properties. The addition of solid sodium chloride to blood fibrin, for example, caused the formation of a precipitate, but did not do so when added to egg white, leading Rudolf Virchow to speculate that different proteins may exist in different states.

The challenge issued by Chevreul and Mulder was taken up by Miescher. And while the precise chemical nature of proteins remained elusive, one thing was clear: to describe the cell in terms of its chemical components, Miescher needed to accurately identify the different types of protein present.

The first hurdle to clear was that the contents of animal cells are enclosed within a membrane made of fats. But by washing his leukocytes in warm alcohol which dissolved this fatty membrane, Miescher made them spill their contents into two immiscible fractions—a lipid portion that was soluble in the alcohol, and an aqueous portion containing the proteins. Having separated the proteins in this way, the next step was to distinguish the different types that were present.

The means to do this were a little blunt and unsophisticated. In essence, they differed little from making scrambled eggs. This is because all proteins, such as albumin in egg white are great long molecular chains made up of individual amino acids bonded together in a very specific order. In living systems, these molecular chains fold up through a kind of molecular origami into a very precise three-dimensional structure. And it is thanks to their 3-D conformation that different proteins can perform specific functions such as, in the case of haemoglobin, the binding and transport of oxygen.

But when a protein is heated, the atoms in its chain gain more energy and start to vibrate more vigorously. Eventually their movement becomes so violent that the precise three-dimensional folding unravels, resulting not only in a loss of function but also of solubility, as occurs when egg albumin is heated and coagulates from a clear, viscous liquid into a white, rubbery solid. The precise folding of a protein molecule can also be unravelled by changes in acidity or alkalinity which disrupt vital electrostatic attractions between charged groups within the molecule, or with the surrounding solvent—which is why, for scrambled eggs with the perfect consistency, a few drops of lemon juice should be added.

By precipitating them using either heat, or dilute solutions of hydrochloric acid and sodium carbonate to alter their pH, Miescher hoped that he would be able to distinguish many different types of cellular protein. But he was to be severely disappointed. Using these methods, he was able to identify only five distinct proteins at the most.

Fig. 3.2 Wilhelm Kühne (1837–1900) was, like Hoppe-Seyler, one of the pioneers of the new discipline of physiological chemistry, but at the same time one of his greatest professional rivals in the field. Portrait taken by Max Kögel. (Reproduced with kind permission of the University of Heidelberg Library)

Not that anyone else had fared any better. Even the distinguished German physiologist Wilhelm Kühne (1837–1900) (Fig. 3.2) who had taken over the position of assistant to Rudolf Virchow when Hoppe-Seyler had left Berlin for Tübingen had recently reported the identification of five different types of cellular protein, one of which was myosin, the contractile component of muscle tissue.

But this must have brought little consolation to Miescher. From the sheer scale of complexity and organisation within the cell, he knew that there simply had to be far more than just five proteins present. But his hunch counted for little without the means to detect them. And although Miescher reflected that 'It would be worthwhile to trace these similarities [with Kühne's work] further', he also knew that this was not enough.[24] If he really wanted to make

[24] (Miescher 1871); p. 447; translation (Hall and Sankaran 2021); p. 23.

a name for himself as a scientist, he needed to do a lot more than simply replicate the findings of another, more well-known researcher such as Kühne. Rather than walk in the shadow of someone else, he needed to come up with a novel discovery that would put his name on the map.

In desperation, he clung to the hope 'that there may yet be small amounts of other proteins that are perhaps easily lost in the successive washings during the extraction process.'[25] But given that he had failed to find even paraglobulin, one of the most readily detectable protein types, in his preparations, his chances of discovering anything more exotic appeared to be diminishing rapidly—as were those of a glowing future career in research.

Miescher was not only frustrated by his failure—he was also at a loss to explain it. The best he could offer was to speculate that this 'may be due to any number of other reasons about which I withhold comment due to an absence of facts'—hardly a level of analysis likely to earn much favour with a PhD examiner today.[26] But if Miescher's thinking was clouded on this matter, perhaps it was simply because, as he complained to his parents in 1869, that he found his work on proteins to be 'a real headache.'[27]

But the history of science is replete with examples of how a change in perspective can often turn what might at first sight appear to be failure into something very different. The drug Tamoxifen, for example, was once considered to be little more than 'a failed contraceptive' until research into its pharmacology revealed how it might be used in the prevention of breast cancer.[28] And for Miescher, even though his grand vision of identifying a rich cytoplasmic cornucopia of proteins now lay in ruins, there nevertheless lay a hidden gem amidst the rubble.

To the untrained eye, the gem in question did not have much sparkle, but was instead just a cloudy precipitate, formed under mind- numbingly boring conditions. This might seem unlikely to set pulses racing, but for Miescher this material had one novel property that left him very excited. So excited in fact, that he felt compelled to write to his Uncle about it (Fig. 3.3):

> *In my experiments with weak alkaline solutions, precipitates formed in the solutions after neutralisation that could not be dissolved in water, acetic acid, highly diluted hydrochloric acid or in a salt solution, and therefore do not belong to any known type of protein.*[29]

[25] Ibid.

[26] (Miescher 1871); p. 448; translation (Hall and Sankaran 2021); p. 23.

[27] Miescher, 11th June 1869, Letter III in (His 1897); p. 39.

[28] (Jordan 2020); R16.

[29] Miescher, 26th Feb 1869, Letter I in (His 1897), p. 34–35.

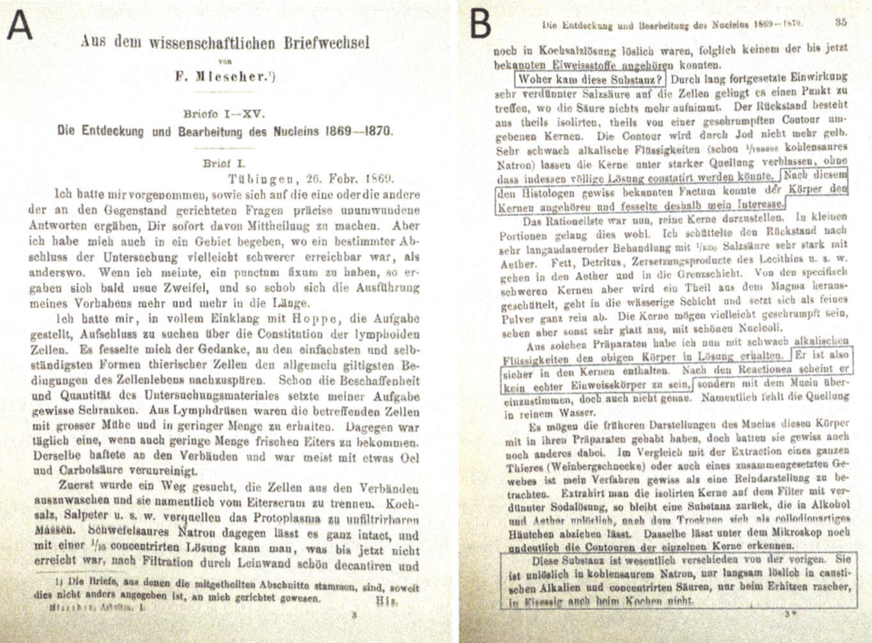

A

Aus dem wissenschaftlichen Briefwechsel

von

F. Miescher.[1])

Briefe I—XV.

Die Entdeckung und Bearbeitung des Nucleins 1869—1870.

Brief I.

Tübingen, 26. Febr. 1869.

Ich hatte mir vorgenommen, sowie sich auf die eine oder die andere der an den Gegenstand gerichteten Fragen präcise unumwundene Antworten ergäben, Dir sofort davon Mittheilung zu machen. Aber ich habe mich auch in ein Gebiet begeben, wo ein bestimmter Abschluss der Untersuchung vielleicht schwerer erreichbar war, als anderswo. Wenn ich meinte, ein punctum fixum zu haben, so ergaben sich bald neue Zweifel, und so schob sich die Ausführung meines Vorhabens mehr und mehr in die Länge.

Ich hatte mir, in vollem Einklang mit Hoppe, die Aufgabe gestellt, Aufschluss zu suchen über die Constitution der lymphoiden Zellen. Es fesselte mich der Gedanke, an den einfachsten und selbständigsten Formen thierischer Zellen den allgemein giltigsten Bedingungen des Zellenlebens nachzuspüren. Schon die Beschaffenheit und Quantität des Untersuchungsmateriales setzte meiner Aufgabe gewisse Schranken. Aus Lymphdrüsen waren die betreffenden Zellen mit grosser Mühe und in geringer Menge zu erhalten. Dagegen war täglich eine, wenn auch geringe Menge frischen Eiters zu bekommen. Derselbe haftete an den Verbänden und war meist mit etwas Oel und Carbolsäure verunreinigt.

Zuerst wurde ein Weg gesucht, die Zellen aus den Verbänden auszuwaschen und sie namentlich vom Eiterserum zu trennen. Kochsalz, Salpeter u. s. w. verquellen das Protoplasma zu unfiltrirbaren Massen. Schwefelsaures Natron dagegen lässt es ganz intact, und mit einer 1/10 concentrirten Lösung kann man, was bis jetzt nicht erreicht war, nach Filtration durch Leinwand schön decantiren und

1) Die Briefe, aus denen die mitgetheilten Abschnitte stammen, sind, soweit dies nicht anders angegeben ist, an mich gerichtet gewesen. His.

Miescher, Arbeiten. I. 3

B

Die Entdeckung und Bearbeitung des Nucleins 1869—1870. 35

noch in Kochsalzlösung löslich waren, folglich keinem der bis jetzt bekannten Eiweissstoffe angehören konnten.

Woher kam diese Substanz? Durch lang fortgesetzte Einwirkung sehr verdünnter Salzsäure auf die Zellen gelingt es einen Punkt zu treffen, wo die Säure nichts mehr aufnimmt. Der Rückstand besteht aus theils isolirten, theils von einer geschrumpften Contour umgebenen Kernen. Die Contour wird durch Jod nicht mehr gelb. Sehr schwach alkalische Flüssigkeiten (schon 1/100000 kohlensaures Natron) lassen die Kerne unter starker Quellung verblassen, ohne dass indessen völlige Lösung constatirt werden könnte. Nach diesem den Histologen gewiss bekannten Factum konnte der Körper den Kernen angehören und fesselte deshalb mein Interesse.

Das Rationellste war nun, reine Kerne darzustellen. In kleinen Portionen gelang dies wohl. Ich schüttelte den Rückstand nach sehr langandauernder Behandlung mit 1/1000 Salzsäure sehr stark mit Aether. Fett, Detritus, Zersetzungsproducte des Lecithins u. s. w. gehen in den Aether und in die Grenzschicht. Von den specifisch schweren Kernen aber wird ein Theil aus dem Magma herausgeschüttelt, geht in die wässerige Schicht und setzt sich als feines Pulver ganz rein ab. Die Kerne mögen vielleicht geschrumpft sein, sehen aber sonst sehr glatt aus, mit schönen Nucleoli.

Aus solchen Präparaten habe ich nun mit schwach alkalischen Flüssigkeiten den obigen Körper in Lösung erhalten. Er ist also sicher in den Kernen enthalten. Nach den Reactionen scheint er kein echter Eiweisskörper zu sein, sondern mit dem Mucin übereinzustimmen, doch auch nicht genau. Namentlich fehlt die Quellung in reinem Wasser.

Es mögen die früheren Darstellungen des Mucins diesen Körper mit in ihren Präparaten gehabt haben, doch hatten sie gewiss auch noch anderes dabei. Im Vergleich mit der Extraction eines ganzen Thieres (Weinbergschnecke) oder auch eines zusammengesetzten Gewebes ist mein Verfahren gewiss als eine Reindarstellung zu betrachten. Extrahirt man die isolirten Kerne auf dem Filter mit verdünnter Sodalösung, so bleibt eine Substanz zurück, die in Alkohol und Aether unlöslich, nach dem Trocknen sich als collodionartiges Häutchen abziehen lässt. Dasselbe lässt unter dem Mikroskop noch undeutlich die Contouren der einzelnen Kerne erkennen.

Diese Substanz ist wesentlich verschieden von der vorigen. Sie ist unlöslich in kohlensaurem Natron, nur langsam löslich in caustischen Alkalien und concentrirten Säuren, nur beim Erhitzen rascher, in Eisessig auch beim Kochen nicht.

3*

Fig. 3.3 Montage of Miescher's Letter of 26th February 1869 in which he describes the discovery of nuclein. (In His 1897; p. 33; Montage compiled and photographed by R. Dahm)

The handful of cytoplasmic proteins that had been identified to date could all be dissolved in weak acid. The substance Miescher had found, by contrast, remained stubbornly insoluble under such conditions. This might seem to be a mere technical distinction (and a rather dull one at that) but it left Miescher feeling like a zoologist who has just discovered not just a new species, but an entire new taxonomic kingdom. His observation of that cloudy precipitate floating in suspension and insoluble in weak acid was a landmark in the history of science that would transform our understanding of the living world.

Not, of course, that Miescher knew any of this at the time. His immediate question was—what might this new substance be? And where in the cell was it found? Two crucial observations suggested an answer. The first of these was that this precipitate could be re-dissolved in weakly alkaline solutions which suggested that it was most likely acidic itself. The second was the curious behaviour of the cell nuclei when they were treated with weak acid and alkali:

> *Through prolonged exposure of the cells to diluted hydrochloric acid it is possible to reach a point when the acid will not take anything up anymore. The residue consists partly of isolated [nuclei] and partly of nuclei surrounded by a shrunken contour.*

> *The contour can no longer be stained yellow with iodine. Very weak alkaline solutions…lead to a strong swelling and fading of the nuclei, without them, however, dissolving into the solution.*[30]

The failure of the nuclei to turn yellow when stained with iodine told Miescher something very important—that treatment with acid had removed all traces of protein from them. So whatever substance remained in the nuclei after treatment with acid, it could not be protein. So, what then, was it?

The swelling of the nuclei in weak alkaline solutions that Miescher mentioned in his letter offered a clue. Could this be because the nuclei contained the same alkaline-soluble material that Miescher had found in the cloudy precipitate about which he had written in such excitement to his uncle? Miescher was convinced that the substance in his cloudy precipitate 'could only belong to the nuclei and therefore captivated my interest.'[31]

The nucleus is crucial to the life of the cell. When the Soviet and American cosmologist George Gamow (1904–1968) turned his attention to solving the genetic code, he once likened the cell to a factory in which proteins are produced. In Gamow's analogy, the sites of protein synthesis in the cytoplasm were the factory floor, while the nucleus was the factory manager's office containing the file cabinets in which the blueprints and instructions for making those proteins were held.[32]

At the time that Miescher began his work however, pretty much nothing was known about what the function of the nucleus might be. Three years earlier, the eminent German biologist Ernst Haeckel (1834–1919) had proposed that it might well contain factors necessary for the transmission of hereditary traits.[33] But with next to nothing known about its chemistry, how it might do this remained a mystery. The next step was now clear to Miescher. His attempt to isolate and describe cellular proteins might well have failed, but now he had a new aim—to 'tackle the question of the chemical constitution of the cell nucleus in earnest.'[34]

But once again, there was a formidable hurdle to overcome. Before their chemical constitution could be studied, the nuclei needed to be removed from their cells. Once again, Miescher's patience proved to be a virtue. By repeatedly washing the cells for prolonged periods in very dilute hydrochloric acid,

---

[30] Ibid., p. 35.

[31] Miescher, 26th Feb, Letter I in (His 1897); p. 35.

[32] (Gamow 1955).

[33] (Haeckel 1866).

[34] (Miescher 1871); p. 452 ; translation (Hall and Sankaran 2021); p. 23.

he was able to burst them open and strip away most of the cytoplasm from their nuclei.

After vigorous shaking with ether and water to remove fats, Miescher found that the nuclei separated into two portions. Those with residual cytoplasm still attached collected at the interface between the ether and the aqueous phase. Fully extracted nuclei however, sank down to collect as a fine powder at the bottom of the aqueous phase. Again, this may sound like a rather dull observation, but Miescher had just become the first person to successfully separate nuclei from the rest of the cell—a technical achievement that would become vital step in the preparation of DNA by countless subsequent generations of PhD students and post-docs to this day.

Moreover, these nuclei behaved exactly as he had predicted. On addition of alkaline solutions, they swelled up; whilst the addition of acid gave rise to the formation of a white 'woolly' precipitate. This confirmed to Miescher that the mysterious precipitate he had originally observed was indeed localised in the nucleus. In recognition of this, he named the novel substance 'Nuclein.'

But what exactly was the chemical nature of this Nuclein? The answer was far from clear. Its behaviour in acids and alkali was quite unlike that of a protein. Nor when boiled did it coagulate as a protein would. Yet when subjected to the xanthoprotein colour assay which tested for the presence of protein, it gave a positive result.

Miescher was not the first to have isolated a substance that perplexed him. A few years earlier, the German-Lithuanian naturalist Eduard Georg Eichwald (1838–1889) had described the properties of mucin, a substance he had isolated from snails. Subjecting the mucin to exhaustive chemical analysis, including treatment with acids and alkali as Miescher would later do with his nuclear material, Eichwald found that when subjected to chemical tests, mucin behaved like a protein, whereas in others it did not. On finding that sugar was released from mucin under certain conditions, Eichwald proposed an answer to the puzzle of its apparently ambiguous nature. He suggested that mucin was 'a conjugated single compound consisting of a moiety with all properties of a genuine protein and a moiety released under certain conditions as a sugar.'[35]

In trying to solve the similarly ambiguous chemical nature of nuclein, Miescher took his inspiration from Eichwald:

[35] (Cabezas 1994); p. 3; (Eichwald 1865); p. 206.

According to these reactions it [nuclein] does not seem to be a real protein, but rather corresponds to mucin, albeit not precisely.[36]

And, just as Eichwald had done, the only way to resolve this ambiguity was to carry out a chemical analysis of Nuclein to determine which elements it contained. But this would require nuclei in far greater quantities and of much higher purity. And effective as his new method of separating nuclei from the cytoplasm had been, Miescher was all too aware of its shortcomings:

The minimal quantities of nuclei obtained by the described methods barely allowed for the determination of the few aforementioned reactions; an analysis of their elements was out of the question.[37]

It was back to the drawing board: Miescher needed to find a new way of isolating nuclei that would get rid of all contaminating cytoplasm. And thanks to the work of Wilhelm Kühne, he soon found one—although it would prove to be only slightly less unsavoury than sniffing discarded bandages and washing pus from them.

In addition to identifying the protein myosin in muscle, Kühne had also been studying the action of digestive juices on proteins. He had found that pepsin, a component of digestive juice secreted into the stomach, could dissolve whole cells to remove cytoplasm and proteins, leaving behind what he described as 'very shrunken nuclei'.[38]

Kühne's method was just what Miescher needed. But first he would have to get hold of some pepsin—a task far from easy at the time. Today, any PhD or post-doctoral researcher in a similar position could just take a quick glance at the online catalogue of several commercial suppliers, before clicking to order a supply of the purified enzyme. Miescher, by contrast, found himself faced with the thankless task of having to rinse out the stomachs of pigs with dilute hydrochloric acid to obtain a crude solution of pepsin.

Once the leukocytes had been broken open with warm alcohol and the lipids removed, treatment with pepsin resulted in a fine, grey sediment containing the nuclei. After shaking these vigorously with ether and warm alcohol to remove any stubborn residual lipids, Miescher washed them in dilute alkaline solutions such as sodium carbonate.

---

[36] Miescher, 26th Feb 1869, Letter I in (His 1897), p. 35.

[37] (Miescher 1871); p. 454 ;translation (Hall and Sankaran 2021); p. 27.

[38] (Kühne 1868); pp. 49–50.

The nuclei isolated using Kühne's method behaved exactly as Miescher expected they should. Treatment with an excess of acetic or hydrochloric acid resulted in a flocculent precipitate that could be redissolved by the addition of alkaline solutions. This is because, being mildly acidic in nature itself, DNA becomes insoluble as acidity increases, but dissolves in more alkaline conditions.

Using this method, Miescher had achieved the very first comparatively clean preparation of DNA ever made (Fig. 3.4). This alone was impressive enough, but what made it even more so were the conditions under which it had been done. Aware that the nuclear substance might well be delicate and prone to degradation at normal room temperatures, Miescher took drastic steps to prevent this from happening. The winters experienced in South-West Germany at that time were particularly cold which Miescher used to his advantage by keeping the windows of his laboratory open throughout this work to maintain what he described as a 'wintry temperature.'

But as he shivered in his laboratory, his only thoughts were on identifying the chemical nature of nuclein. Using a method of incineration similar to that developed by Raspail, Miescher burned his samples of Nuclein in the presence of various chemicals that reacted selectively with the different constituent elements. He then weighed the various compounds resulting from this combustion to determine the amount of each element present.

The process was tedious, time consuming and was later derided by Miescher as mere 'factory-work'.[39] But despite this, it proved to be worthwhile. Because alongside the presence of carbon, hydrogen, oxygen and nitrogen, all of which are abundant in proteins, there were a couple of very surprising findings. Firstly, Miescher found that Nuclein contained no sulphur—an element known to be present in proteins. Secondly, he found something that astounded him so much he felt compelled to write and proudly tell his parents:

> *I have been engaged with the definitive analysis of the nuclear substance. From this it emerges that it is comparatively very rich in phosphorus—and not in the form of lecithin.*[40]

Lecithins are a group of fats that contain phosphorus and were one of the few organic compounds known to do so at the time. But its high phosphorus content suggested that Nuclein was something truly novel:

---

[39] (Miescher, 29th June 1876, Letter XLII in (His 1897), p. 87.

[40] Miescher, 21st Aug 1869, Letter IV in (His 1897), p.39.

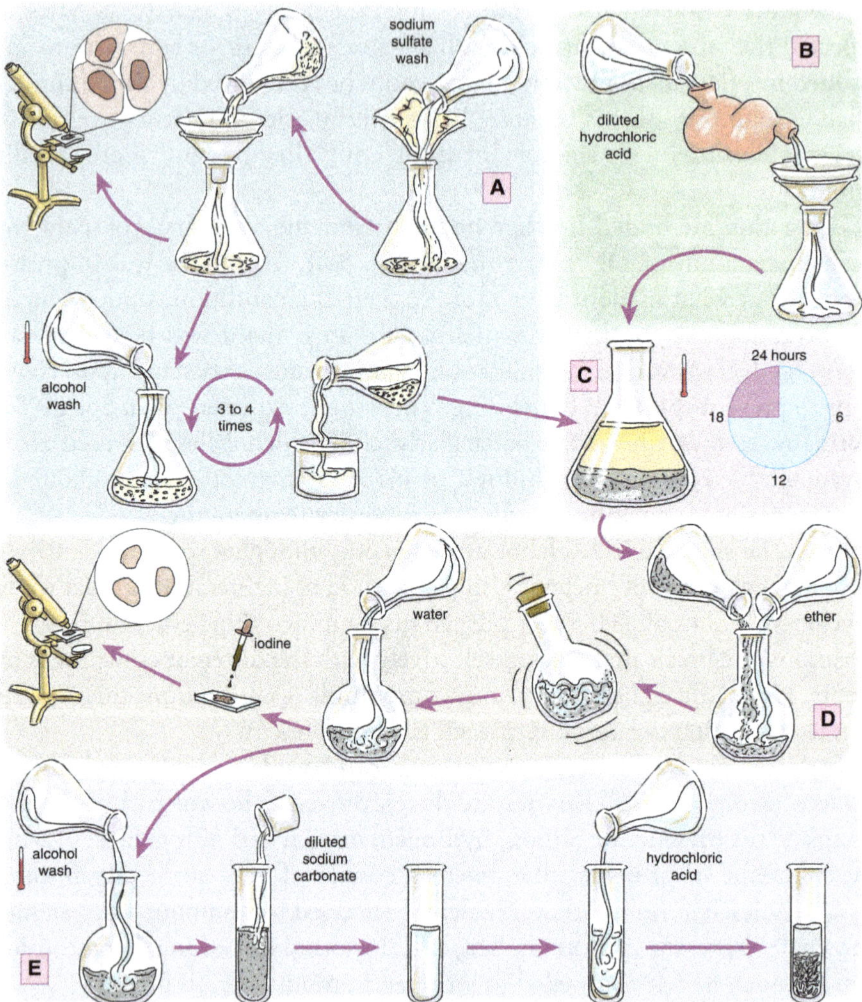

Fig. 3.4 Friedrich Miescher's protocol for isolating nuclein was meticulous, tedious, and often had to be carried out in uncomfortably cold temperatures to prevent degradation of the samples. It was Miescher's insight as a scientist and extreme dedication that led to his success. First, he had to wash pus-covered bandages with a dilute solution of sodium sulfate to extract the white blood cells, which he then filtered to remove stray cotton fibres. After the cells settled to the bottom of the container for several hours, he examined them under a microscope to make sure they were intact. He then washed the cells with warm alcohol several times, which broke them open and removed most of the lipids (**a**). Miescher also washed pig stomachs out with dilute hydrochloric acid to get a solution containg the protein-digesting enzyme pepsin (**b**), in which he repeatedly soaked the cells for 18 to 24 hours at warm temperatures. At that point, the cells turned into a fine, gray, powdery sediment of isolated nuclei, separated from a faintly yellow liquid (**c**). He shook the nuclei with ether several times to remove any residual lipids. After washing them with water and staining them with iodine, Miescher

> *I think that the given analyses — as incomplete as they might be—show that we are not working with some random mixture, but ... with a chemical individual or a mixture of very closely related entities... We are dealing with an entity sui generis not comparable to any hitherto known group.*[41]

This last sentence would prove to be something of an understatement, to put it mildly. Not content to have merely isolated a cloudy precipitate with curious chemical properties, Miescher had delved deeper. And in doing so he had become the first to discover a novel cellular substance with properties unlike any other. It was a striking example of the observation often ascribed to the prolific science fiction writer Isaac Asimov that 'The most exciting phrase to hear in science, the one that heralds new discoveries, is not 'Eureka!' but 'That's funny..."[42]

But what might the function of this novel substance be? Miescher was quick to offer some suggestions—all of which pointed to a central role in the cell. Given its high content of phosphorus, might it be a metabolic precursor of compounds such as lecithin which also contained this element? 'I cannot ignore the thought,' speculated Miescher, 'that it is here that the essential physiological function of phosphorus in the organism is revealed.'[43]

But there was another, even more exciting possibility. Miescher found it remarkable that in plants at least, phosphorus was confined 'almost exclusively' to regions of rapid growth. And was it just coincidence, he wondered, that such regions also contained large amounts of nuclei? Might there be some intimate connection between nuclein and the proliferation of cells and nuclei during tissue growth?

Miescher was starting to get the dizzying feeling that he might well be on the verge of a whole new understanding of cells. Writing to his uncle Wilhelm His in December 1869, he already began to envisage practical applications for

Fig. 3.4 (continued) checked the nuclei under a microscope. A lack of staining indicated that the proteins had been successfully removed (d). After another series of warm alcohol washes, Miescher began his testing. When he added dilute sodium carbonate, an alkaline solution, to the nuclei, they would swell and become translucent. By adding acid, he again obtained an insoluble, flaky precipitate—nuclein, or DNA (e). (From 'The First Discovery of DNA' by Ralf Dahm, American Scientist (2008) 96: pp. 320–327; p. 324). (Reproduced with kind permission of Sigma Xi, The Scientific Research Honor Society)

[41] (Miescher 1871); p. 458; translation (Hall and Sankaran 2021); p. 23.

[42] There is some debate about whether it was actually Asimov who first said this. But regardless of who said it, the observation is certainly an astute one.

[43] (Miescher 1871); p. 459; translation (Hall and Sankaran 2021); p. 36.

his discovery. Could, for example, measurements of the relative amounts of Nuclein and protein be used to characterise and distinguish different pathological conditions?

Miescher was confident that this approach would be far more reliable and informative than current methods which relied upon investigation of pathological tissue by microscopy—an instrument which, as he complained to Wilhelm His, 'does often let one down.'[44] From Miescher's perspective, there were two major problems with microscopy. Firstly, it revealed only the appearances of cellular structures and was thus restricted to mere phenomenology. By contrast, Miescher's chemical approach offered more than simply a visual description of these structures: it opened up the possibility of delving beneath the mere appearance of cellular structures to reveal their molecular composition. Secondly, structures observed under the microscope might not be genuine cellular components but simply artefacts arising from the staining and preparation of samples.

By quantifying the amount of nuclein present and observing how it changed, Miescher hoped to provide a more reliable method of studying the cell than simply relying on microscopy. This approach might, for example, reveal physiological processes to be fundamentally different at a chemical level, rather than being similar—as had been assumed by microscopical investigations. Comparing the relative amounts of nuclein, proteins, and products of metabolism might offer a more accurate means of identifying the precise physiological condition of a cell, such as whether it was in a state of growth or death. Moreover, in rapidly proliferating tissues such as those in a tumour, Miescher predicted 'an increase in nuclear substances [as] a preliminary phase to cell division.'[45] In the light of what we know today, this makes perfect sense as, prior to undergoing division, a cell needs first to replicate its entire DNA content.

In his excitement, Miescher was eager to see whether nuclein could be found in cell types other than leukocytes, such as yeast, kidney, liver and nucleated red blood cells.[46] But at the same time he was also keen to share his discovery with the scientific world and began writing up his work. In a letter to his parents, he modestly described his manuscript as being 'restricted in

[44] Miescher, 20th Dec 1869, Letter V in (His 1897); p. 41. In the original collection of letters by Wilhelm His, this particular document is dated 1890 and not 1869, but the content of the letter suggests that Miescher is at the start, and not the end, of his career and that the date of 1890 must therefore be a printing error.

[45] Miescher, 20th Dec 1869, Letter V in (His 1897); p. 41.

[46] Miescher, 11th June 1869, Letter III in (His 1897), p. 39.

scope, somewhat like a moderate dissertation.'[47] Yet even as he settled down to write what would become his first scientific publication, he clearly had a sense that its ultimate scope would be neither restricted, nor moderate:

> *An entire family of such phosphorus-containing substances, which differ slightly from one another, will reveal itself, and that this family of nuclein bodies will prove tantamount in importance to the proteins…Knowledge of the relationship between nuclear substances, proteins, and their closest conversion products will gradually help to lift the veil which still utterly conceals the inner processes of cell growth.*[48]

And when it came to articulating the grander aims of physiological chemistry, Miescher was anything but modest in the scale of his ambition. The task of the physiological chemist, he declared, was nothing less than to:

> *…take the path of seeking out beneath the most simple external conditions of living organisms, the basic theme of chemical movements, perhaps also the principle of the divergence of plants and animals.*[49]

First, however, Miescher needed to make his discovery known. But if he was hoping that his mentor Hoppe-Seyler would share his enthusiasm for this, he was to be very disappointed. With Germany and France about to be plunged into war, Miescher's eagerness to get his work into print was running ahead of events in the wider world. Added to which, there was always the fear gnawing away at him that, if he did not publish this groundbreaking work, then someone else might…

[47] Miescher, 21st August 1869, Letter IV in (His 1897), p. 39.

[48] (Miescher 1871); p. 460. translation (Hall and Sankaran 2021); p. 23.

[49] Miescher, 20th Dec 1869, Letter XX in (His 1897), p. 61.

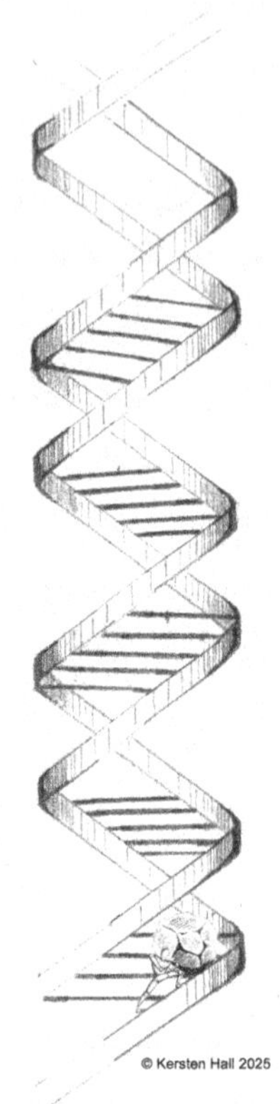

Drawing by Kersten Hall

References

Cabezas, J. A. 1994. The origins of glycobiology. *Biochemical Education* 22:3–7.

Eichwald, E. 1865. Beiträge Zur Chemie Der Gewebbildenen Substanzen Und Ihre Abkömmlinge: Ueber Das Mucin Besonders Der Weinbergschnecke. *Justus Liebigs Annalen Der Chemie* 134:177–211.

Fruton, J. 1999. *Proteins, enzymes, genes: The interplay of chemistry and biology*. New Haven, CT: Yale University Press.

Gamow, G. 1955. Information transfer in the living Cell. *Scientific American* 193:70–78.

Haeckel, E. 1866. *Generelle Morphologie Der Organismen*. Berlin: Reimer.

Haeckel, E. 1868. Monographie Der Monoren. *Jenaische Zeitschrift für Medizin und Naturwissenschaft* 4:64–137.

Hall, K. T., and N. Sankaran. 2021. DNA translated: Friedrich Miescher's discovery of Nuclein in it's original context. *The British Journal for the History of Science* 54:1–9. https://doi.org/10.1017/S000708742000062X.

His, W. 1897. *Die Histochemischen Und Physiologischen Arbeiten von Friedrich Miescher*. Leipzig: F. C. W. Vogel.

Klose, A. 2007. Victor von Bruns Und Die Sterile Verbandwatte. In *Hin Und Weg: Tübingen in Aller Welt. Tübinger Katalogue 77*, 35–45. Tübingen: Tübingen Stadtmuseum.

Kühne, W. 1868. *Lehrbuch Der Physiologischen Chemie*. https://archive.org/details/bub_gb_DxNu12EEKh4C

Lister, J. 1867. On the antiseptic principle in the practice of surgery. *The Lancet* 90:353–356.

Miescher, F. 1871. Ueber Die Chemische Zusammensetzung Der Eiterzellen. *Medicinisch-Chemische Untersuchungen Aus Dem Laboratorium Für Angewandte Chemie Zu Tübingen* 4:441–461.

Miescher, F., and O. Schmiedeberg. 1896. Physiologische-Chemisch Untersuchungen Über Die Lachsmilch. *Naunyn-Schmiedeberg's Archives of Pharmacology* 37:100–155.

Saltzwedel, G. 1977. Victor von Bruns (1812-1883): Leben Und Werk. *Beiträge zur Geschichte der Eberhard-Karls Universität Tübingen* 13:1–214.

Waller, J. 2002. *Fabulous science: Fact and fiction in the history of scientific discovery*. Oxford: Oxford University Press.

Welch, G. R. 1995. T.H. Huxley and the protoplasmic theory of life: 100 years later. *Trends in Biochemical Sciences* 20:481–485.

Young-Simpson, J. 1867. Carbolic acid and its compounds in surgery. *The Lancet* 90 (November): 546–549.

4

Blood Feud

Contents

Abstract Hoppe-Seyler has good reason to be cautious about Miescher's claim of having isolated a new type of biomolecule. When the recent discovery made by Oskar Liebreich, another of Hoppe-Seyler's students, of a novel compound called protagon had turned out to be nothing more than an artefact of purification, Hoppe-Seyler's reputation had suffered a blow. To avoid a similar embarrassment over nuclein, Hoppe-Seyler insists that he personally repeat Miescher's work. To complicate matters further, Hoppe-Seyler is embroiled in an ongoing feud with Ludwig Thudichum, another eminent figure in the emerging field of physiological chemistry. Thudichum was a German political émigré who had ascended to the heights of the British medical establishment and with whom Hoppe-Seyler had conducted a sometimes bad tempered dispute on the pages of various journals. This had begun with disagreement over the properties of the blood protein haemoglobin but would also influence how Miescher's paper on nuclein was received when it was finally published in 1871. By this

K. Hall, R. Dahm, *The Dawn Fisherman*, Copernicus Books,
https://doi.org/10.1007/978-3-032-14219-1_4

time Miescher has returned to his home city of Basel where he not only takes up a professorship but also, thanks to an anatomical quirk of the Rhine salmon that come to spawn there, finds a rich new supply of nuclein in their sperm.

Keywords Miescher • DNA • Nuclein • Discovery • Hoppe-Seyler • Pus • White blood cells • Leukocytes • Physiology • Chemistry • Biology • History of Science • Research • Experimentation • Verification • Replication • Mentorship • Academia • Switzerland • University of Basel • University of Leipzig • Embryology • Cellular biology • Biochemistry

Miescher wants to share his discovery with the world – but Hoppe-Seyler is cautious…and with good reason

In December 1869 and with Christmas approaching, Miescher found himself busy sealing and addressing a parcel. But this was no festive gift left until the last minute to be hastily packed and posted. Writing to his parents on the day before Christmas Eve, he believed it to be something far more exciting:

> *On my table lies a sealed and addressed packet. It is a manuscript, for whose shipment I have already made all necessary arrangements. I will now send it to Hoppe-Seyler in Tübingen. So, the first step into the public is done, given that Hoppe-Seyler does not refuse it.*[1]

By now Miescher had left Tübingen and, after a short holiday, had come to Leipzig to spend a year working with the distinguished German physiologist Carl Ludwig (1816–1895) (Fig. 4.1). Like Hoppe-Seyler, Ludwig was convinced that living systems could be understood in terms of physical and chemical processes and his research covered a diverse range of subjects including urinary excretion, anaesthesia, blood pressure and blood gases. In the course of this work, he also developed several instruments such as the kymograph for the written recording of blood pressure variations, and the blood mercurial pump for the analysis of blood gases.

This had a powerful influence on Miescher, leaving him convinced of the important role that instrumentation played in scientific research. He was also

[1] Miescher 23rd December 1869, cited in (His 1897); p. 9.

Fig. 4.1 Carl Ludwig (1826–1895) with whom Miescher worked at the University of Leipzig for a short while after leaving Hoppe-Seyler's laboratory in Tübingen, was another important mentor of Miescher. (Photographs reproduced with kind permission of the archive of the University of Leipzig)

impressed not only by the diversity of Ludwig's research, but also that of his research team. This included scientists from at least nine different countries such as The Netherlands, Norway, Hungary, Germany, Britain, Italy, Switzerland, Russia, Egypt and the USA, who as Miescher recalled were:

> *...crowded around the experimentation tables... You can see that the most important schools of physiology were represented...Every one brings something with him, and even to my humble self it was permitted to give wise counsels in matters of hemoglobin....*[2]

[2] Miescher 1870, Letter VIII, in (His 1897); p. 47.

The cosmopolitan make-up of Ludwig's lab was unusual for the time and Miescher forged several lifelong friendships during his time there whilst conducting research into the nerves of the spinal cord. But although Miescher carried out this research with what his uncle Wilhelm His later described as his usual diligence, his work on the nervous system failed to excite him in the way that nuclein had done.[3]

Nuclein was still very much on his mind and, eager to tell the world about this new substance, he spent his first few months in Leipzig putting the finishing touches on a manuscript describing his discovery. At last, on 23rd December 1869, he placed his manuscript into the package on his table, sealed and addressed it, and sent it off to Hoppe-Seyler in Tübingen.

Then he held his breath. For any PhD student or post-doctoral researcher, getting their first piece of work published is a milestone, and things were no different in Miescher's day. The publication of his discovery in an academic journal would not only make nuclein known to the scientific world but would establish Miescher as an independent researcher and give him the crucial foothold on which to build the rest of his career.

But as weeks turned into months with still no reply, Miescher's excitement and eager anticipation soured into anxiety. 'I have become doubtful whether the work that I sent to Hoppe arrived at its address,' he confessed to his parents, 'as I have received no signs of life from Tübingen. I have therefore written again to Hoppe.'[4]

When a reply from Hoppe-Seyler finally arrived in February of 1870, Miescher was relieved to read that both his letters had been received, but perhaps not so pleased at what followed:

> *I have safely received your two friendly letters regarding your work and must apologise that I have not replied sooner. The main reason for my silence lay in two doubts– the first of these relates to your work and the accuracy of the phosphorus containing nuclear material that is supposedly insoluble in ether and alcohol, and the second relates to whether I can yet release this to the medical and chemical journals….*[5]

This was not the reply for which Miescher had been hoping. Especially when Hoppe-Seyler explained that publication would be delayed until he had repeated Miescher's work to his own satisfaction.

[3] (His 1897); p.11.

[4] Miescher, 19th Feb 1870, Letter in (His 1897); p. 9.

[5] Hoppe-Seyler to Miescher, 24th February 1870, Letter VI in (His 1897); p. 42.

One of Hoppe-Seyler's concerns was that nuclein might simply be an artefact arising from the breakdown of proteins. Through its digestive action on proteins, pepsin had been found to give rise to several different substances that could be dialysed through a semi-permeable membrane. This suggested that these products were significantly smaller than the proteins from which they had been derived and, pursuing this work further, Carl Gottshelf Lehman (1812–1863), professor of physiological chemistry at the University of Leipzig, gave them the name 'peptones.'[6] The discovery by one of his other students, Nikolai Nikolaevich Lubavin (1845–1918), that peptones were produced by treatment of the milk protein casein with acid gave Hoppe-Seyler reason to be cautious about Miescher's claim. Might Miescher's nuclein be nothing more than a type of peptone that had somehow incorporated cellular phosphorus?

Phosphorus was a subject about which Hoppe-Seyler felt particularly sensitive. When Hoppe-Seyler carried out his own isolation and analysis of nuclein, he found that its phosphorus content differed significantly from that obtained by Miescher.[7] This may have been because, when Hoppe-Seyler repeated Miescher's work, he used a slightly different purification protocol that did not involve Kühne's method of using pepsin to remove proteins.[8] Perhaps his ongoing dispute with Kühne had left him reluctant to use his rival's method, but whatever the reason, the disparity that he found in the phosphorus content between his own findings and those of Miescher set alarm bells ringing—and not without good reason.

Recent bitter experience had taught him to be particularly cautious regarding claims by students to have found novel cellular compounds. And especially so if these novel compounds were alleged to contain phosphorus. Only a few years earlier, another of Hoppe-Seyler's students, Oskar Liebreich (1838–1908), had made a similar claim that had caused lasting damage to Hoppe-Seyler's reputation. In time, it would also cast a long shadow over Miescher's work on nuclein.

---

[6] (Fruton 2002); p. 131.

[7] Ibid.

[8] (Hoppe-Seyler 1871); p. 488.

4.1 Oskar Liebreich's 'Discovery' of Protagon: Once Bitten, Twice Shy

By the time that Liebreich had begun his work, several different lipids such as lecithin, all of which contained phosphorus, had been found in the brain. But, in an address to the 39th meeting of the Association of German Naturalists held in Giessen, Liebreich claimed to have discovered that all these compounds were derived from the breakdown of a single phosphorus-containing substance that he christened 'protagon.'[9]

According to Liebreich, protagon was the main lipid component of brain tissue, and as his supervisor, Hoppe-Seyler must have felt an initial glow of pride to see that his student's discovery was covered by the prestigious British Medical Journal in a report on the meeting in Giessen.[10] Convinced that protagon was the substance by which plants incorporated the phosphorus that was essential for their growth, Hoppe-Seyler declared it to be 'a substance that is, in so many respects, of physiological importance.'[11] Hoping to show that it was of similar importance in animal tissues he began trying to detect its presence in red blood cells.

But his confidence was to be short-lived. When J.L. Parke, another of Hoppe-Seyler's students tried to determine the amount of protagon in egg yolk he found the amount of phosphorus to be so high that it could not possibly be accounted for by the presence of protagon alone. This led both him and Hoppe-Seyler to suggest that there must be a second phosphorus-rich substance present.[12,13] Things got worse when yet another of Hoppe-Seyler's students, Konstantin Diakonow (1839–1868), found that the treatment of lecithin with barium hydroxide, gave the very same breakdown products as Liebreich's protagon.[14]

Perhaps out of respect for his supervisor, Diakonow took a diplomatic tone at the start of his paper, saying that despite this result 'it would be hasty to conclude that lecithin is nothing more than protagon contaminated with fats.'[15] But only a few pages further on, having described a detailed chemical analysis, he balanced tact with blunt honesty:

---

[9] (Liebreich 1865).

[10] 'Thirty-Ninth Meeting of the Association of German Naturalists', British Medical Journal, 22nd October 1864.

[11] (Hoppe-Seyler 1866b); p. 142.

[12] (Parke 1866).

[13] (Hoppe-Seyler 1866a); p. 215.

[14] (Diakonow 1867).

[15] (Diakonow 1867); p. 223.

Fig. 4.2 Ludwig Thudichum (1829–1901) who, like Wilhelm Kühne, was another of Hoppe-Seyler's most eminent—and vociferous—rivals in the field of physiological chemistry. National Library of Medicine.

> *The qualitative determinations of phosphoric acid in alcoholic or etheric extracts from various animal organisms and membranes are not sufficient to prove the existence of Protagon*[16]

If protagon even existed at all, Diakonow said that it was most likely a phosphorus-free substance that had simply co-purified with lecithin in Liebreich's preparations, thus giving the mistaken impression that it contained phosphorus.[17]

Protagon was rapidly becoming a kind of biochemical Loch Ness monster, and Diakonow was far from alone in harbouring doubts over its existence. As Liebreich gave his presentation to the audience in Giessen, the German clinician and biochemist Johann Ludwig Wilhelm Thudichum (1829–1901) (Fig. 4.2) sat listening intently and taking the odd note for a report that he was writing on the meeting. When his article first appeared in the British

[16] (Diakonow 1867); p. 227.

[17] (Diakonow 1868); p. 98.

Medical Journal, Thudichum did not pass judgement on Liebreich's claim to have found a novel substance: he simply reported what had been presented at the meeting. But this was soon to change.

Thudichum would not only very quickly become one of the most outspoken critics of Liebreich, but also of Miescher and his claim to have discovered nuclein. And aside from making scientific claims that he found to be dubious, Liebreich and Miescher shared something in common that was guaranteed to incur Thudichum's hostility: they had both been students of Hoppe-Seyler.

4.2 Ludwig Thudichum, 'The First British Biochemist'

Perhaps some of Thudichum's belligerence was born out of an urgent sense that he had to make a name for himself. As a young man with a politically liberal outlook, he had supported the revolution of 1848 that challenged the ruling autocratic governments in the Confederation of German States. But his progressive politics had come at a cost and in the aftermath of the failed uprising he found that his career options were limited. After being rejected for a post at Giessen University, he emigrated to London in the hope of starting afresh.

As an outsider trying to find his place within the British medical establishment, Thudichum fought to earn recognition and his ferocious ethic of 'work, work, and again, work' paid off.[18] By the time that he had travelled to Giessen to hear Liebreich's paper on protagon, he had been made a member of the Royal College of Physicians and in the following year was appointed director of the newly founded chemical pathology laboratory at St. Thomas' Hospital.[19] Here his work made such a favourable impression on Sir John Simon (1816–1904), the Principal Medical Officer to the Privy Council,[20] that Thudichum was awarded generous government grants and publications of his work were included as appendices to Simon's annual reports.

Thudichum's biographer David Drabkin called him 'the first British biochemist.'[21] His work was characterised by stressing the importance of precise chemical analysis of the composition of his subjects—whether these were

[18] Ibid.; p. 68.

[19] (Breathnach 2001); p. 286/284.

[20] A body composed mainly of senior politicians from the House of Commons and the House of Lords who act as advisors to the monarch of the United Kingdom.

[21] (Drabkin 1958); p. 32.

urine, bile, pigments or blood. But his research interests were by no means confined to the lab bench. As a lover of wine, he mixed business with pleasure through a novel approach to studying the metabolism of alcohol which, although it may have left his students delighted, would be unlikely to meet with the approval of a research ethics committee today:

> *We drank, from 2 o'clock in the afternoon till 7 in the evening 44 bottles of wine, consisting of white and red Hungarian, Burgundy and Sauterne. The alcoholic contents of total was an aggregate of 4000 grammes of absolute alcohol. All the urine passed from 2 o'clock until 6 next morning was collected and distilled—only 10 grammes of alcohol were collected. The rest was burned away in the system.*[22]

Outside of science, Thudichum was reputed to be skilled in fencing although he was said to have little love for the sport. He was not, however, averse to fighting duels over scientific disagreements. And the adversary with whom he crossed swords most often, was Felix Hoppe-Seyler.

4.3 Thudichum and Hoppe-Seyler Cross Swords

Their dispute began as a blood feud. Or more accurately, a disagreement over the nature of haemoglobin, the iron-containing protein that transports oxygen around the blood, and upon which Hoppe-Seyler had built his early career. In 1862, Hoppe-Seyler had described the spectrum of oxygenated haemoglobin only to feel rather put out when George Gabriel Stokes (1819–1903), Lucasian Professor of Mathematics at the University of Cambridge, took this work further by showing how both the binding and release of oxygen by haemoglobin cause changes in its spectrum that account for the difference in colour between venous and arterial blood.

Feeling that he had a territorial claim on haemoglobin, Hoppe-Seyler was not best pleased to find that Chief Medical Officer Sir John Simon's 1867 report to the Privy Council contained an appendix in which Thudichum described his own studies of what he called haematocrystalline, but what would today be recognised as haematin—the chemical group within haemoglobin that contains iron.

Despite Hoppe-Seyler having already published research into the chemical composition of haemoglobin, Thudichum considered this to be a question that was far from settled. In his own paper, he drew attention to the fact that

[22] (Drabkin 1958); p. 63, 64.

five other researchers, Hoppe-Seyler included, had all published significantly different values for the iron content of haemoglobin.[23] Of these five studies, it was Hoppe-Seyler's work that he singled out for comment—and what he wrote was hardly likely to foster a cordial relationship.

By making a point of informing his readers that Hoppe-Seyler 'has just retracted his statements regarding hematin' Thudichum may have believed he was simply promoting an open and honest scientific discussion.[24] If so, he was naïve in the extreme. Because Hoppe-Seyler saw things very differently. With his work on haemoglobin, Thudichum had encroached on intellectual territory that Hoppe-Seyler considered to be his own. But now, with the remark about his retraction, Thudichum had added insult to injury by attempting to humiliate him and undermine his scientific authority.

Thudichum had fired the opening salvo in a bitter feud that would go on to cast a shadow over Miescher's own work. And, skilled as he was in fencing, Thudichum must have known that when an opponent as eminent as Hoppe-Seyler came under attack, a riposte would surely follow. The following year Hoppe-Seyler delivered his counterattack and, writing in a major journal, he did not pull his punches:

> *Thudichum's results from the bile dyes, the blood pigment and its decomposition products, dyes which he has obtained from the urine, show so much deviation from the previous investigations and are so often obviously wrong that it becomes impossible to reproduce his work in summary here.*[25]

Thudichum was quick to fire back. Writing to the editor of the journal, Rudolf Virchow, Thudichum threw Hoppe-Seyler's own phrase 'obviously wrong' straight back at him. It was indeed true, admitted Thudichum in his letter that many of his results were original and some disagreed with what had previously been accepted. 'But this very fact,' he wrote, 'ought to make the presentation of these new results by one of your colleagues [Hoppe-Seyler] an act of duty, even if they are not to their liking. But for reasons that are not clear to me, Mr. Seyler has neglected this duty, has chosen to conceal what he could, and should have reported, and to bring into suspicion an unspecified section without offering the slightest evidence that it contains error.'[26]

---

[23] (Thudichum 1867); p. 226.

[24] Ibid.

[25] (Hoppe-Seyler 1868); p. 85. https://www.digitale-sammlungen.de/en/view/bsb10480499?page=94,95&q=thudichum

[26] Thudichum to Virchow, 1st Aug 1869, in (Drabkin 1958); pp. 249–251; p. 250.

There was a lot at stake here for Thudichum. Having left his native Germany to make a new start in London, he had fought tooth and nail to make his way in the British medical establishment. He was not about to have this struggle placed in jeopardy by Hoppe-Seyler's aspersions:

> *Although some recognition of my many years of effort by my German compatriots brings me some certain satisfaction and encouragement, and I would have seen it as some recompense for the numerous and great sacrifices which I have made through love of my research, I would have been less disturbed by the treatment contained in your report had I not held a public position, alongside my personal one here in London and England, which is essentially characterized and conditional upon my occupation in the field of medicinal chemistry.*[27]

Fearing perhaps that Hoppe-Seyler's dismissal of his work might cause him to lose the respect of his British colleagues and put his professional position in danger, Thudichum felt that he could not 'allow the suspicions contained in your report of my investigations to go in silence.'[28]

He demanded that an independent researcher be allowed to repeat his work and to publish their results in a future edition of Virchow's Archive journal. Furthermore, he proposed that the costs for this be paid for by himself and Virchow, that he himself would supply the materials, and that Professor Justus von Liebig—one of Germany's most eminent scientists at the time—be assigned as an adjudicator.[29]

Virchow replied within only a matter of days but not with the news that Thudichum was hoping to hear.[30] Although he conceded that 'I would have thought it more appropriate if Mr Hoppe had referred to the main points of your work', he was unwilling to follow up the proposals that Thudichum had suggested in his letter.

After being told by Virchow about the situation, Hoppe-Seyler wrote to Thudichum saying that although the phrase he had used still expressed his opinion, he would have had it corrected had he known it would cause such offence. But at the same time, he made clear to Thudichum that 'the discussion in your report of the blood pigment and its decomposition products, something with which I believe I am well acquainted through many years of

[27] Ibid.

[28] Ibid.

[29] (Drabkin 1958); p. 250.

[30] Virchow to Thudichum, 6th Aug 1869, in (Drabkin 1958); pp. 251–252.

almost uninterrupted work, contains some things for which I would have reason to feel offended by.'[31]

He dismissed Thudichum's proposal of having the work carried out by an independent researcher on the grounds that 'I can never trust the work of others as I do my own' and gave Thudichum a warning:

> *the judgment of later researchers will decide which of us is right, but if, therefore, you continue to treat my work in this way as I assume you will, I will in the most extreme case, oppose you in public as I had to do against Lehmann and more recently against Pflüger.*[32]

He did however offer to write a new report on Thudichum's work in the next edition of the Annual Report and concluded his letter in a conciliatory tone with the reassurance that it was his 'sincere wish to make amends as far as possible for an injustice done to you by a possibly one-sided and overly critical judgment.'[33]

Five days later, Thudichum replied in the same spirit to thank Hoppe-Seyler for his offer of writing another report and expressing gratitude that he had 'been kind enough to address it in a way that will enable you and me to immediately eliminate the difference between us.' Extending the olive branch, he went on to write:

> *I have always regarded your work with much admiration, and if I have somehow disregarded your achievements, this can only be due to the fact that my sources were inadequate.*[34]

Thudichum even offered to send Hoppe-Seyler some samples that he had prepared but despite this apparently new-found spirit of reconciliation, there were other forces driving Thudichum. After fleeing the hostile political climate in his native Germany, and then having to make his own way in London, he clearly felt like an eternal outsider seeking recognition:

> *And as everything becomes clear over the course of time with energy and sacrifice on my part, your German compatriots will be convinced that it is your duty to no longer neglect my work.*[35]

---

[31] Hoppe-Seyler to Thudichum 7th Aug 1869, in (Drabkin 1958); pp. 255–256; p. 255.

[32] Ibid. p. 255.

[33] Ibid.

[34] Thudichum to Hoppe-Seyler 12th Aug 1869, in (Drabkin 1958); pp. 256–257; p. 256.

[35] Ibid. p. 257.

A few months later, Thudichum wrote to Virchow, and was pleased to tell him 'that Mr. Hoppe has given me explanations with which I have declared myself satisfied and which I hope will lead to a lasting good understanding.'[36]

But this hope was to prove sorely misplaced. The feuding between Hoppe-Seyler and Thudichum was by no means confined to their dispute over haemoglobin. Thanks to his study of the cerebral symptoms of infectious diseases, Thudichum had become fascinated by the chemistry of the brain, which he described as 'the most diversified chemical laboratory of the animal body.'[37] A striking example of such chemical diversity was the variety of phosphorus-containing lipid compounds such as Liebreich's protagon that had been isolated from brain tissue. But Thudichum also knew that the discovery of these substances needed to be greeted with a degree of caution.

In 1864, Thudichum had been present in the audience at Giessen where Liebreich announced his discovery of protagon. As he listened with interest and jotted a few notes, a question began to form in his mind—was it not possible that some of these apparently novel phosphorus-containing lipids were nothing more than mere artefacts? Three years later, Thudichum resolved to address this question once and for all:

> *Even the existence of cerebric acid which had been discovered by Fremy had been disputed by Liebreich; and the researches [sic] of Liebreich had been controverted just a few months before by Köhler who had spent no less a time than three years over his subject. It was therefore necessary for me to institute for my own guidance some preliminary experiments on the question of the constitution of the brain matter. These researches [sic] are far from complete, but every step of them has yielded a new and reliable result.*[38]

He soon believed he had found an answer. In one of his appendices to the Report of the Chief Medical Officer of the Privy Council, he made his scepticism plain: 'Lately, Liebreich, having examined the same substance gave it the name protagon, but without any valid reason.'[39] Moreover, his own chemical analyses of this material led him to a blunt and pretty damning conclusion: 'between Liebreich's protagon and Fremy's cerebric acid [another lipid containing phosphorus] the difference is more in name than in substance.'[40]

---

[36] Thudichum to Virchow 19th Oct 1869, in (Drabkin 1958); pp. 252–254; p. 254.

[37] (Thudichum 1884); p. 27.

[38] (Thudichum 1867); p. 184.

[39] (Thudichum 1867); p. 175.

[40] Ibid., p. 184.

Thanks to his meteoric rise through the British medical establishment, when Thudichum spoke, his peers listened. So too did his rivals. With the existence of protagon, and by implication the credibility of his own research group, called into question, it was little wonder that Hoppe-Seyler insisted on first repeating and confirming Miescher's isolation of nuclein to his own satisfaction before going ahead with publication. The controversy that had erupted over protagon was not one that he wished to repeat with nuclein. Nor could his reputation afford such a repeat.

Miescher took the news that the publication of his work would be delayed, with good grace. But he also had some concerns. Confident that he had made an important discovery and fearing that any delay in publication might cost him his priority in claiming it, he asked Hoppe-Seyler to make clear that the work had been done in 1869.[41] He was also very accommodating about Hoppe-Seyler's reservations. 'Your doubts regarding the phosphorus-containing nuclear material were not unexpected,' he wrote to his former mentor, before adding that 'I am in suspense to learn the results of your investigations—whether heard from your mouth, or in print in the journal.'[42]

But he would have to remain in suspense for a while longer. While the ongoing feud with Thudichum caused Hoppe-Seyler to delay the publication of Miescher's work, a conflict of a very different nature was about to intrude upon all their lives.

On 19th July 1870, the French emperor Napoleon III declared war on Prussia. He was convinced that Otto von Bismarck, the Prussian statesman and chancellor of the North German Confederation, was deliberately trying to weaken the status of France, and was equally confident that technical innovations on the battlefield would guarantee a swift and decisive victory. But when Prussian troops lay siege to Paris in September of that year, this confidence proved to have been catastrophically misplaced.

With war having broken out between two major European powers, time, energy and resources were, understandably diverted away from basic scientific research. Keen to submit his habilitation (a kind of post-doctoral thesis) at the University of Basel, in the hope of being appointed to a professorship there, Miescher was eager to hear from Hoppe-Seyler about the progress of his publication. Moreover, he had a nagging fear that if publication was delayed for much longer, other scientists might well publish something on nuclein before him.

---

[41] Miescher 1870, letter VII, in (His 1897); p. 45.

[42] Ibid.

Writing to Hoppe-Seyler in August 1870, Miescher tactfully wondered whether the further delay was due to the war:

Being unable to find a favourably inclined ear, trusting in your consistent friendship, I knock once more upon your door. I am currently being asked so often and urgently about the fate of my work in Tübingen that whether I like it or not, I must make enquiries about the health of my first-born offspring. Since February, I have been somewhat unclear as to how things stand. At that time and also until the summer, there was no urgency. Now however, I wish to ensure the presentation of this publication for my intended post-doctoral qualification before the New Year—even if not in the form of a doctoral paper, from which essential benefit can be made. In so far as this, I would wish for it to appear in your journal very soon.

I can well imagine that the reason for the delay is that, following the withdrawal of material in July, the outbreak of the war suddenly threw everything in the air that did not belong to journals.

It is therefore important for me to know definitely, and particularly whether the matter is at the current moment in press, or whether it has been further put to one side for an indefinite period. In the first case, I would like to repeat my request to send the revision. If it has not yet gone to press, I would like in the coming time to give a lecture on it to the local natural science society in order to sufficiently comply with the requirements of the local conditions.[43]

In his letter, Miescher went on to say that, if Hoppe-Seyler envisaged any further delay, then it might be best if he simply returned the manuscript so that Miescher could take it elsewhere. If this was a ploy by Miescher to spur Hoppe-Seyler into action, it seems to have worked. When Miescher finally received a reply on sixth October—the delay no doubt caused by the war—Hoppe-Seyler offered him some reassurance:

I hope that my work will help to put the importance of your discovery in the correct light. Some other 12 pieces of work are lying equally ready for the journal. Because of the connection between the aforementioned work, I beg you not to withdraw yours. I will attach an apology as an editorial note to your work for the late publication.[44]

[43] Miescher, August 1870, Letter IX in (His 1897); pp. 48–49.

[44] Hoppe-Seyler to Miescher, 6th Oct 1870, Letter X in (His 1897); pp. 49–50.

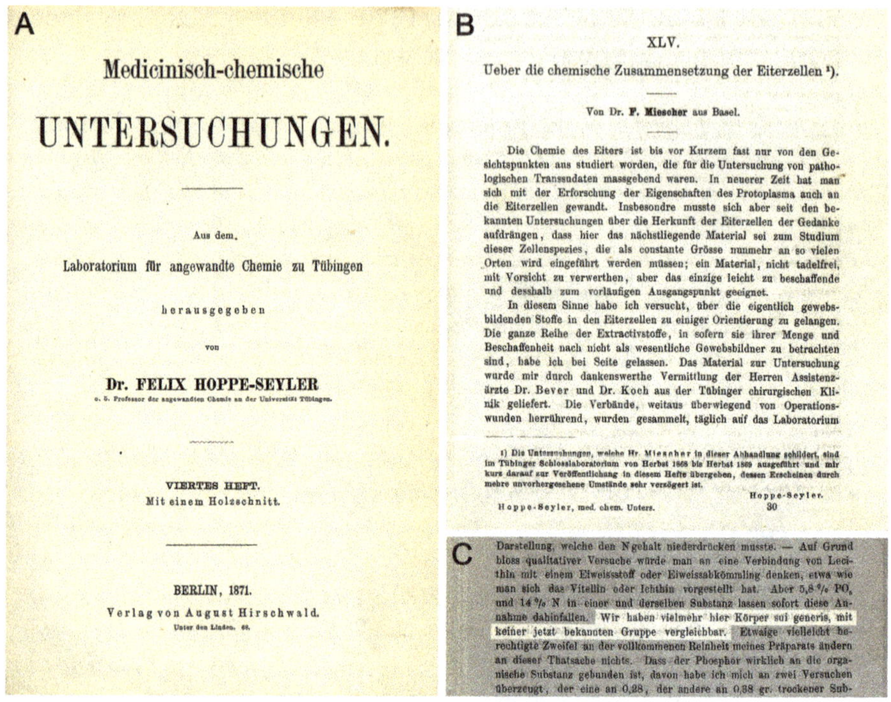
A

Medicinisch-chemische

UNTERSUCHUNGEN.

Aus dem.

Laboratorium für angewandte Chemie zu Tübingen

herausgegeben

von

Dr. FELIX HOPPE-SEYLER

o. ö. Professor der angewandten Chemie an der Universität Tübingen.

VIERTES HEFT.

Mit einem Holzschnitt.

BERLIN, 1871.

Verlag von August Hirschwald.

Unter den Linden. 68.

B

XLV.

Ueber die chemische Zusammensetzung der Eiterzellen [1]).

Von Dr. F. Miescher aus Basel.

Die Chemie des Eiters ist bis vor Kurzem fast nur von den Gesichtspunkten aus studiert worden, die für die Untersuchung von pathologischen Transsudaten massgebend waren. In neuerer Zeit hat man sich mit der Erforschung der Eigenschaften des Protoplasma auch an die Eiterzellen gewandt. Insbesondre musste sich aber seit den bekannten Untersuchungen über die Herkunft der Eiterzellen der Gedanke aufdrängen, dass hier das nächstliegende Material sei zum Studium dieser Zellenspezies, die als constante Grösse nunmehr an so vielen Orten wird eingeführt werden müssen; ein Material, nicht tadelfrei, mit Vorsicht zu verwerthen, aber das einzige leicht zu beschaffende und desshalb zum vorläufigen Ausgangspunkt geeignet.

In diesem Sinne habe ich versucht, über die eigentlich gewebsbildenden Stoffe in den Eiterzellen zu einiger Orientierung zu gelangen. Die ganze Reihe der Extractivstoffe, in sofern sie ihrer Menge und Beschaffenheit nach nicht als wesentliche Gewebsbildner zu betrachten sind, habe ich bei Seite gelassen. Das Material zur Untersuchung wurde mir durch dankenswerthe Vermittlung der Herren Assistenzärzte Dr. Bever und Dr. Koch aus der Tübinger chirurgischen Klinik geliefert. Die Verbände, weitaus überwiegend von Operationswunden herrührend, wurden gesammelt, täglich auf das Laboratorium

1) Die Untersuchungen, welche Hr. Miescher in dieser Abhandlung schildert, sind im Tübinger Schlosslaboratorium von Herbst 1868 bis Herbst 1869 ausgeführt und mir kurz darauf zur Veröffentlichung in diesem Hefte übergeben, dessen Erscheinen durch mehre unvorhergesehene Umstände sehr verzögert ist.

Hoppe-Seyler.

Hoppe-Seyler, med. chem. Unters. 30

C

Darstellung, welche den Ngehalt niederdrücken musste. — Auf Grund bloss qualitativer Versuche würde man an eine Verbindung von Lecithin mit einem Eiweissstoff oder Eiweissabkömmling denken, etwa wie man sich das Vitellin oder Ichthin vorgestellt hat. Aber 5,8 % PO_5 und 14 % N in einer und derselben Substanz lassen sofort diese Annahme dahinfallen. Wir haben vielmehr hier Körper sui generis, mit keiner jetzt bekannten Gruppe vergleichbar. Etwaige vielleicht berechtigte Zweifel an der vollkommenen Reinheit meines Präparats ändern an dieser Thatsache nichts. Dass der Phosphor wirklich an die organische Substanz gebunden ist, davon habe ich mich an zwei Versuchen überzeugt, der eine an 0,28, der andere an 0,38 gr. trockener Sub-

Fig. 4.3 Montage of Miescher's original 1871 paper in which he first described the discovery of nuclein (DNA). (Montage compiled and photographed by R. Dahm)

About a week later, Hoppe-Seyler wrote once more, reassuring Miescher that the publication was in the final stages, and that the war all seemed rather distant:

> *The war will once again have drawn a little close to you, hopefully without causing significant disturbance to the neighbouring region. Here we are in the deepest piece and the deepest holidays for perhaps a long time. A Russian in my laboratory worked mostly throughout the entire holiday. Only the newspapers and the wounded, whom we have to nurse, remind us of the war.*[45]

When Paris fell to the besieging Prussian forces in January 1871, an armistice was signed and in May that year, the war was formally ended with the signing of the Treaty of Frankfurt. But for Miescher, the dawning of 1871 brought another good reason to feel a sense of relief.

After two long years of waiting, his paper on nuclein was finally published (Fig. 4.3). The title, 'Ueber die chemische Zusammensetzung der Eiterzellen'

[45] Hoppe-Seyler to Miescher, 14th Oct 1870, Letter XI in (His 1897); pp. 50–51.

(On the chemical composition of pus cells) was unlikely to set the scientific world on fire, but today it stands as a landmark in the story of DNA. Alongside it was a paper by Pal Plosz (1844–1902), another of Hoppe-Seyler's students, who, building on Miescher's work, had examined red blood cells from different species for the presence of nuclein. In those with nuclei, such as birds and snakes, Plosz found nuclein to be abundant, but in cells lacking nuclei such as red blood cells from cows, nuclein was absent.

Accompanying the papers by Miescher and Plosz was also one by Hoppe-Seyler in which he described his own attempts to replicate and verify Miescher's findings.

Having originally questioned the values Miescher had obtained, Hoppe-Seyler now sang the praises of his student:

> *The analyses by Mr. F. Miescher presented here have not only enhanced our understanding of the composition of pus more than has been achieved in the past decades; for the first time they have also allowed insights into the chemical constitution of simple cells and above all their nuclei. Although I am well acquainted with Dr. Miescher's conscientious proceeding, I could not suppress some doubts about the accuracy of the results, which are of such great importance; I have therefore repeated parts of his experiments, mainly the ones concerning the nuclear substance, which he has termed nuclein; I can only confirm that I have to fully confirm all of Miescher's statements that I have verified.*[46]

When he read Hoppe-Seyler's words, Miescher must have breathed a sigh of relief and felt a warm glow of pride that his discovery of nuclein had finally been confirmed by his very own mentor. For Oskar Liebreich and protagon, however, things did not look so rosy. With his judgement that 'Liebreich's Protagon has been called into question, especially by Diakonov', Hoppe-Seyler had hammered yet another nail into the coffin of a discovery made by one of his very own students.[47]

Not that Miescher's work was entirely free from error either. Along with the paper on the isolation of nuclein from white blood cells, Hoppe-Seyler's collection contained a second paper by Miescher in which he described how the granular bodies found in the yolk of hens' eggs contained a substance that appeared to be nuclein.

Miescher had been encouraged to study these granular bodies by his uncle, Wilhelm His who, along with other histologists and embryologists had long been interested by them. His suspected that egg yolk might well play a far

[46] (Hoppe-Seyler 1871); p. 486.
[47] Ibid.; p. 487.

Fig. 4.4 Karl Martin Leonhard Albrecht Kossel (1853–1927). (https://wellcomecollection.org/works/vpu74v8k). (Reproduced with kind permission of the Wellcome Collection)

greater role in the development of the embryo than merely providing nutrition and that these structures might be nuclei. Struck by the high phosphorus content found in the yolk of hens' eggs, Miescher was confident that detecting the presence of nuclein would confirm his uncle's theory.[48]

His confidence appeared to be well founded when, using the same extraction procedure with which he had isolated nuclein from leukocytes in pus, he found a phosphorus-rich substance in the yolk. Admittedly its phosphorus content differed somewhat from the nuclein found in leukocytes from pus, but he nevertheless concluded this to be evidence of 'an entire group of nucleins that will certainly increase with other members.'[49]

Alas, on this occasion however, Miescher was wrong. Perhaps his enthusiasm for nuclein was skewing his judgement, leading him to infer its presence wherever he looked. It fell to another of Hoppe-Seyler's students—Albrecht Kossel (1853–1927) (Fig. 4.4) to show that, in this instance at least, Miescher had been mistaken and that the phosphorus-rich substance he had found in the granular bodies of egg yolk was not in fact, nuclein, but the phospholipid

[48] Miescher to Hoppe-Seyler, 31st Oct 1870, Letter XIII in (His 1897); p. 53.

[49] (Miescher 1871); p. 507.

lecithin. It was an auspicious start to Kossel's career for in time he too would go on to play a crucial role in the story of nucleic acids.

Along with his papers on the isolation of nuclein from pus and the yolk of hen's eggs, Miescher submitted a third manuscript for inclusion in Hoppe-Seyler's publication. In it he offered what he felt to be some important reflections about his work on nuclein and physiological chemistry in general. Hoppe-Seyler, however, felt differently and chose not to include this paper on the grounds that it did not contain any new experimental results—a decision which Miescher accepted with good grace:

> *I understand fully that you did not include my subsequent remarks; the further details therein are a definition of my position in opposition to the histologists; I still hold this to be necessary, but perhaps however they do not belong in your chemical journal; Following your advice, I will save these matters for a later opportunity.*[50]

But while Miescher did not present any new experimental data in this manuscript, he certainly made clear in it his conviction that nuclein played some crucial role in biology. Excited by Hoppe-Seyler's identification of a substance in yeast cells that strongly resembled nuclein, Miescher proclaimed the discovery of 'a new factor in the life of both the lowest as well as the highest organisms has been obtained, and one which raises a whole range of new questions for all of physiology.'[51]

He also used this manuscript as an opportunity to make a bold claim for the superiority of physiological chemistry over microscopy. According to Miescher, microscopy was, by its very nature, merely empirical and descriptive. It could reveal nothing more about structures within the cell other than their appearance—and this itself was prone to misinterpretation due to artefacts from staining. Physiological chemistry on the other hand promised much more. It had the power to penetrate beyond the veneer of stained organelles and reveal their molecular nature. Using this approach, Miescher looked forward to a time when '…the understanding of the nucleus will be freed from mere appearances…'.[52]

Hoppe-Seyler reassured Miescher that a later opportunity would arise for him to express these thoughts in a publication, but he would have to wait in vain. In the end, his paper entitled 'Subsequent Remarks' was not published

[50] Miescher to Hoppe-Seyler, 1870, Letter XV, in (His 1897); p. 55.

[51] (Miescher, Nachträgliche Bemerkungen) in (His 1897); p. 33.

[52] (Miescher, Nachträgliche Bemerkungen) In (His 1897); p. 34.

until after his death when it appeared in a volume of his works collected and edited by his uncle Wilhelm His.

At around this time, Miescher had also completed another important piece of work. This was his habilitation—the highest formal academic qualification in universities of German-speaking countries—and an essential step in climbing the rungs of the career ladder. Obtained whilst working as a post-doctoral researcher, the habilitation confirms that the holder has demonstrated independent research, and in Miescher's time was a prerequisite for a professorial position. Nor would Miescher have to wait very long at all before this most coveted accolade was bestowed upon him.

4.4 Homecoming

Having submitted his habilitation on the exchange of gases, Miescher now returned to his home city of Basel where he took up the post of Privatdozent[53] and began working with his Uncle Wilhelm. As an embryologist, Wilhelm His was fascinated by the question of development—that is, how does a single fertilised egg cell divide and differentiate to give rise to all the different specialised cell types that make up a complete multicellular organism. When Wilhelm His left Basel to continue his research in this area at Leipzig, Miescher—aged only 28 years old—was given the professorial post vacated by his Uncle (Fig. 4.5).

Gaining a professorial chair at such a young age was a meteoric ascent, but the return to Basel had brought something even more valuable. In one of his letters to Hoppe-Seyler about his studies of hen egg yolk, Miescher also mentioned that he had just purchased three pounds of fish eggs for nine Kreuzer.[54] This was not an insight into Miescher's culinary tastes, but rather a development that was to be of great significance for his scientific work—all thanks to the bulging sex organs of the Rhine salmon. These were to occupy Miescher for years to come, but although they would prove a great source of nuclein for his ongoing research, in time they would prove to be as much of a curse as a blessing.

[53] An academic post at universities in German-speaking countries that often precedes a professorship.

[54] A unit of currency in Austria, Switzerland and the southern German states prior to the introduction of the German goldmark in 1871.

Fig. 4.5 Photograph of Miescher's Uncle, Wilhelm His taken by Carl August Teich circa 1882, by which time he had vacated his post as a Professor in Basel making way for his nephew Friedrich Miescher to take up this position in his place. The impressive attire worn by His in this photograph conveys the eminence that he had achieved by this time—as also does the bust of him photographed in 1950 and shown in Fig. 5.1b. Images reproduced with kind permission of the University of Basel Library

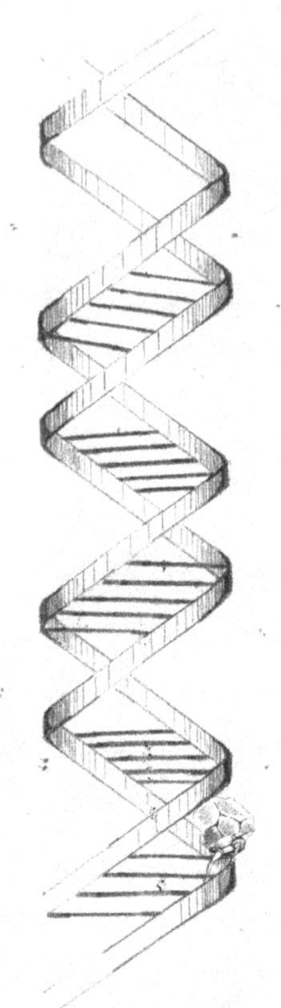

Drawing by Kersten Hall

References

Breathnach, C. S. 2001. Johann Ludwig Wilhelm Thudichum 1829-1901, bane of the Protagonisers. *History of Psychiatry* 7:283–296.

Diakonow, C. 1867. Ueber Die Phosphorhaltigen Körper Der Hühner- Und Störeier. *Medicinisch-Chemische Untersuchungen Aus Dem Laboratorium Für Angewandte Chemie Zu Tübingen* 2:221–227.

Diakonow, C. 1868. Das lecithin Im Gehirn. *Centralblatt Für Medzinische Wissenschaft* 7:97–99.

Drabkin, D. 1958. *Thudichum: Chemist of the brain*. Philadelphia, PA: University of Pennsylvania Press.

Fruton, J. 2002. A history of pepsin and related enzymes. *The Quarterly Review of Biology* 77:127–147.

His, W. 1897. *Die Histochemischen Und Physiologischen Arbeiten von Friedrich Miescher*. Leipzig: F. C. W. Vogel.

Hoppe-Seyler, F. 1866a *Medicinisch-Chemische Untersuchungen: Aus Dem Laboratorium Für Angewandte Chemie Zu Tübingen.*

Hoppe-Seyler, F. 1866b. Ueber Das Vitellin, Ichthin Und Ihre Beziehung Zu Den Eiweissstoffen. *Medicinisch-Chemische Untersuchungen Aus Dem Laboratorium Für Angewandte Chemie Zu Tübingen* 1:215–220.

Hoppe-Seyler, F. 1868. Physiologische Chemie. *Jahresbericht über die Leistungen und Fortschritte in der gesamten Medizin* 1:67–101.

Hoppe-Seyler, F. 1871. Ueber die Chemische Zusammensetzung des Eiters' [on the Chemical composition of pus]. *Medicinisch-Chemische Untersuchungen* 4:486–501.

Miescher, F. 1871. Ueber Die Chemische Zusammensetzung Der Eiterzellen. *Medicinisch-Chemische Untersuchungen Aus Dem Laboratorium Für Angewandte Chemie Zu Tübingen* 4:441–461.

Parke, J. L. 1866. Ueber Die Chemische Constitution Des Eidotters. *Medicinisch-Chemische Untersuchungen Aus Dem Laboratorium Für Angewandte Chemie Zu Tubingen* 1:209–214.

Thudichum, J. L. W. 1867. *Dr. Thudichum on researches intended to promote an improved treatment of disease*, 152–294. Report of Chief Medical Officer of Privy Council Appendix 7.

Thudichum, J. L. W. 1884. *Physiological chemistry of the brain*. London: Bailliere, Tindall and Cox.

5

The Sperm Campaign

Contents

Abstract Rising early in the morning during winter, Miescher makes trips down to the banks of the Rhine river in order to catch the salmon which, in the months that they spend there, show a disproportionate increase in the mass of their sex organs in preparation for breeding. Embarking on what he later called his 'sperm campaign', Miescher finds that the sperm of the fish is a particularly rich source of nuclein (DNA) and is able to refine his studies of this substance. Through his work with the Rhine salmon he becomes ever more interested in the process of fertilisation as well as how the fish are able to break down muscle tissue and redirect these nutrients towards the growth of their sex organs—a metabolic feat which allows them to remain reproductively active for months in their breeding ground without feeding. All of this is carried out in working conditions that were far from state-of-the-art: Miescher complains of having to work in a corridor, and on one occasion when short of laboratory glassware resorts to using Sevre porcelain from his own home. But while Miescher remains so immersed in his work that he even

K. Hall, R. Dahm, *The Dawn Fisherman*, Copernicus Books,
https://doi.org/10.1007/978-3-032-14219-1_5

has to be dragged from the lab bench to attend his own wedding, a battle is raging over nuclein (DNA).

Keywords Miescher • Nuclein • DNA • Fertilization • Development • Sperm • Egg • Cell • Nervous system • Mechanical • Stimulus • Chemical • Protamine • Salmon • Biology • Chemistry • Histology • Microscope • Morphology • Physiology • Genetics • Reproduction • Embryology • Biochemistry

Thanks to the reproductive organs of the Rhine salmon, Miescher finds a welcome alternative source of nuclein to pus-soaked bandages

On 21st March 1878, Miescher's bride Maria Anna Rüsch stood at the altar in her wedding dress nervously watching the hands of the clock and wondered why her future husband was nowhere to be seen (Fig. 5.1). Guests in the congregation were no doubt muttering under their breath and casting knowing glances at each other. Thanks to a hastily organised search party, Miescher was eventually found and brought to the church. But if Maria Anna breathed a sigh of relief, it was short-lived. Because it was now obvious that her intended husband was already married.

Not that Miescher was a bigamist. It was simply that he was wedded to his scientific work. For the search party of friends who had eventually brought Miescher to the church had been able to do so only by dragging him away from his laboratory where they had found him so immersed in his work that he had been oblivious to the fact that his own wedding service was about to begin.[1]

The warning signs that Maria Anna would face stiff competition for her husband's attention during their marriage were evident in his curious habit of rising before dawn during winter to go and catch fish from the river Rhine.

This was not done as a pleasant diversion before beginning work at the laboratory bench but rather because Miescher had become fascinated by the sexual organs of the Rhine salmon.

Starting in the North Sea, these salmon swim 500 miles upstream to their spawning grounds in the upper Rhine, by which time they are sexually mature. Throughout the 8–10 months that they spend in fresh water the fish remain highly active but neither feed nor even secrete digestive juices. Yet despite this

[1] (Suter 1944); p. 10.

Fig. 5.1 Portrait of Miescher with his wife, Maria-Anna Rüsch taken around 1870 by Jakob Höflinger. (Image reproduced with kind permission of the University of Basel Library)

lack of nutritional intake, their bodies undergo a seismic change in metabolism and anatomy, the biggest of which is in their sex organs, which undergo a huge increase in size. Miescher found, for example, that when female salmon leave the sea, their ovaries account for just 0.5% of their body weight, yet by the time they had reached Basel, this figure had increased to a startling 26%—a more than 50-fold growth.

This made salmon eggs the ideal material for the studies into the nature of fertilisation that Miescher had begun with his Uncle. But he also realised that this anatomical quirk of the salmon might be a gift to physiological chemistry. In a letter written in early 1871, his old friend the German pharmacologist Rudolf Boehm, received the honour of being the first person with whom Miescher shared his idea[2]:

[2] Miescher to Boehm, 23rd Sept 1871, Letter XXV in (His 1897); p. 64.

At the moment, I use the recurring fish season to work on sperm. It contains colossal heaps of a substance with an extremely high phosphorus content. If you want to have nucleic bodies by the hundredweight, you collect ripe salmon testes here in November… I expect to obtain from these objects of investigation, information that is of more general significance than merely for the physiology of sperm. The current view that the spermatozoid is a complete cell with a predominant nuclear mass seems to coincide very smoothly with my analyses so far. I can get the organs at any time, and at any stage of maturation. The mature organ consists almost exclusively of spermatozoa... If there is any place where one can trace the chemical process of nuclear formation with its main and secondary links, it is here….[3]

So began what Miescher would later call his 'sperm campaign.'[4] With 90% of their total cell mass accounted for by the nucleus, salmon sperm were a far more abundant source of nuclein than leukocytes.[5] And as it involved neither discarded surgical bandages nor pus, isolating nuclein from them was a far less unpleasant procedure.

But the task was nevertheless still far from straightforward. As with leukocytes, the procedure had to be carried out at low temperatures to prevent degradation of the nuclein and, without the benefit of fridges or cold rooms at the time, this meant that it could only be done during the winter months.

When nuclein is to be prepared, I go at five o'clock in the morning to the laboratory and work in an unheated room. No solution can stand for more than five minutes, no precipitate more than one hour before being placed under absolute alcohol. Often it goes on until late in the night. Only in this way do I finally get products of constant phosphorus composition.[6]

But having to work in freezing cold temperatures was perhaps the least of the discomforts that Miescher experienced on taking up his new position in Basel. Despite having been made a professor, his working conditions were hardly what might be described as 'state-of-the-art'. Miescher's uncle Wilhelm His later recalled how, having been relegated to working in a corner of the chemical laboratory, which belonged to the Professor of Chemistry and was already

[3] Miescher to Boehm, 23rd Sept 1871, Letter XXV in (His 1897); p. 63.

[4] Miescher, 2nd March 1891, Letter LXIX in (His 1897); p. 108.

[5] Miescher's discovery that salmon sperm was an abundant source of DNA has left another legacy familiar to molecular biologists although few are likely to know of its origins. To this day, researchers in molecular biology laboratories who are studying physical interactions between specific DNA-binding proteins and their target DNA sequences, still use commercial preparations of salmon sperm DNA as an agent to block non-specific binding factors that might otherwise interfere with the experiment. (With thanks to Professor Pete McHugh, Institute of Molecular Biology, Oxford, UK for reminding us of this.)

[6] Miescher, 16th January 1874, Letter XXXV in (His 1897); p. 76–77.

overflowing with students, his nephew 'longed for the fleshpots of the laboratory in Tübingen castle.'[7] Writing to Hoppe-Seyler in the summer of 1872, Miescher expressed his despair at how he was supposed to achieve anything of worth under his current working conditions:

> *How and when, by the way, this whole multifaceted investigation will be completed is a mystery to me. I am now about to take on eight hours of lecturing. But there is no mention of chemical work in the room assigned to me, neither space nor furnishings. In the chemical laboratory, in which I maintain a corner with difficulty and effort, the occasions when I am able to seize the opportunity for a proper analysis amidst the manifold obstacles - lack of space, apparatus and time - seem like a cause for celebration.*[8]

Perhaps the university authorities in Basel were wary of making life too easy for the young professor. When Miescher had first taken up the Chair of Physiology, previously occupied by both his father and uncle, there had been rumours of nepotism. It may have been to refute such allegations that Miescher worked himself to the point of exhaustion, for as one of Miescher's few students, F. Suter, recalled: 'There was no relenting…his demon impelled him to overcome all obstacles that stood in the way of scientific knowledge of the problems that occupied him.'[9] One such obstacle was a lack of sufficient glassware in the laboratory, which Miescher solved simply by using Sevre porcelain from his own home. Whether Maria Anna was impressed by her husband's display of initiative remains unknown. Even the location of their family home may well have been motivated by Miescher's passion for his work. The house at Augustinergasse 21 was not only situated in the shadow of Basel's cathedral, but more importantly it overlooked the Rhine, making it much easier for Miescher's nocturnal and dawn expeditions to go fishing for salmon.

Whilst despairing over his working conditions, Miescher nevertheless had some exciting news for Hoppe-Seyler when he wrote to him in the summer of 1872. 'It is a complete mystery to me,' he began 'how it can be that I have not written to you for almost one and a half years.'[10] Yet he had good reasons for having been too preoccupied to write. As he went on to explain, he had now isolated a second novel substance from sperm nuclei. Giving it the name 'protamine', he believed that it was bound with nuclein in a complex to form a salt. Moreover, although protamine did not contain phosphorus it was

[7] In (His 1897); p. 17

[8] Miescher to Hoppe-Seyler, 20th July 1872, Letter XXVII in (His 1897); p. 69.

[9] (Suter 1944); p. 10.

[10] Miescher, 1872, Letter XXVI in (His 1897); p. 64.

highly basic in nature which confirmed his earlier hypothesis based on his first observations made in Tübingen that nuclein must be acidic.[11,12]

Excited as he was by these findings, Miescher conceded that much remained yet to be done. 'From what I have shared with you in confidence,' he admitted, 'you will see that this story is in no way yet ready for publication.'[13]

One problem was that it was not easy to separate and purify nuclein from the complex in which it was bound with protamine.[14] A bigger headache however was that when Miescher extracted nuclein from other species such as ox, frog, carp and hen, he could find no protamine.[15,16] Added to these frustrations was the ongoing issue of his working conditions. Having successfully prepared a crystalline salt of protamine, he was eager to determine its chemical composition but as he complained to Hoppe-Seyler, 'the laboratory is only open to me at night and on Sundays due to overcrowding with students.'[17]

His analysis of the chemical composition of nuclein however was proving to be less frustrating. Building on the work he had done in Tübingen, Miescher determined the relative proportions of the chemical elements present in nuclein, from which he proposed that it had the formula $C_{29}H_{49}N_9P_3O_{22}$.[18] He also deduced that, as nuclein could not pass through a parchment filter, it must be a very large molecule indeed.[19]

As for what the function of nuclein might be, Miescher had a suspicion that it might play a role in the fertilisation of egg cells. There had been much debate over whether the egg was a cell but by the mid-nineteenth century, thanks to observations such as how the surface of fertilised eggs of frogs and toads became segmented, it was becoming clear that the embryo arose by the division of a single fertilised cell through binary fission to produce daughter cells.[20]

But what was the mechanism of fertilisation? In 1868, Wilhelm Kühne had published a rudimentary analysis of the chemical composition of sperm cells and urged researchers to look within semen for the presence of substances

[11] Ibid., p. 66.

[12] (Miescher 1874); p. 153.

[13] Miescher to Hoppe-Seyler, 1872, Letter XXVI in (His 1897); p. 67.

[14] Ibid.; (Miescher 1874); p. 159.

[15] Miescher, Basel 29th Jan 1873, Letter XXIX in (His 1897); p. 73.

[16] Miescher, 20th Sept 1873, Letter XXX in (His 1897); p. 73.

[17] Miescher to Hoppe-Seyler, Summer 1872, Letter XXVI in (His 1897); p. 67.

[18] (Miescher 1874); pp. 165–166.

[19] (Miescher 1874); p. 158.

[20] Harris (1999); p. 127.

known as 'ferments'.[21] The term 'ferments' was a broad one, referring to agents involved in well known chemical transformations such as the formation of ammonia from urinary urea, of lactic and butyric acids in the souring of milk and bread respectively, the action of digestive juices from the pancreas and stomach on proteins, and extracts from germinating barley that could turn starch into sugar. Heeding Kühne's call to identify the agent in semen responsible for fertilisation, Miescher believed his work on salmon sperm offered the ideal candidate:

> *In so far as we are to suppose that a single substance is the specific cause of fertilisation, perhaps acting as a ferment [enzyme] or in any other way, such as a chemical stimulus, then one would undoubtedly have to think above all of nuclein.*[22]

If this were indeed the case, and nuclein provided the chemical stimulus for fertilisation, did this mean that nuclein in the egg and sperm played equal roles in the process? Or was one more important than the other? In addressing this question, Miescher was unfortunately about to be misled up a blind alley, and one largely of his own making.

5.1 A Wrong Turn

This arose from Miescher's own mistaken belief that he had found nuclein in the yolk of hens' eggs. Miescher had observed a substance in egg yolk which he believed to be nuclein but which was present in vastly greater amounts than in sperm cells. This left him struggling to see how sperm cells could play any significant role in fertilisation. Unaware that what he had found in the egg yolk was not nuclein, but lecithin, (as would later be shown by the German biochemist Albrecht Kossel (1835–1927)) Miescher therefore concluded that, due to their higher nuclein content, egg cells must play the main role in fertilisation, while sperm had only a minor walk-on part.

As for the mechanism by which nuclein might cause fertilisation, Miescher recognised that he faced a daunting explanatory challenge. How was it possible that a single substance could not only trigger fertilisation, but also the growth and development of multicellular organisms with all their diverse traits? Whilst acknowledging the possibility that at least some of this biological variation might arise from differences in the chemical structure of nuclein,

[21] (Kühne 1868); p. 558.

[22] (Miescher 1874); p. 192.

Miescher failed to see how this alone could account for the enormous diversity of living systems. 'If we think of the sperm only as a carrier of a specific fertilisation substance,' wrote Miescher as he struggled with this question, 'how do we explain the changes, the variations of the effect, from species to species, from race to race, from individual to individual?... differences in chemical structure of the molecules will occur, but only in limited diversity.'[23]

His mistaken belief that he had detected nuclein in the yolk grains of hen's eggs muddied the picture even further. Especially since it seemed to be no different in its chemical composition to that from sperm. 'There is nothing,' he wrote, 'to suggest that the sperm nucleins possess any special characters compared to the egg nucleins.'[24] This, together with his failure to detect protamine in the sperm of carp, a close relative of the salmon, led him to seriously call into question his original conviction that a single chemical substance might be responsible for fertilisation[25]:

> *...the mystery of fertilization cannot reside in a particular substance; this can already be said with great probability. Not one part, but the whole as such is effective, through the interaction of all its parts.*[26]

Fertilisation then, was the result not of a single substance, but of an interaction between multiple components within the system. And of these components, it was the egg, and its internal conditions, that played a crucial role in shaping the future growth and development of the organism. Sperm cells by contrast, played only a minor part in this process, limited only to:

> *the weakening or strengthening of a few individual properties of the germ. Thus with the exception of certain modifications, which recur again and again in a similar way, the entire wealth of individual imprints would pass from the mother to the offspring as an inheritance.*[27]

But how did these two components of the system, egg and sperm, interact to result in fertilisation? In considering this question, Miescher had arrived at an answer that was not without a certain irony:

---

[23] Ibid., p. 193–194.

[24] Ibid., p. 193.

[25] Ibid., p. 195.

[26] (Miescher 1874); p. 193. https://www.biodiversitylibrary.org/item/42692#page/146/mode/1up

[27] (Miescher 1874); p. 194.

A range of reasons suggest that chemical entities alone cannot be the deciding factors [in fertilisation]... There is no single fertilisation substance.[28]

Ever since the days when he had first washed pus from old bandages in his kitchen laboratory in Tübingen castle, Miescher had been confident that the secret to understanding living systems would be revealed through their chemistry. But when it came to fertilisation at least, he seemed now to have conceded that chemistry alone was not enough to account for this phenomenon. 'The chemical facts are of secondary significance,' he declared, 'They are subordinate to a higher perspective.'[29]

As for what such a 'higher perspective' might be, Miescher believed he knew the answer:

Even clearer than all chemical analyses, of course, are the effects of the laws of procreation, which are revealed to us daily through the inheritance of paternal characteristics.[30]

5.2 A Higher Perspective: The Search for Laws

In his search for laws that lay beneath the apparent messiness of biology, Miescher had the company of someone whose name was to become far more well-known than his own. When not attending matins and evensong services, Gregor Mendel (1822–1884) (Fig. 5.2), an abbot in Brünn (now Brno), Moravia, had been busy in the garden of the St. Thomas monastery searching for law-like behaviour in biology.

Having studied at the University of Vienna, Austria, where he had studied combinatorial mathematics, probability theory, and systematic botany, he applied what he had learned to a meticulously planned series of experiments carried out over several years. These involved crossing pea plants, which exhibited discrete binary traits such as having flowers that were either white or violet, or seeds that were either green or yellow, and counting the number of plants that showed these traits in successive generations.

Thanks to his observation that these traits recurred in precise numerical ratios over successive generations, Mendel's name is today known to every student of biology from high school to undergraduate level for being the

[28] Ibid., pp. 195–196.

[29] Ibid.

[30] Ibid., p. 193.

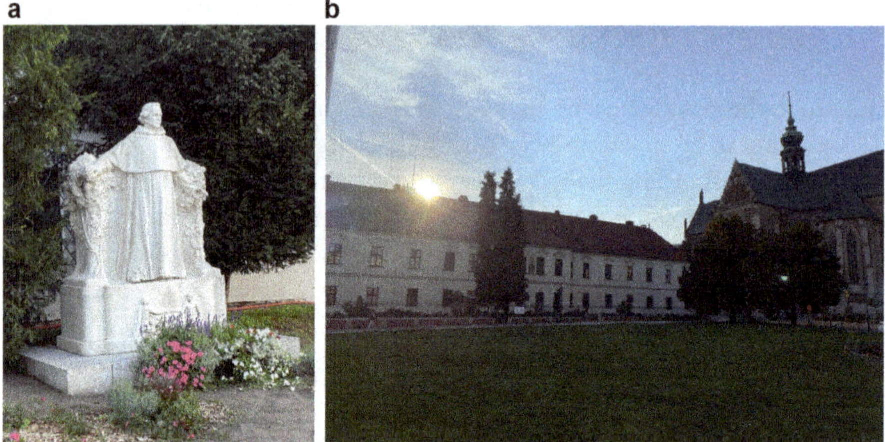

Fig. 5.2 (**a**) Statue of Gregor Mendel (1822–1884) and (**b**) the St. Thomas Abbey in Brno, Czech Republic (formerly Brünn) where he conducted his famous experiments with peas. Mendel's former home was in one of the houses shown next to the abbey. (Photographs by R. Dahm)

discoverer of genes and the laws of heredity that describe their transmission. Yet if Mendel could somehow be whisked from his mid-nineteenth century monastery garden and shown the biology textbooks of the early twenty-first century, these accolades would probably leave him feeling flattered but at the same time utterly bewildered. For the laws which are today named after him, are not quite the same as those he believed he had found. The title of Mendel's original paper offers a clue as to why this should be so. Called 'Experiments on Plant Hybrids' it describes Mendel's search for a universal law describing the formation and development of hybrids.[31] This was an important question at the time, because it promised to reveal whether species were fixed entities or whether new species emerged through the hybridisation of pre-existing ones.

The fruitless search for a law describing hybrid formation had so far left naturalists in a state of frustration but with the discovery of his numerical ratios, Mendel believed he had finally succeeded.[32,33] And his success in this regard may shed important light on something that, at first sight, seems to be a puzzle of stunning proportions.

[31] For a recent translation of Mendel's paper, accompanied by a commentary, see Gregor Mendel 'Experiments on Plant Hybrids (1866)' Translation and commentary by Staffan Müller-Wille and Kersten Hall (2020) Masaryk University Press, Brno.

[32] Ibid.; p. 23.

[33] Ibid.; p. 29; p. 53.

The textbook accounts of Mendel, which portray him as the discoverer of the gene, often add extra drama to his story by claiming that his work on heredity was so far ahead of its time that it lay neglected in the scientific wilderness for nearly four decades until its apparent rediscovery at the start of the twentieth century. But more recent scholarship suggests that, far from languishing in obscurity, Mendel's paper was being circulated and read at the time. Far from being isolated as is often claimed, the Brno naturalists' association to whom Mendel had first presented his work in 1865, had contacts with over 130 other societies in Europe and North America.[34]

If Mendel's paper was known to the scientific community, why did none of its readers recognise Mendel as having discovered the gene? The answer is most likely that Mendel's paper was understood and interpreted exactly for what it claimed to be—yet another study (albeit a highly original and innovative one) in a well-established tradition of research into hybridisation.

In 1901, the English biologist William Bateson pondered what might have happened had Charles Darwin been able to incorporate Mendel's work into his own. Darwin was struggling at the time to explain how traits which conferred a survival advantage could be passed on to successive generations and thus allow evolution by natural selection to take place. To Bateson it was clear that Mendel's paper was evidence for a particulate theory of heredity and that this would have given Darwin exactly what he had needed. Darwin had in his possession a book that contained a reference to Mendel's experiments but even had he read it, his most likely reaction, according to historian Pablo Lorenzano, would have been to conclude that Mendel was 'just another "hybridist", though perhaps a very good one on account of his experimental skills.'[35]

With his discussion of fertilisation in hybrid formation and his search for laws describing biological processes, Mendel appears to have had at least some common ground with Miescher. Whether Miescher was aware of Mendel's work is not known, but even had he been, his response too would probably have been that Mendel was 'just another "hybridist." After all, a study on crossing pea plants would hardly be likely to catch the eye of someone who was trying to pioneer a chemical approach to cells and the mystery of living systems.

But Miescher nevertheless shared Mendel's conviction that biological processes, whether concerning the formation of hybrids or the fertilisation and subsequent development of an organism, unfolded according to deterministic

[34] (Mielewczik et al. 2017).

[35] (Lorenzano 2011); p. 41.

laws. The question for Miescher to answer was—if such a law existed, how was it manifest in physical terms? He had already conceded that chemistry alone was insufficient to account for fertilisation and development. But if not chemistry, then what?

5.3 The Missing Screw

The nervous system offered an answer. Struck by the way in which electrical impulses travelling along nerves were able to excite contraction in muscle, Miescher asked:

> *Why should such fundamental properties of organized substances also be a privilege of the nervous system, which, after all, with all other organs, has sprung from the mass of furrowing balls? Nowhere than in this field do we know such great effects induced by impulses of such an undetectably small amount of living force.*[36]

Here, it seemed, was a solution to the conundrum of fertilisation. In the nervous system, 'impulses of such an undetectably small amount of living force' travelling along the neurons could result in 'great effects.' What if sperm cells, despite their apparently much lower content of nuclein, were to elicit a similar excitation in the egg? But whereas the chemical and physical alterations resulting from the stimulation of muscle by nerves were only temporary, those induced in the egg by the mechanical stimulus from sperm would initiate a 'chain of processes' resulting in the formation of a new organism. The unfertilised egg was like a ball sitting at the top of a hill laden with gravitational potential energy and needing only the impetus from the physical impact of a sperm cell to send it rolling on its way to trigger the development of a new organism.

Miescher likened the egg to 'a machine that generates or transforms a movement of some kind. A composite whole, effective not by any individual, but by its composition. The motion produced is capable of countless modifications in the finest gradations and of great variety depending on substance, shape, size, mutual position of the parts.'[37]

Elsewhere, he likened the unfertilised egg to an unwound clock that has stopped and is dependent on the mechanical stimulus from a sperm cell to get

[36] Ibid., p. 196.
[37] Ibid., p. 196.

it moving again.[38] Imagining the egg cell to be a complex mechanical assembly, which is unable to operate due to the lack of a single vital screw, Miescher suggested that the crucial role of sperm cells was to provide the missing component.

> *The spermatozoon inserts the screw again at one point and completes the active arrangement. This is all it takes. At this point, the chemical-physical pause is now disturbed, the machine starts working again, each cell supplies protamine to its neighbouring cells and so the movement spreads according to certain laws.*[39]

He also believed that he had identified what this missing screw might be. 'Could it ultimately be,' Miescher asked his old friend Rudolf Boehm, 'that nucleic acid-protamine (can you think of a better name?) or a related, and therefore physiologically equivalent, molecule could be a crucial link in the system of cell multiplication?[40]

A casual reader might be forgiven for thinking that, by 1874 when his paper on protamine and nuclein in salmon sperm was finally published, Miescher had turned his back on nuclein and, having previously championed its importance he was now casting doubt on its role. But this was not quite the case. Instead, he envisaged a model of fertilisation in which chemical agents and a mechanical stimulus might work in concert:

> *However, the two possibilities discussed about the nature of the fertilization process are by no means mutually exclusive. It could, for example, be the specific stimulus of a molecular nature, whereas the peculiarities of the locomotion movement could have an influence on the place where the stimulus acts on the germ*[41]

In fact, far from playing down the role of chemical agents, Miescher concluded his paper with a more general call-to-arms for the cell to be understood in terms of chemistry. The interpretation of the role played by sperm heads in fertilisation would ultimately be dependent on a clear understanding of the nucleus. And, as far as Miescher was concerned, the understanding of the nucleus had so far been limited by histological approaches which studied morphology and appearance under the microscope. For Miescher, this was an approach that was riddled with error:

---

[38] Miescher, 25th Aug 1872 (first Dec 1895), Letter XXXVIII in (His 1897); p. 80.

[39] Miescher to Boehm, 2nd May 1872, Letter XXVIII in (His 1897); p. 71.

[40] Ibid., p. 70.

[41] (Miescher 1874); in (His 1897); p. 99.

> *One will have to admit that the main criterion for the recognition of a body as a nucleus was now not its nature, but the place where it is found; any roundish structure in the protoplasm of a cell, which is not a drop of fat, a crystal, a chlorophyll, starch, or glycogen grain or an undoubted vacuole, is immediately identified as being nuclear in nature.*[42]

In place of arguments over what might well be artefacts observed down the microscope, Miescher argued that understanding the nucleus needed to be rooted in a firm foundation that was grounded in chemistry:

> *It is therefore time to finally detach the concept of the cell nucleus from these fluctuating external appearances and to link it to such properties which must be closely related to its general physiological function. This includes, in particular, its chemical composition.*[43]

And chemical analysis of the nucleus might well unlock one of the greatest secrets in nature:

> *Should it be possible to provide strict proof that the entry of the seed to the egg is mainly equivalent to the addition of a physiologically complete nucleus to the mass of the egg, the mystery of fertilization will miraculously merge with the general elementary problem of cell life; we would have in it an inimitable natural experiment which would allow us a deep insight into the role of the nucleus in general, and the study of the seed would have a significance far beyond the question of procreation; for here Nature offers us, in a form accessible to decomposition, one of those simple fundamental active arrangements which are able to convert elastic forces into the specific form of vital movement processes.*[44]

Confident that the sperm of the Rhine salmon had given him a key with which to unlock the mysteries of fertilisation and development, Miescher published this work in 1874. The fears and frustration that had plagued him only a few years earlier as he had sat waiting patiently for Hoppe-Seyler to repeat his experiments, all the while fearing that someone else might scoop his discovery of nuclein were now well and truly in the past. On the pages of respected journals, he was recognised as the undisputed discoverer of nuclein.

---

[42] (Miescher 1874); p. 204.

[43] Ibid., p. 205.

[44] Ibid., p.207.

This was an accolade for which he had long waited. But unfortunately, having achieved it, Miescher was soon to learn the painful truth of the proverb, 'Take care what you wish for…'.

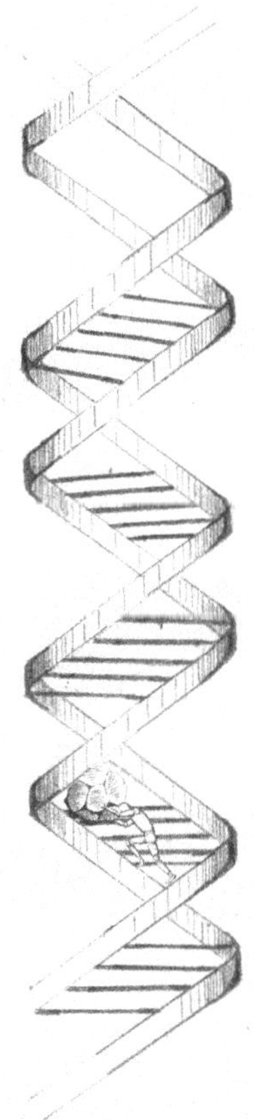

Drawing by Kersten Hall

References

Harris, H. 1999. *The birth of the cell.* New Haven, CT: Yale University Press.

His, W. 1897. *Die Histochemischen Und Physiologischen Arbeiten von Friedrich Miescher*. Leipzig: F. C. W. Vogel.

Kühne, W. 1868. *Lehrbuch Der Physiologischen Chemie.* https://archive.org/details/bub_gb_DxNu12EEKh4C

Lorenzano, P. 2011. What would have happened if Darwin had known Mendel (or Mendel's work)? *History and Philosophy of the Life Sciences* 33:3–48.

Mielewczik, M., D. P. Francis, B. Studer, M. V. Simunek, and U. Hossfelder. 2017. Die Rezeption von Gregor Mendels Hybridisierungsversuchen Im 19. Jahrhundert—Eine Bio-Bibliographische Studie. In *150 Jahre Mendelsche Regeln: Vom Erbsenzählen Zum Gen-Editieren. Nova Acta Leopoldina*, vol. 143, 83–134. Stuttgart: Wissenschaftliche Verlagsgesellschaft.

Miescher, F. 1874. Die Spermatozoen Einiger Wirbeltiere. Ein Beitrag Zur Histochemie. *Verhandlungen Der Naturforschenden Gesellschaft in Basel* 6:138–208.

Suter, F. 1944. Prof. F. Miescher. Persönlichkeit Und Lehrer. *Helvetica Physiologica et Pharmacologia Acta Supplementa* 2:5–17.

6

The Great Unwashed

Contents

Abstract While Miescher busies himself with the sex organs of the Rhine salmon and procuring domestic china ware for his laboratory, his work on DNA has been received against the backdrop of an often acrimonious dispute between Felix Hoppe-Seyler and his rivals in the field, Ludwig Thudichum and Wilhelm Kühne. Now, Miescher's discovery of DNA becomes yet another subject over which they hurl insults at each other. One British article, written by authors supporting Thudichum, dismisses nuclein (DNA) as a contaminating artefact in the purification protocol, describing it as 'the Great Unwashed.' Miescher meanwhile makes little comment, instead keeping himself immersed in his work. But this is not without its own challenges. For despite being an abundant source of DNA, the Rhine salmon would prove to be as much a curse as a blessing.

K. Hall, R. Dahm, *The Dawn Fisherman*, Copernicus Books,
https://doi.org/10.1007/978-3-032-14219-1_6

Keywords Nuclein • DNA • Friedrich Miescher • Felix Hoppe-Seyler • Ludwig Thudichum • Physiological chemistry • Protagon • Bile pigments • Bilirubin • Rhine salmon • Cell nuclei • Chemical composition • Molecular structure • Scientific controversy • Criticism • Skepticism • Nineteenth century science • Biochemistry • Genetics • Molecular biology • Historical figures • Scientific rivalry • Academic feuds • Laboratory research • Discovery

Miescher finds that his discovery of nuclein meets with a mixed – and not always friendly – reception

With a second paper now published on nuclein, Miescher might have had cause to feel that its existence would no longer be in question. If so, he was in for a shock. As an emerging field, physiological chemistry faced many challenges. Much remained unknown and what few facts had been gleaned so far often seemed disconnected and incoherent. Added to which, the new discipline was fraught with disputes. Some of these were scientific in nature, rooted in arguments over a lack of clear definitions for concepts; others were not.

Science is often perceived as being the disinterested pursuit of objective truth, but the reality is that scientists never work in a vacuum: their research always takes place within a wider social, historical, political and cultural context. All of which can be important factors in influencing the reception, direction and adoption of scientific work. For Miescher, the cultural landscape into which he introduced his discovery of nuclein was hardly a hospitable one. By the time his paper on the sperm of the Rhine salmon was published in 1874, the field of physiological chemistry was starting to resemble an intellectual bare-knuckle cage fight. Disagreements that had begun as differences over scientific matters spilled over into acrimonious feuding and often very personal attacks conducted on the pages of journals as eminent figures within the new discipline jostled to establish themselves at its head. And it was amidst this raging maelstrom of scientific egos as they smashed into each other, that Miescher and his discovery of nuclein were caught.

6.1 Trouble Ferments

One such clash was a long running feud between Felix Hoppe-Seyler and Wilhelm Kühne. This was fought on many fronts, one of which was the question of whether haemoglobin contained sulphur. Hoppe-Seyler claimed to have detected its presence in the blood protein, whilst Kühne replied that

Hoppe-Seyler's results were simply due to contamination. Other sources of discord included the mechanism of vision. When Hoppe-Seyler published a historical account of research in this field, Kühne was quick to challenge it. Describing Hoppe-Seyler's interest in the subject as mere 'excursions', Kühne accused him of having made historical errors 'that cannot be passed over with silence.'[1] And as far as Kühne was concerned, this was not the only field in which Hoppe-Seyler had made errors of presentation. Writing in the journal 'Investigations from the University of Heidelberg' in 1877, Kühne took pains to point out how his own research into the effects of acids on the digestive enzymes pepsin and trypsin had been misrepresented by Hoppe-Seyler. It was, he said, an error 'which cannot remain without correction.'[2]

But even these were just minor skirmishes. Where Hoppe-Seyler and Kühne really clashed was over the nature of fermentation. At the time, this was a broad term that encompassed not just the conversion of sugars to alcohol as is understood by the modern meaning of the word, but many other processes including the production of lactic acid when milk goes sour, the formation of butyric acid in bread, and the release of ammonia from urea in the urine. What remained unclear however was whether such processes occurred only in the presence of living cells such as micro-organisms, or whether they were in essence no different from any other chemical transformation that might be carried out in the test tube?

For many distinguished scientists such as Louis Pasteur, and the Swiss scientist Carl Wilhelm von Nägeli (1817–1891) the answer was clear - 'The agent of fermentation is inseparable from the substance of the living cell.'[3] But for Hoppe-Seyler, this placed the whole future of physiological chemistry as an independent discipline in danger. 'It is easy to see that these methods virtually negate physiological chemistry,' he warned, 'and make a biological-botanical or zoological consideration alone the dominant one.'[4]

He had an ally in the French chemist Marcelin Berthelot (1827–1907) who sought to show that 'in every fermentation, one must try to reproduce the same phenomena by chemical methods and to interpret them by exclusively mechanical considerations. To banish life from all explanations relative to organic chemistry, that is the aim of our studies.'[5] Berthelot's words could easily have been adopted as a mission statement for the discipline of

[1] (Kühne 1882); p. 488.
[2] (Kühne 1877); pp. 325–326.
[3] (Fruton 1999); p. 143.
[4] (Hoppe-Seyler 1876); p. 5.
[5] Cited in (Fruton 1999); p. 133.

physiological chemistry, and Miescher's own ambition for a new description of the cell in chemical terms.

A big part of the problem was one of definition. Processes such as the conversion of sugar to alcohol, or lactic acid which Pasteur maintained were dependent on micro-organisms were known as 'formed', or 'organised' ferments, while those which could apparently proceed independently of micro-organisms, were called 'unformed', or 'unorganised' ferments. Examples included the digestion of proteins by pepsin, or of starch by diastase, and were confined only to processes that resulted in the degradation and breakdown of a substance.

For Hoppe-Seyler, this distinction was misplaced. Convinced that all such processes were ultimately chemical in nature and could therefore take place independently of a living system, he argued that a far more accurate distinction was between that of organisms and the ferments that they contained.

Unsurprisingly, Kühne disagreed. In 1878, he proposed to settle the debate once and for all by coining a new term for the agents involved in fermentation:

> *I have only taken the occasion…to denote some of the better-known substances, called by many, "unformed" ferments, as enzymes.*[6]

Today, Kühne's term 'enzyme' refers to the biological catalysts without which the vast array of metabolic reactions within the cell could never occur. But at the time that Kühne coined this term, it had a much more restricted meaning, confined only to agents such as pepsin or diastase which acted outside the cell and were involved in the breakdown of substances.[7]

And although Kühne's term has today come to be at the heart of modern biochemistry, it left Hoppe-Seyler distinctly unimpressed:

> *Recently, Kühne has issued the demand that the distinction I made be opposed; however, since he offers no reason worthy of consideration, I do not find it necessary to say anything in reply. The new word enzyme may be added to the numerous names Kühne has proposed, insofar as they are designations of substances that are still unknown.*[8]

But the growing enmity between Hoppe-Seyler and Kühne went much deeper than a mere disagreement over nomenclature. In the same article that he

[6] (Kühne 1878); p. 293.

[7] (Köhler 1973); pp. 181–196; pp. 187–188.

[8] (Hoppe-Seyler 1878); pp. 2–4.

poured cold water on the term 'enzyme', Hoppe-Seyler accused Kühne of throwing into question the originality of work done by one of his students, Nicholas Lubavin. Lubavin had been studying the action of the gastric enzyme pepsin on proteins and found that this gave rise to several products including the amino acids leucine and tyrosine. Kühne meanwhile, had shown that pancreatic juice had a very similar degradative effect on proteins and seemed, according to Hoppe-Seyler at least, to be insinuating that Lubavin's findings were little more than a repetition of this work. Rising to the defence of his student, Hoppe-Seyler landed what he hoped would be a knockout blow. He argued that Kühne's own work was hardly original and that several other researchers had already demonstrated the action of pancreatic juice on proteins. 'There is not one correct thought in this passage…Everything is untrue,' he fumed, before going on to add:

> *I do not think it is necessary to respond to all the further attacks directed against me by Mr Kühne; They are meaningless. If you take away the foreign ideas and the grandiloquence of the phases, there is hardly anything left of the work.*[9]

Yet despite seeking to demolish Kühne's scientific credibility, Hoppe-Seyler was willing to make one concession to him. He admitted to having made an error when he had misread some details of the concentrations and types of acid used in Kühne's work. It was an error that Kühne had himself (most likely with some satisfaction) already taken pains to point out in one of his own publications that year, insisting that 'it cannot go without correction.'[10] But such was Hoppe-Seyler's animosity towards Kühne that even his mea culpa now became twisted into an insult:

> *I regret my mistake all the more since it now turns out that the statement in question by Kühne contained nothing worthy of note and that I could have ignored it altogether.*[11]

Objective disagreements over the interpretation of experiments and concepts are essential to science. Without them it cannot evolve and grow. But the situation between Hoppe-Seyler and Kühne had crossed a boundary. It was no longer merely a professional dispute but was rapidly becoming a personal one. Both Hoppe-Seyler and Kühne were no longer just playing the ball, but rather

---

[9] Ibid; p. 3.
[10] (Kühne 1877); pp. 325–326.
[11] (Hoppe-Seyler 1878); p. 3.

the man—as was evident from a paper published by Kühne with the title 'Response to an Attack by Mr. Hoppe-Seyler':

> *Obviously, this can only mean two things: either the matter itself was not worth mentioning and Hoppe-Seyler only did so because he considered it an opportunity to contradict me, or the fact (prevention of trypsin digestion by very dilute acids) was to be discussed, and then he would have found me, who found it or confirmed and expanded it according to Danilewsky, not mentioned. It is sad that a writer took this opportunity to give such an insight into his methods.*[12]

6.2 Physiological Chemistry in the Cross-Hairs

Some historians have suggested that the dispute between Hoppe-Seyler and Kühne was rooted in a tension over what exactly the role of chemistry in the biological sciences should be. Kühne considered himself primarily a physiologist for whom chemistry served a subordinate role, while for Hoppe-Seyler chemistry was central.[13] Throwing his weight behind Kühne was the eminent German physiologist Eduard Pflüger (1829–1910) who warned that with his emphasis on chemistry, Hoppe-Seyler was placing the future of physiology as a unified discipline under threat.[14]

While Kühne and Hoppe-Seyler crossed swords at seemingly every opportunity, the discipline of physiological chemistry was itself coming under attack from several directions. By the mid-nineteenth century, Germany had a rapidly growing chemical industry that led the world in the production of substances such as synthetic dyes. This was thanks largely to German expertise in the field of organic chemistry—a discipline that was built on solid theoretical foundations which allowed the precise synthesis of defined compounds by known chemical pathways. Organic chemists proficient in these methods, looked down upon physiological chemistry as a shabbily dressed, down-at-heel relative with no hope of ever achieving the prestige and precision of their own discipline. Deriding it as nothing more than a chaotic mess of poorly defined chemical entities, they dismissed it on the grounds that 'Tierchemie ist Schmierchemie.' (Animal chemistry is smudged chemistry).[15]

For Miescher, however, organic chemistry too was not without its own limitations. He felt that in trying to understand the processes by which

[12] (Kühne 1878); p. 68.
[13] Fruton (1990), pp. 90–92.
[14] (Pflüger 1877); pp. 597–598; (Pflüger 1878); pp. 361–365.
[15] (Fruton 1999); p. 57.

biological molecules were broken down or assembled, 'organic chemistry for all its prestige is of little help.'[16] Its reagents were so harsh and indiscriminate that they 'destroy everything together at once. Wherever the precise work of the watchmaker is to be found, they take an axe and a sledgehammer to it.'[17]

Miescher recognised that the sheer complexity of the cell and the fragility of its chemical components, meant a chemical approach to studying it would require methods that were more subtle and gentle than the blunt tools of organic chemistry. 'Somehow, physiologists must find a way of helping themselves,' he declared, whilst recognising that since 'physiological chemistry emerges from such a pile of unconnected facts to which it makes little sense to add yet more shreds,' the task would be far from easy.[18,19]

The challenge for physiological chemists was to be able to weave all these apparently disparate shreds together so that they offered a coherent account of the structure and function of living systems. Miescher was confident that nuclein would prove to be an exemplar of how this might be achieved but others remained unconvinced.

6.3 Miescher: 'The Contaminator'[20]

The physiologist Jacob Worm-Müller (1834–1889) hailed Miescher's demonstration that nuclei could be isolated and purified by digesting cells with pepsin, as 'an estimable investigation.'[21] But although he had no reservations about praising Miescher's practical methods, Worm-Müller was much more cautious about what he claimed to have discovered with them. Based on his own analyses of nuclein, Worm-Müller made three criticisms of Miescher's claim to have discovered this novel substance. Firstly, he pointed out that

[16] Miescher, 16th Jan 1874, Letter XXXV in (His 1897); p. 76.

[17] Ibid. Miescher's use of the term 'watchmaker' is worth noting here. Thanks to the use of this term by the English theologian William Paley (1743-1805) in his 1802 book 'Natural Theology or Evidences of the Existence and Attributes of the Deity' it had come to be synonymous in the anglophone world with the idea that the apparent design found in the natural world was due to the hand of a divine creator. It was in rebuttal of this idea that biologist Richard Dawkins used the term in his 1986 book 'The Blind Watchmaker: Why the Evidence of Evolution Reveals a Universe Without Design.' Whether 'Natural Theology' was widely read in continental Europe however is unclear and Miescher may have been unfamiliar with Paley's term for the divine. Miescher may instead have been simply reaching for a striking visual metaphor of a complex mechanical system and which, given his Swiss background, the one that sprang most easily to mind was that of a watch.

[18] Ibid.

[19] Miescher, 27th Feb 1874, Letter XXXVI in (His 1897); p. 77.

[20] (Veigl et al. 2020); pp. 463–464.

[21] (Worm-Müller 1874); p. 190.

results for the chemical composition of nuclein were highly variable and often appeared to be dependent on the technical details of whatever extraction protocol had been used. This led him to believe that nuclein was a mixture of substances. Secondly, he argued that nuclein was almost certainly associated not just with a mixture of proteins but also of organic compounds such as lecithin which might well be the source of the phosphorus Miescher had detected. Finally, he felt there to be no clear evidence yet that nuclein was exclusively located in the nucleus. 'My preliminary work on the chemistry of the muscle, the liver and the egg yolk,' he concluded, 'have convinced me that it would take several years to sufficiently sift through the material to reach even a preliminary conclusion with regard to nuclein.'[22]

Others were far less diplomatic in their tone. To Hoppe-Seyler's adversaries, nuclein was a gift, for it gave them a new source of ammunition with which to attack him. The pages of scientific journals and books became the battleground on which this engagement was fought. In 'Animal Chemistry, or the Relations of Chemistry to Physiology and Pathology', published in 1878, the British clinician Charles Thomas Kingzett (1852–1935) complained that in a recent book by the Austrian chemist and clinician Karl B. Hoffmann (1842–1922) 'more space is devoted to 'nuclein' – an impure ordinary albuminoid than to the whole brain chemistry'.[23] Referring dismissively to 'the substance termed by Miescher 'nuclein'', Kingzett felt that the attention being given to this material was utterly without foundation.

Nuclein was a particularly sharp thorn in Kingzett's side. A year earlier he and his co-worker, Henry Wilson Hake, had published a paper entitled 'Physiology and its Chemistry at Home and Abroad' in which they lamented the state of physiological chemistry. Although they felt the new discipline had got off to a promising start, they feared it had quickly fallen into a state of deterioration and was in dire need of reform if it was to survive. As evidence of this decline, they offered a few examples of low-quality work that they felt besmirched the field of physiological chemistry. At the top of their list was a name that must have made Miescher's heart sink:

> *We will commence by considering the alleged discovery of a substance to which has been given the name "nuclein".*[24]

[22] Ibid., p. 194.

[23] (Kingzett 1878); p. 280.

[24] (Kingzett and Hake 1877); p. 92.

Kingzett and Hake compared the results of chemical analyses of nuclein carried out by Miescher and Hoppe-Seyler with the values obtained for several different albumins, or what would today be called proteins. Noting these values to be so similar, they concluded that nuclein was not a novel cellular substance at all, but simply another type of protein. As for its high phosphorus content, they pointed out that this had varied significantly between researchers and was therefore most likely to have arisen from an impurity in the extraction protocol.[25]

The repeated misspelling of Miescher's name by Kingzett and Hake may well have been an innocent typo. But given their disparaging tone in assessing his work, it was more likely a bit of cheap gamesmanship. Claiming that errors in Miescher's analysis of the chemical composition of nuclein and protamine had led him to give incorrect formulae for them, Kingzett and Hake suggested that both these substances 'may not unfitly be classified, together with certain human beings, under one comprehensive heading—the Great Unwashed.'[26] Their final verdict on Miescher's work was damning: 'More seriously, we feel that, as a result to Science, whatever might have been valuable in these researches is, for the time being, absolutely valueless.'[27]

Another of Hoppe-Seyler's students, the Austro-Bohemian clinician Rudolf von Jaksch (1855–1947), received only a slightly less harsh treatment. Two years after Miescher published his work on nuclein in salmon sperm, Jaksch reported the isolation of 'a substance in human brain that is closely analogous in its chemical and physical properties to Miescher's nuclein.'[28] Moreover, Jaksch was confident that the substance he had found 'was a nuclein-like body in the Miescherian sense' because its high phosphorus content could not have arisen from impurities.[29]

Kingzett and Hake were unimpressed. 'Without claiming absolute purity for his preparation,' they scoffed, 'He [Jaksch] entertains no doubt "that it is a nuclein kind of body in a Miescher" (why not "Pickwickian"?) "sense."[30] Their suggestion that nuclein would more accurately be described as 'Pickwickian' was not intended as a compliment. It refers to Mr. Samuel Pickwick, the main character in Charles Dickens' first novel, 'The Pickwick Papers', published in serial form between 1836–1837. Samuel Pickwick is something of a comedic, jovial and inoffensive figure and the definition of

[25] Ibid., p. 93.
[26] Ibid., p. 95.
[27] Ibid.
[28] (von Jaksch 1876); p. 469.
[29] Ibid., p. 473.
[30] (Kingzett and Hake 1877); p. 94.

"Pickwickian' offered by the Oxford English Dictionary is: 'Frequently humorous. Of a word, expression, etc.: not literally meant; (sometimes) interpreted in such a way as to avoid unpleasantness, difficulty, etc.' But Kingzett and Hake may well have had another meaning in mind when using this word. For in the opening chapter of Dickens' novel, Mr. Pickwick delivers a scientific paper with the grand sounding title of "Speculations on the Source of the Hampstead Ponds, with some Observations on the Theory of Tittlebats[31]"to the London club that takes its name from him. Dickens uses the scene to gently poke fun at Pickwick's mild pomposity in coming up with such a grand title for his paper while at the same time failing to recognise that the obscure and parochial nature of its subject make it unlikely to set the scientific world on fire. And while most of Pickwick's immediate audience deliver uncritical praise for his address, one lone member, Mr. Blotton dismisses him as a 'humbug.' Perhaps it was to this opening scene of the novel that Kingzett and Hake were making a sly reference—casting themselves in the role of Blotton to Miescher's Pickwick, when they suggested that nuclein might well be better described as 'Pickwickian' than 'Miescherian.'

6.4 Thudichum Enters the Ring

Writing to Hoppe-Seyler, Miescher regretted that his discovery of nuclein was being called into question by what he described as 'a rabble.'[32] But even though he distanced himself from the fray and tried to focus on his work, the 'rabble' grew ever louder. As far as Kingzett and Hake were concerned, Jaksch's claim to have isolated nuclein was not his only crime. Equally grievous was that, in studying the chemical composition of brain tissue, he had violated someone else's intellectual territory: 'The fact that the above [Jaksch's paper] was printed after the publication of Thudichum's researches in brain chemistry,' they sneered, 'intensifies rather than palliates the ignorance it displays.[33]

Kingzett and Hake's enthusiastic adoption of Ludwig Thudichum as a champion, together with the vitriol they directed towards Miescher may well have been rooted in the shifting geopolitics of the time.With Otto von Bismarck having launched a succession of wars in a relatively short period of time, first against Denmark in 1864, then with Austria 2 years later, and

[31] Tittlebat is another name for a stickleback fish.

[32] Miescher to Hoppe-Seyler, Summer 1872, Letter XXVI in His 1897, pp. 64–68; p. 68.

[33] (Kingzett and Hake 1877); p. 94.

finally against France in 1871, Britain was becoming nervous about the military power of the newly unified German state and its intentions.

This wariness of all things German also cast a shadow over the emerging field of physiological chemistry. Kingzett and Hake's 1877 paper in which they attacked nuclein was titled 'Physiological Chemistry at Home and Abroad' but it was clear that when making this comparison, they had only one foreign country in mind: Germany. And it was also clear that their intention was to show Britain to be the better of the two when it came to physiological chemistry. 'In any comparison between Germany and England, we are by no means disposed to concede the laurel to Germany,' they declared, before going on to argue that in many German universities, 'the enormous mass of work published represents also Russian, Polish, Austrian, Hungarian, and Belgian, nay Italian workers.'[34] Miescher might well have been Swiss, but as part of the German-speaking world, it seemed this was enough to make him guilty in the eyes of Kingzett and Hake.

Dismissing much German work in physiological chemistry as being of 'inferior character', Kingzett and Hake offered a diagnosis of the problem. It was the result, they said, of students being taught by professors of medical chemistry—a situation which they claimed could never end well:

> *When it is therefore considered that the medical men are far behind Englishmen in their acquaintance with Materia Medica…it is not surprising that the results in many cases should be so deplorable.*[35]

Who better therefore to adopt as the posterboy for British physiological chemistry than Ludwig Thudichum? Here was a German clinician whose publications had been rejected by the scientific establishment of his homeland and become an honorary Englishman. For Kingzett and Hake, Thudichum must have seemed to be the living embodiment of the superiority of the British system.

Thudichum himself was no doubt delighted to receive such vocal support on the pages of respected journals. Despite having achieved what he described as 'moderate successes' in London, he still felt that his own contribution to this field had been insufficiently recognised. Nowhere did he feel this was truer than in his homeland of Germany where both Rudolf Virchow and Eduard Pflüger had declined offers from Thudichum to submit papers for publication in their respective journals. Undeterred—and determined at all

---

[34] (Kingzett and Hake 1877); p. 105.

[35] Ibid.; p. 106.

costs—to get his work into print by whatever means necessary—Thudichum responded by setting up his own journal.

When the first volume of his 'Annals of Chemical Medicine' appeared in 1879, Thudichum envisaged that it would become the very first British journal dedicated to physiological chemistry. But his hopes were soon dashed when it received a savage review in the leading medical journal *The Lancet* which pointed out that every one of the 23 articles that it contained appeared to have been written by Thudichum himself.[36]

Despite this far from spectacular debut, a second volume followed 2 years later but the damage was done and not even a mildly favourable review in the British Medical Journal could save Thudichum's flagship from going under.[37] But although it had failed to become the trailblazing publication for physiological chemistry that Thudichum had hoped, it had nevertheless succeeded in serving another purpose—one that had already been discerned by a perceptive reviewer at the journal *Nature* who warned readers—'Those who open this work expecting to find it adequately fulfilling the promise of its title will be disappointed. Had they read the initial preface they would have been prepared for this, for it indicates very clearly the intention of the promised series… '.[38]

And it was clear from the preface to the first volume that Thudichum's intention was as much about settling scores with rivals than the promotion of physiological chemistry:

> *Some, nevertheless, men of externally high position, have persistently done all they could, sometimes by silence, sometimes by innuendo, sometimes by obstruction to thwart the endeavours which others have made to advance medical science by original chemical researches [sic].*[39]

According to Thudichum, the emerging field of physiological chemistry was dominated by an elite group of self-appointed autocrats who were hell-bent on excluding him and his work from their cabal. Worse still, one of their number had now revived the question of protagon—a subject that Thudichum believed he had buried over a decade ago.[40]

---

[36] 'Reviews and Notices of Books,' The Lancet 31st Jan 1880, p. 175.

[37] 'Reviews and Notices: Annals of Chemical Medicine by J.L.W. Thudichum M.D.,' The British Medical Journal, 5th June 1880, p. 853–854.

[38] 'Thudichum's Annals of Chemical Medicine', Nature 19th Jan 1882, p. 263.

[39] (Thudichum 1879a); p. ix.

[40] (Thudichum 1879b); p. 254.

This was Arthur Gamgee (1841–1909) who, aged only 32 had been appointed Brackenbury Professor of Physiology at Owens College, Manchester. But before this he had done postgraduate work with Carl Ludwig in Leipzig and with Wilhelm Kühne in Heidelberg where the importance of physiological chemistry had left a powerful impression on him. When Gamgee died in 1909, *The Lancet* hailed his 'Textbook of the Physiological Chemistry of the Animal Body, Including an Account of the Chemical Changes Occurring in Disease', as being 'beyond doubt' his greatest accomplishment.[41] But following the book's publication in 1880, Thudichum had very different thoughts about its worth, deriding it as being an 'impediment' to progress in the field of physiological chemistry and 'humiliating to scientific literature.'[42] His criticism of the book also presented a golden opportunity to attack two old rivals, for in one particular sentence that dripped with caustic sarcasm, he noted that it contained:

> *...the united wisdom of Professors Kühne, of Heidelberg, and Hoppe-Seyler, of Strasburg. The strangeness of the combination, in view of the well-known divergence of these two authors on many matters...the reader might at once arrive at the conclusion that the work of Dr. Gamgee, although it might contain much that was good, yet would not contain much that was new. However, in this respect also expectations are surpassed not only by the complete absence of anything exhibiting any advance upon previous knowledge, but by the extraordinary way in which the facts of science and the data of literature have been treated.*[43]

Thudichum brought a lengthy charge sheet of crimes against Gamgee's work—the histological content was unoriginal, the section on proteins too short, key researchers had not been cited. But of all these alleged crimes, Gamgee's greatest transgression had been to champion the existence of protagon.[44] As far as Thudichum was concerned, he had passed final sentence on this matter back in 1874 when, in an exhaustive report into the chemical constituents of the brain, he had concluded that protagon was not a novel, distinct chemical entity, but rather a mixture of fatty substances containing phosphorus.[45] It was, he said, nothing more than an artefact arising from variations in technicalities such as the particular solvent used for extraction

[41] 'Obituary: Arthur Gamgee M.D. Edin., F.R.C.P. Lond., F.R.S.' The Lancet 17th April 1909, pp. 1141–1142.

[42] (Thudichum 1881); p. 183.

[43] Ibid., p. 184.

[44] Ibid., p. 187.

[45] (Thudichum 1874); p. 209;

that gave rise to 'a delusive appearance of definiteness'.[46] 'The doctrine of 'protagon'' he declared in conclusion, 'has been properly rejected by all physiological chemists who know their business.'[47]

Hoppe-Seyler was one of those who had now, perhaps with some reluctance, come to doubt the existence of protagon. But this did not mean that his relationship with Thudichum had improved. In the 10 years since Hoppe-Seyler had curtly dismissed his work as being 'obviously wrong' a simmering resentment had festered within Thudichum. This finally erupted into outright hostility when he published a paper with a title that was hardly likely to win favour with Hoppe-Seyler. Called 'Erroneous Statements of Hoppe-Seyler Concerning the Colouring Matters of Bile', Thudichum used it to pour scorn upon Hoppe-Seyler's research into the bile pigment bilirubin which he ridiculed as having 'no better foundation than the imagination of Hoppe-Seyler', before passing a damning verdict[48]:

> *Towards the end of his description of the matters in question, the confusion of Hoppe-Seyler rises to a wild conflict of errors…In short, the entire representation of this subject given by Hoppe-Seyler is unworthy of literature and of science.*[49]

Miescher would quickly find himself caught in this crossfire. For of all the work to come out of Hoppe-Seyler's laboratory, Thudichum's greatest scorn was reserved not for claims about bile pigments, but Miescher's discovery of nuclein. How could anyone take this claim seriously, asked Thudichum, when 'Like other choice things, protagon amongst them, it came from Hoppe-Seyler's school, which it was then the pleasure of the physiological aspirants of this country to declare, perhaps to believe, to be infallible. Whosoever did not join that ring, or do their bidding, was by them sequestrated in the monastic sense, his existence was affectedly ignored, and his work treated with incomparable naïveté, in these days of printing, as non-extant.'[50]

Citing Kingzett and Hake, Thudichum declared nuclein to have been 'exploded.' It was a damning verdict and arriving amid such acrimonious

[46] (Thudichum 1879b); p. 263.

[47] Ibid.

[48] Ibid; p. 332.

[49] Ibid.

[50] (Thudichum 1881); p. 186.

feuding and toxic career rivalry, Miescher's discovery of nuclein stood little chance of being evaluated from a purely objective perspective.

But what was Miescher's response to the storm that raged around nuclein? In a letter of December 1869 written as he was preparing his manuscript on nuclein, he appears clearly to have recognised that his work might well ruffle a few feathers—and that this in turn might have a detrimental effect on him:

> *The more that I lay out the plans for my next work, the more the need presses itself upon me to keep disturbances away from a number of points of action and to avoid involvement of discussion in the journals for a while longer. The way in which the journals are currently driven would put me in such a feverish haste which, as I know from experience, would cause the path before me to become as obscured as that behind me.*[51]

Perhaps the peace and quiet of those early morning trips down to the banks of the Rhine to catch salmon for his experiments helped to bring Miescher some much needed solace. Whilst Hoppe-Seyler traded blows with his critics over nuclein on the pages of the journals, Miescher was becoming preoccupied with a new avenue of research.

In offering a rich and abundant source of nuclein, the bulging testes of the Rhine salmon had been a blessing for Miescher and his work on this substance. Now this anatomical curiosity was about to open a second, whole new avenue of scientific enquiry for him. But this time, what had begun as a blessing would turn out to be a curse.

[51] Miescher, 20th Dec 1869, Letter V in (His 1897); p. 39.

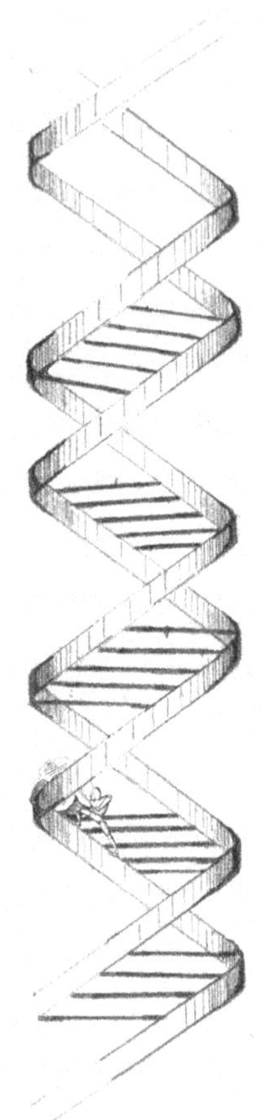

Drawing by Kersten Hall

References

Fruton, J. Contrasts in scientific style: Research groups in the chemical and biochemical sciences. American Philosophical Society, Philadelphia, PA 1990.

Fruton, J. 1999. *Proteins, enzymes, genes: The interplay of chemistry and biology*. New Haven, CT: Yale University Press.

His, W. 1897. *Die Histochemischen Und Physiologischen Arbeiten von Friedrich Miescher*. Leipzig: F. C. W. Vogel.

Hoppe-Seyler, F. 1876. Ueber Die Processe Der Gährungen Und Ihre Beziehungen Zum Leben Der Organismen. *E Pflüger Archiv Für Physiologie* 12:1–17.

Kingzett, C. T. 1878. *Animal chemistry, or the relations of chemistry to physiology and pathology*. London: Longman and Green.

Kingzett, C. T., and H. W. Hake. 1877. Physiology and its chemistry at home and abroad. *Quarterly Journal of Science* 14:91–109.
Köhler, R. E., Jr. 1973. The enzyme theory and the origin of biochemistry. *Isis* 64:181–196.
Kühne, W. 1877. Nachtrag Zur Geschichte Des Trypsins. *Untersuchungen aus dem Physiologischen Institut der Universität Heidelberg* 1:325–326.
Kühne, W. 1878. Erfahrungen Und Bemerkungen Über Enzyme Und Fermente. *Untersuchungen an Dem Physiologischen Institut Der Universität Heidelberg* 1:291–324.
Kühne, W. 1882. Bemerkungen Zu Herrn Hoppe-Seyler's Darstellung Der Optochemie. *Untersuchungen Aus Dem Physiologischen Institut Der Universität Heidelberg* 2:488–492.
Pflüger, E. 1877. Die Physiologie Und Ihre Zukunft. *Deutsche Medizinische Wochenschrift*597–598.
Pflüger, E. 1878. Die Physiologie Und Ihre Zukunft. *Archiv Für Physiologie* 15:361–365.
Thudichum, J. L. W. 1874. *Researches on the Chemical constituents of the brain*, 113–247. Reports of the Chief Medical Officer of the Privy Council.
Thudichum, J. L. W. 1879a. Note and experiments on the alleged existence in the brain of a body termed "Protagon". *Annals of Chemical Medicine—Including the Application of Chemistry to Physiology, Pathology, Therapeutics, Pharmacy, Toxicology and Hygiene* 1:254–263.
Thudichum, J. L. W. 1879b. Preface. *Annals of Chemical Medicine—Including the application of Chemistry to Physiology, Pathology, Therapeutics, Pharmacy, Toxicology and Hygiene* 1:v–x.
Thudichum, J. L. W. 1881. On modern text-books as impediments to the progress of animal chemistry. (A deduction). *Annals of Chemical Medicine* 2:183–189.
Veigl, S. J., O. Harman, and E. Lamm. 2020. Friedrich Miescher's discovery in the historiography of genetics: From contamination to confusion, from Nuclein to DNA. *Journal of the History of Biology* 53:451–484. https://doi.org/10.1007/s10739-020-09608-3.
von Jaksch, R. 1876. Ueber Das Vorkommen von Nuclein Im Menschengehirn. *Pflüger's Archive: European Journal of Physiology* 13:469–473.
Worm-Müller, J. 1874. Zur Kentniss Der Nuclein. *Pflügers Archiv: European Journal of Physiology* 8:190–194.

7

Hacking the Heads Off the Hydra

Contents

Abstract Thanks to his work with salmon sperm, Miescher publishes a second paper on DNA in 1874 but now finds himself growing ever more distracted. His interest in how the Rhine salmon was able to redirect muscle tissue to the growth of the sex organs leads him to undertake exhaustive wider investigations of its diet, ecology and behaviour for the benefit of the fishing industry. This in turn results in him being recruited by the Swiss government to undertake a study of the diet of prison inmates. On top of this he now holds the post of director of a new institute for physiological research, which brings with it the burden of administrative duties. All of this leaves Miescher feeling ever more frustrated at the lack of time and attention he is giving to the question of what the function of DNA might be. Reaching to classical mythology to express his despair, he complains that his never ending

K. Hall, R. Dahm, *The Dawn Fisherman*, Copernicus Books,
https://doi.org/10.1007/978-3-032-14219-1_7

workload felt like the heads of the Hydra, or likens himself to Sisyphus, cursed by the gods of Olympus to spend eternity rolling a boulder up a mountain. In the meantime however, others have turned their attention to nuclein (DNA). Another of Hoppe-Seyler's students, the chemist Albrecht Kossel shows it to contain four nitrogenous bases which he names adenine, cytosine, guanine, and thymine, whilst one of Miescher's own former students, Richard Altmann gives nuclein a new name by which is still known today—nucleic acid.

By the time that Miescher returns to nuclein (DNA), in 1887, it is too late. Physically and mentally exhausted, he develops tuberculosis and dies just a few years later in a sanatorium in the Swiss resort of Davos in 1895.

Keywords Miescher • Nuclein • DNA • RNA • Phosphorus • Cell nuclei • Kossel • Nitrogenous bases • Adenine • Guanine • Cytosine • Thymine • Biochemistry • Molecular biology • Genetics • Physiology • Chemistry • Research • Discovery • Science • History • Biology • Medicine • Nobel Prize • Nucleic acids

Miescher is distracted from nuclein by ever increasing work commitments—all of which takes a heavy toll on his health

By 1887, Miescher was burdened by a growing sense of unease—and guilt. In May of that year, he complained 'I feel like one who is in debtor's prison. And so long as my old debts remain unpaid, I am unable to get on with any new work.'[1] The debt in question was the work on nuclein that he had initially begun with such enthusiasm, but which he now felt he had allowed to fall into neglect. From the publication of his paper in 1874 on salmon sperm nuclein and until 1887, Miescher did no more work on this subject. But throughout those 13 years an uncomfortable sense of an unpaid debt had nagged away at him. Now at last he was determined that the time had finally come for it to be paid off in full.

The cause of this 13-year period of neglect is not without a certain irony. Although the bulging gonads of the Rhine salmon had been a gift in providing Miescher with a rich source of nuclein, they had also set Miescher on a path that would distract and divert his attention away from further research into its properties for over a decade following the publication of his paper on salmon sperm nuclein in 1874.

---

[1] Miescher, 16th May 1887, Letter LXI, in (His 1897); p. 105.

7.1 Prison Food

The annual trek made by salmon from the North Sea to their breeding grounds in the Rhine is impressive enough, but it was their curious behaviour on arrival there that fascinated Miescher. For the next 6–9 months that they spend in freshwater, the fish eat nothing. Yet despite their lack of nutritional intake, they undergo a seismic anatomical transformation during this period. The muscles in their trunk and tail shrivel away, whilst at the same time their gonads grow to such a size that they can account for up to a quarter of the body mass of the fish.

Miescher found out the metabolic secret behind this phenomenon. The nutrients needed to fuel the massive growth of the reproductive organs were derived from the breakdown of proteins in the trunk and tail muscles. 'It is the merit of the Rhine salmon,' he wrote in triumph, 'to have revealed to us a previously unknown factor of animal economics the migration of substances that make up tissues from organ to organ in a truly magnificent way.'[2]

But, ever curious, Miescher found that he could not confine his interest to this anatomical remodelling of the salmon. With the help of Mr. Glaser Sohn, the most prominent fish merchant in Basel at the time, he studied not only the dramatic physiological changes in the Rhine salmon, but also their ecology and migratory behaviour. These first investigations were presented not in an academic journal, but rather in a technical appendix to a contribution by the Swiss section of the International Fisheries Catalogue in Berlin in 1880. Miescher had recognised that studies of the physiology and behaviour of migrating salmon might not only have scientific benefits but also more immediate practical benefits for the local economy. 'It is not here the place to go into more detail about the purely scientific results as such: this will soon be done in another place,' he wrote in the introduction to his work, 'but since a more precise knowledge of the living conditions of the Rhine salmon can also be advantageous in practical terms, especially for the creation of a rational fishing policy, I have compiled among my results those which I deemed to be of any practical interest.'[3]

Miescher presented reams of data such as tables showing exhaustive analyses of the average weight of salmon caught each month in the years from 1872–1879, comparisons of the ratio of average weights of Rhine salmon to those caught in Holland, or the increase in percentage body mass of their eggs. He made comparisons between different types of salmon, drawing

[2] (Miescher 1880); p. 116.

[3] Ibid., p. 117.

attention to variations in their anatomy or behaviour, producing graphs of variation in length or weight, analysing their diet and even turning to questions such as at what age do salmon first begin to migrate?

But the problem of course was that whilst Miescher's detailed studies may well have helped the fisheries of Basel, it sapped his focus and energy away from work on nuclein. And although his sense of public duty in wishing to provide practical help for the local economy was commendable, it was not only leading him ever further away from his passion but slowly wearing him out. Writing to his friend Rudolf Boehm in May 1878, Miescher was disappointed not yet to be able to give any details on the progress of his work and explained that 'so many different things are gnawing away at me, it becomes impossible to achieve one's goals…I am sinking under 12 h of weekly lectures, all without any assistants so that I am hardly able to come to my senses.'[4]

One of the main reasons for his increased workload was that by now Miescher's exhaustive research into the dietary requirements of the Rhine salmon had come to the attention of the Swiss authorities. They too now laid a heavy claim upon Miescher's sense of civic obligation. In the autumn of 1876, the Swiss government asked him to produce a report into the diet of inmates at a Basel prison—a task which he took on, albeit begrudgingly. By spring the following year he complained that for the duration of March and April he had 'been bound in chains…with the most arduous and thankless task of my entire life.'[5] He nevertheless completed his task with such characteristic and customary diligence that he soon received requests to produce similar reports for other prisons each of which, according to his uncle Wilhelm His, wanted its own specific menu.[6]

But all these commitments came with a price. 'I have made myself too green, and now the goats are eating me,' he exclaimed in despair, 'Inquiry on the diet of the Swiss people, cookbook for the workers, nutrient tables for the national exhibition, controversies with the Chalmer milk company—in brief, I am on the best way to becoming the guardian for the stomachs of all three million Swiss.'[7]

Not that the sex organs of salmon and formulating prison diets were the only sources of distraction for Miescher. In 1879, a year after getting married, his first child—a daughter—was born, followed by a son in 1881. Four years

[4] Miescher, 6th May 1878, Letter L, in (His 1897); p. 97.

[5] Miescher, 25th May 1877, Letter LXXXV, in (His 1897); p. 129.

[6] In (His 1897); p. 27.

[7] Ibid.

later, another daughter was born in what would turn out to be an eventful year for other reasons.

7.2 The Lernaean Hydra

In 1885, Miescher founded Basel's first anatomical-physiological institute which, 4 years later, hosted the First International Congress of Physiology. The institute was housed in the 'Vesalianum', a building named in honour of the sixteenth century physician Andreas Vesalius. Had Miescher only known that, over a 100 years later, he too would have research institutes named in his honour (Fig. 7.1), this might well have brought him some much-needed consolation. For although his position as director of the institute brought prestige, it was also something of a poisoned chalice. The demands of his new position gave Miescher little time to carry out research and left him instead consumed with administrative tasks (a situation with which many in academia will no doubt readily identify). One of these was the recruitment of technicians for the construction of new instruments with which to make precise physiological measurements—such as for the composition of blood gases. This had been the subject of his original Habilitation and on returning to it

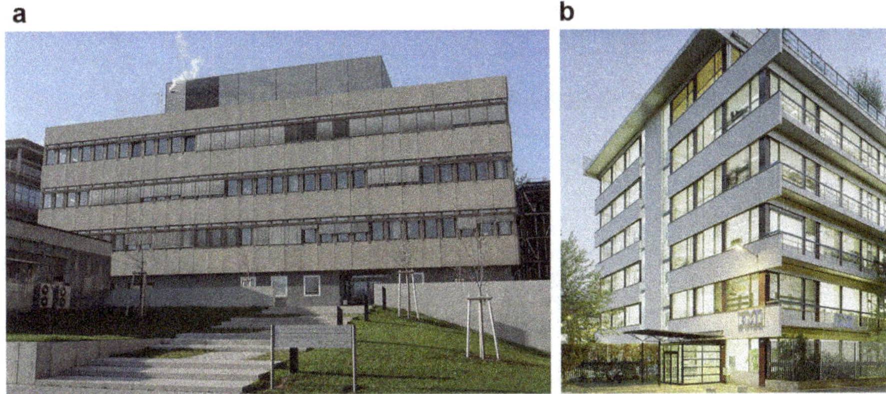

Fig. 7.1 Over a century after founding the first physiological institute in Switzerland at the Vesalianum, Miescher now has the Friedrich Miescher Laboratory (FML) at the Tübingen Max Planck Institute (**a**), and the Friedrich Miescher Institute in Basel (**b**) named in his honour. (© Friedrich-Miescher-Labor des Max-Planck-Instituts für Biologie Tübingen Credit: MPI für Biologie, Tübingen, Max Planck Society; Photograph reproduced with kind permission of the Friedrich Miescher Institute for Biomedical Research)

after all these years, he now investigated how the composition of blood varies at differing altitudes and showed that it is the concentration of carbon dioxide, not oxygen, in the blood that regulates breathing.

Miescher's return to nuclein might have been expected to bring him some relief, especially since, thanks to his new post as director of the institute, he now had new equipment such as a centrifuge at his disposal, which made the preparation of nuclein much easier.[8] But no amount of new equipment could make up for the immense frustration and despair that Miescher felt as he surveyed the vast amount of work on which he needed to catch up after a 13-year hiatus. The relentless grind was taking its toll on him. He was becoming tired and starting to show signs of depression. Alongside comparing his situation with that of Sisyphus, Greek mythology offered a ready supply of other powerful metaphors for his frustration and despair. 'I need fifty hands – not just two,' he cried in exasperation, 'Year after year, I am busy from early in the morning until midnight and there will be few of my colleagues who take any less rest. But the task is like the Lernaean Hydra—for each head that is struck off, six more grow in its place.'[9]

Not that Miescher was entirely alone in hacking away at those ever-multiplying heads. Because during the years that Miescher had been preoccupied with the sex life and migration patterns of salmon, or the diets of Swiss prisoners, another protégé of Hoppe-Seyler had turned his attention to nuclein.

7.3 Albrecht Kossel and the 'Building Bricks' of Life[10]

This was Albrecht Kossel (1853–1927), who was once described as 'a real scientific man, modest, kindly, simple, sincere, with a brilliant imagination and indefatigably at work in the laboratory even up to the time of his death.'[11] These admirable traits served him well, for in 1910 he received the Nobel Prize for Physiology or Medicine, having been nominated for the award eight

[8] Miescher, 16th May 1887, Letter LXI, in (His 1897); p. 105.

[9] Miescher, 2nd March 1891, Letter LXIX, in (His 1897); p. 108–109.

[10] Readers who would like to know more about Kossel's life and work might be interested in 'Albrecht Kossel: Ein Nobelpreisträger aus Mecklenburg' (Albrecht Kossel: A Nobel Prize Winner from Mecklenburg) - a biography of him by proud fellow Mecklenburgers, Edith and Joachim Framm (Wismar 2024).

[11] Ibid.

times since 1902.[12] The award was given "in recognition of the contributions to our knowledge of cell chemistry made through his work on proteins, including the nucleic substances", which sounds almost as if his research into those 'nucleic substances' had been included in the nomination as a mere afterthought.[13] But far from being eclipsed by his work on proteins, Kossel's insights into nuclein were to be of profound importance.

In his lecture to the Nobel Academy, Kossel articulated a simple, but powerful idea. This was that huge, complex biological molecules such as proteins were made up of smaller chemical building blocks—a kind of chemical Lego brick from which larger molecules could be assembled. Kossel gave these basic units the name 'Bausteine' which, in his native German, means 'building bricks' and by the time he gave his Nobel lecture it was becoming evident that the particular 'building bricks' from which proteins were built belonged to the group of chemicals known as amino acids. But what Kossel showed was that nuclein—(or 'nucleic acid' as it had become known by the time he was awarded the Nobel Prize) was also made of chemical building bricks—albeit of a different nature to those found in proteins.

Kossel grew up in the Prussian city of Rostock on the Baltic coast and was described as having 'a simple, friendly, affectionate and generous nature. He had nothing of the insolence, conceit and arrogance so often associated with the Prussian…'[14] In 1872, he headed south to study medicine at the newly established University of Strasburg where Hoppe-Seyler was now a professor. Following the Prussian victory in the war of 1870–71 with France, Strasburg and the surrounding region of Alsace-Lorraine had become German territory again and, with the founding of a new university there, Hoppe-Seyler had taken the opportunity to establish the first German department of biochemistry. The year that Kossel spent in Strasburg before returning to his native Rostock where he continued his studies, was long enough for him to fall under the spell of Hoppe-Seyler's courses in physiological and pathological chemistry. After completing his course in medicine at Rostock, Kossel rushed straight back to Strasburg to work as Hoppe-Seyler's assistant and pick up where Miescher had left off.

In his own work on nuclein, Miescher had shown that it contained carbon, nitrogen, oxygen, hydrogen and phosphorus, but Kossel wanted to know how these elements were arranged to build the molecule. To do this, he isolated

---

[12] https://www.nobelprize.org/nomination/archive/show_people.php?id=4995. Accessed 4/7/24.

[13] The Nobel Prize in Physiology or Medicine 1910. NobelPrize.org. Nobel Prize Outreach AB 2024. Thu. 4 Jul 2024. <https://www.nobelprize.org/prizes/medicine/1910/summary/>

[14] (Mathews 1927); p. 293.

nuclein from yeast before subjecting it to boiling or treatment with barium salts. This harsh treatment resulted in breakdown of the nuclein with the release of what Kossel described as 'a not insignificant amount of hypoxanthine.'[15] Hypoxanthine is an example of a purine, a type of organic molecule containing carbon and nitrogen atoms bonded together into two rings—one with five atoms, and the other with six. Together with pyrimidines, which contain carbon and nitrogen bonded into a single six membered ring, purines make up a chemical family called the nitrogenous bases. Kossel's detection of hypoxanthine showed that these nitrogenous bases (or nucleobases as they are sometimes known today) were a basic building block of nuclein, and their recent detection in meteorites and an asteroid has caused much interest and speculation about the origins of life on Earth (Fig. 7.2).[16]

This inspired him to write a book entitled 'Studies of Nuclein and its Breakdown Products', which was published in 1881 and established Kossel as an authority in this field. As a result of the book's success, he became Director of Chemistry at the Institute of Physiology in Berlin. It was a move which brought not only prestige, but also a rich new source of nuclein.

Using pancreatic tissue obtained from a local slaughterhouse, Kossel extracted nuclein and heated it with acid to discover that it contained two bases.[17] One of these had first been identified in 1844 by the German chemist Julius Unger (1819–1885) who had given it the name 'guanine' on account of having isolated it from bird excrement, or 'guano.' Guanine had already been found by several other researchers in a wide range of tissues including the pancreas, the retina and salmon sperm and was believed to be a product resulting from the breakdown of proteins.[18] But Kossel showed that this was not the case. He found that the high levels of guanine found in rapidly growing, nuclei-rich, tissues, such as leukaemic blood or embryonal muscle cells, were derived from the chemical decomposition of nuclein.[19,20]

The other nitrogenous base that Kossel had found in pancreatic nuclein was, however, a novel discovery and in reflection of its origins he named it 'adenine', after the Greek word for gland. Only a few years later Kossel and his PhD student, Albert Neumann, identified two more bases. Again, this was thanks to tissue obtained from the slaughterhouse. This time however, it was not pancreatic tissue but the thymus glands of calves which are stuffed full of

[15] (Kossel 1879); p. 291.

[16] See for example (Callahan et al. 2011; Witze, 2025).

[17] (Kossel 1885).

[18] (Kossel 1883).

[19] Ibid.

[20] (Kossel 1882); p. 16.

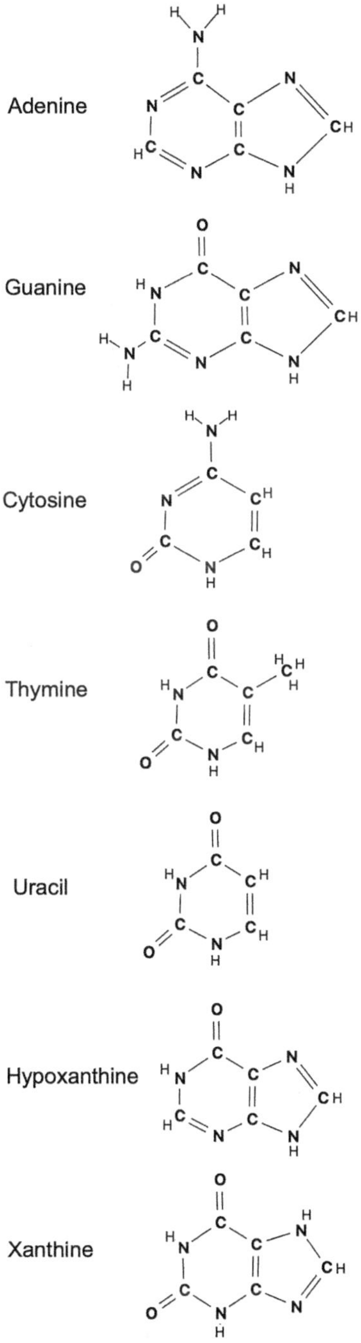

Adenine

Guanine

Cytosine

Thymine

Uracil

Hypoxanthine

Xanthine

Fig. 7.2 DNA contains four different types of nitrogenous base. These are adenine and guanine, which belong to the purine group of nitrogenous bases, and cytosine and thymine, which belong to the pyrimidine group. RNA meanwhile also contains adenine, guanine and cytosine but in place of thymine contains uracil. Hypoxanthine, which Albrecht Kossel detected in significant amounts in nuclein, is a breakdown product of guanine, and can be further oxidized to produce the base xanthine

white blood cells with huge nuclei, and thus a rich source of nuclein.[21] Kossel and Neumann christened one of these two new bases, 'cytosine' from the Greek word for cell and the other, 'thymine' since it had been found in nuclein from thymus tissue.

In a discussion of Kossel's work for a podcast recorded in 2022, biochemist Mark Lorch pointed out that generations of post-docs and PhD students the world over owe a debt of gratitude to Kossel for choosing these particular names for the bases.[22] Today, scientists working with sequences of genetic information abbreviate the bases to single letter—A, C, G, or T—based on the first letter in their name. Had Kossel instead given any two of these bases a name that began with the same letter, scientists today might well have been cursing him for having made the task of deciphering genetic sequences far more formidable than it already is (and the 1997 sci-fi film GATTACA might have had a far less catchy title).

Whilst analysing the chemical composition of nuclein, Kossel also speculated as to what its function might be, and he was clear that one possibility could be ruled out. Miescher had proposed that nuclein might act as a store of phosphorus to serve an organism's metabolic needs. But from his own experiments, which had shown that the quantity of nuclein remained unchanged regardless of whether an animal was starved, Kossel rejected this suggestion as improbable.[23,24]

Kossel's analyses of the bases and his careful quantification of their amounts also revealed why Miescher had been mistaken when he claimed to have discovered that the yolk grains of hen's eggs contained large amounts of nuclein. Believing egg cells to contain vastly more nuclein than sperm, Miescher had been misled by this finding (initially at least) into dismissing the nuclein content of sperm as playing any significant role in fertilisation. Kossel reasoned

[21] (Kossel and Neumann 1893); (Kossel and Neumann 1894).

[22] Professor Mark Lorch, University of Hull in 'Historical Perspectives on Contemporary Issues: The DNA Papers—Episode 2—Albrecht Kossel', Consortium for History of Science, Technology and Medicine. https://soundcloud.com/user-287120959/dna-papers-episode-2-albrecht-kossel

[23] Ibid., p. 15.

[24] Kossel's view was shared by John James Rickard Macleod (1876–1935) a young researcher who had come from his native Aberdeen to spend a few years working with Carl Ludwig in Leipzig. Citing Miescher and Kossel's work, Macleod observed that the high levels of inorganic phosphorus found in muscle could not possibly originate from nuclein because there was simply not enough of it (Macleod 1899). Macleod then went on to show that, during periods of intense muscle activity, levels of another phosphorus-containing compound that was known at the time as 'Nucleon' showed a sharp drop. What Macleod was most likely describing here was adenosine triphosphate—better known today by its acronym, ATP—which is the major currency of energy in the cell. Impressed by the couple of years that he had spent in the German school of physiological chemistry, Macleod returned to Scotland, and then to Canada where in 1923 he shared the Nobel Prize in Physiology or Medicine with Frederick Banting for the discovery of insulin.

however that if egg yolk contained as much nuclein as Miescher claimed, then its degradation should result in significant amounts of the four bases. But when Kossel tested egg yolk for their presence, he found nothing. His conclusion was that the high levels of phosphorus which Miescher had reported in egg yolk arose not from nuclein but from other compounds such as lecithin.

As for what the function of nuclein might be, Kossel noted that an observation made by Louis Pasteur in 1858 might offer a clue. Pasteur had observed that fermenting yeast cells showed a massive increase in a nitrogen-containing substance that he found to be insoluble in weak sulphuric acid. Speculating about what this substance might be, Kossel wrote one simple word—'Nuclein?'[25]

Writing to his uncle in 1869, Miescher had expressed an interest in finding out whether tissues which were undergoing rapid division, such as those in a tumour, contained higher levels of nuclear material.[26] This was also a question which intrigued Kossel. He found that cells rich in nuclei, such as those of the spleen, liver and pancreas also contained high amounts of nuclein as did those undergoing rapid growth such as in an embryo or in leukaemic blood.[27,28] Could it be that nuclein was somehow involved with the multiplication of cells and the formation of tissues?

Kossel certainly thought so, but he was also intrigued to know whether all nuclein, whatever its source, was made up of the four bases he had found so far. To address this question, he returned once more to yeast nuclein and found that, like nuclein from pancreatic tissue or the thymus gland, it too contained adenine, cytosine and guanine. But when he set another of his students, Alfredo Ascoli, the task of finding out whether it contained thymine, he got a surprise.

Ascoli could not detect any thymine in the nuclein of yeast but instead found a new base that he called 'uracil' on account of it being identical to a compound that had been synthesized in the laboratory from uric acid. But this was the first time uracil had been found to occur naturally, and even more puzzling was that it was not found in nuclein from pancreas or thymus tissue.

By the time that Ascoli made this discovery, nuclein had acquired a new name. This had first been coined in 1889 by the German pathologist and histologist Richard Altmann (1852–1900) (who had been one of the few students that Miescher had ever taken on) and, for reasons that will become clear

[25] (Kossel 1879); p. 285. Kossel, 1879; 1880.

[26] Miescher to Wilhelm His, 20th December 1869, Letter V,in (His 1897); pp. 39–41.

[27] (Kossel 1882); p. 12.

[28] Ibid; p. 15, 21, 22.

shortly, it had left Miescher distinctly unimpressed. Altmann had isolated a substance from the nuclei of yeast, calf thymus glands and sperm cells that shared very similar properties to Miescher's. It too had a high content of phosphorus, was precipitated by acids but dissolved by weak alkaline solutions. Yet despite these similarities, Altmann insisted that what he had found was a novel substance distinct from that isolated by Miescher. He called this substance 'nucleic acid', and the name was quickly adopted—perhaps because simply by including the word 'acid', Altmann's new name was perceived by the scientific community as having a greater precision than Miescher's 'nuclein.'

A year after Ascoli had found that yeast nucleic acids contained uracil, Thomas Osborne (1859–1929) and Isaac Harris at the Laboratory of Agricultural Research, Connecticut found that uracil was also present in nucleic acid isolated from wheat embryos. This led them to suggest that the presence or absence of uracil might be indicative of 'a significant difference between animal and plant[29] nucleic acids' with the added caveat that 'further investigation will show how far this applies.'[30,31]

In time, further investigation showed that although nucleic acids could indeed be divided into two types as Osborne and Harris had suggested, this distinction had nothing to do with whether they were of plant or animal origin. Instead, it depended on whether they contained uracil and one other chemical building block that Kossel had identified. In his analysis of yeast nucleic acid Kossel had shown that it contained a type of sugar known as a pentose because it contained five carbon atoms. This pentose sugar component was later identified as ribose, which led to what had once been called yeast nucleic acids becoming known as ribonucleic acid, or RNA. The nucleic acid that Kossel had isolated from pancreas and calf thymus tissue and which did not contain uracil was later shown to also contain a pentose sugar. This pentose sugar differed in structure from the ribose found in RNA only by the lack of a single oxygen atom for which reason it became known as deoxyribose. And the nucleic acids which contained this sugar therefore became known as deoxyribonucleic acid—or DNA.

---

[29] Osborne and Harris appear to have considered yeast as being some primitive type of plant cell. In their paper, they state: 'To our knowledge this is the first nucleic acid which has hitherto been isolated from higher plants, for the one nucleic acid of vegetable origin, which has hitherto been preserved in an almost pure state, has been found in yeast.' (Osborne and Harris 1902; p. 85). In his 'Physiologische Process der Athmung: Akademische Habilitationsrede 1871' Miescher describes yeast as 'our prototype, the yeast cell, itself again a plant organism with all its secrets.', p. 53).

[30] (Osborne and Harris 1902).

[31] Ibid. p. 121.

We now know that it is not the case that DNA is found only in animal cells while RNA is found only in plant cells as Osborne and Harris had proposed. The cells of all living organisms—plants, animals, funghi, archaea and bacteria—contain both DNA and RNA. But until the COVID-19 pandemic and the development of RNA-based vaccines against SARS-CoV 2, although most people grasped that DNA did something very important in biology, few outside a molecular biology lab would have heard of RNA. Yet RNA is a highly versatile and very important molecule which, as is becoming ever more apparent, deserves to share some of the spotlight with its more famous molecular relative.

7.4 'A Lucky Catch'

Long before the realisation that nucleic acids could be classified as DNA or RNA, a distinction of a different kind had left Miescher fuming. This was Richard Altmann's claim that nucleic acid, which he claimed to have discovered, was distinct from Miescher's nuclein. Yet while this irked Miescher, it was a distinction that others, including Kossel appeared to accept:

> *Nucleic acids are the components of young, developmentally capable, cells throughout the animal and plant kingdoms. They are found either freely, or in association with proteins as 'Nuclein' in the material of the nucleus*[32]

For Miescher, the terms 'nuclein' and 'nucleic acid' were much more than a question of mere semantics. Altmann was suggesting that what Miescher called 'nuclein' was not a pure substance, but rather a mixture of what Altmann had christened 'nucleic acid' together with proteins. True, Altmann conceded that the nuclein Miescher had isolated from salmon sperm (which Altmann referred to as 'spermanuclein') was most likely analogous, if not identical to the substance he called nucleic acid.[33] 'Should the identity of Miescher's spermanuclein prove to be the same as my preparation,' he wrote, 'I would plead for the former to be called nucleic acid.'[34]

But Altmann did not think this was the case for the nuclein that Miescher had originally isolated from leukocytes in pus as described in his 1871 paper. He drew attention to the fact that whereas Miescher had shown

[32] (Kossel and Neumann 1894); p. 2753.

[33] (Altmann 1889); p. 532.

[34] Ibid. p. 533.

'spermanuclein' to contain no sulphur, his original preparation of nuclein from pus had contained traces of this element. As sulphur was known to be found in proteins, Altmann concluded that Miescher's nuclein was a mixture of proteins together with an acid that contained high levels of phosphorus which he called 'nucleic acid.'[35]

Altmann believed that his success in separating nucleic acid from proteins had been thanks to the use of a protocol for the preparation of his material that differed slightly from the one used by Miescher. Unlike Miescher, Altmann had first precipitated the nuclear material with acid—a step which he believed had removed all traces of protein to leave only the nucleic acid component. Miescher however, was not quite so impressed. As far as he was concerned, nuclein and nucleic acid were one and the same. He described 'nucleic acid' as being little more than 'a lucky catch' made thanks only slight differences in Altmann's purification protocol.[36] And he noted (perhaps not without a little schadenfreude) that the yield from Altmann's procedure was rather low. His verdict was final—'My salmon nuclein is of course the same as his nucleic acid and was certainly the purest of them.'[37]

In one sense at least, Altmann may have been right about Miescher's nuclein being a mixture. In 2021, researchers in Tübingen carried out a reconstruction of Miescher's original isolation of nuclein. Using protocols that were as close as possible to the ones Miescher had described in his paper of 1871, they found not only DNA, but also RNA in their preparations. When they then analysed a sample of the original nuclein which Miescher had isolated from salmon sperm and which today, stands on display in the museum in Tübingen, they found traces of the base uracil which is only found in RNA. Fittingly, this work was reported in the same journal (by now renamed) in which Miescher's original article had been published.[38]

The criticism that nuclein might not be a novel substance but a mixture must nevertheless have stung. This was the very same criticism that had been levelled at Miescher by detractors such as Kingzett and Hake, but to hear it come from Altmann of all people must have rubbed salt into the wound. For although he was now Professor of Anatomy at the University of Leipzig, Altmann had once been one of the handful of students who had trained under Miescher. Now it seemed that the former student was trying to usurp his own mentor with his claim to have discovered 'nucleic acid.' Little wonder then

[35] (Altmann 1889); p. 524.

[36] Miescher to Hoppe-Seyler, 20th July 1872, Letter XXVII, in (His 1897); pp. 68–69.

[37] Miescher, 2nd March 1891, Letter LXIX, in (His 1897); p. 108–109.

[38] (Thess et al. 2021).

that, in the words of historian Sophie Veigl, Miescher was 'super-pissed' at Altmann.[39]

To add to Miescher's annoyance, he did even not consider this name to be particularly novel.[40] In 1872, he had already described the acidic properties of nuclein to Hoppe-Seyler and had himself used the term 'nucleic acid' in a letter to his friend Rudolf Boehm.[41] Miescher therefore made a point of deliberately referring to 'Altmann's nucleic acid' as if to emphasize that his former student had not discovered a novel substance, but simply coined a new term for nuclein. But in a paper on salmon sperm, published posthumously with his friend and colleague, the German pharmacologist Oswald Schmiedeberg (1838–1921), he nevertheless tried to retain a tone of magnanimity:

> *Since I myself emphasized the strongly acidic properties of the substance obtained from salmon sperm, I have no objection to the term 'nucleic acids' proposed by Altmann.*[42]

There was more at stake here than just an argument about terminology. By coining the term 'nucleic acid', Altmann secured his place in the history books. When the eminent US biologist Edmund Beecher Wilson published the first edition of his influential book 'The Cell in Development and Heredity', Miescher's claim to have already identified the acidic component of nuclein was given little regard:

> *Altmann (89) opened the way to an understanding of the matter by showing that "nuclein" may be split up into two substances; namely (1) an organic acid rich in phosphorus to which he gave the name nucleic acid, and (2) a form of albumin.*[43]

With Altmann's rebranding of nuclein as nucleic acid, the charges originally levelled by Kingzett and Hake that nuclein was 'the great Unwashed', appeared to have stuck once and for all. Historian Sophie Veigl has described the perception of Miescher by his scientific peers as being that of 'a contaminator' who had succeeded only in isolating a poorly characterised mixture, in contrast to Altmann who had obtained a pure substance.[44] Critics who wished to

---

[39] Sophie Veigl, in 'Historical Perspectives on Contemporary Issues: The DNA Papers—Episode 2—Albrecht Kossel', Consortium for History of Science, Technology and Medicine - https://soundcloud.com/user-287120959/dna-papers-episode-2-albrecht-kossel

[40] (Dahm 2008); p. 575.

[41] Miescher, 2nd May 1872, Letter XXVIII, in (His 1897); p. 70–73.

[42] (Miescher and Schmiedeberg 1896) in (His 1897) pp. 359–414; p. 369.

[43] (Wilson 1896); p. 240.

[44] (Veigl et al. 2020).

portray Miescher as such could also point to the fact that when he and his co-worker Jules Piccard had carried out chemical analyses of protamine they had found that it appeared to contain the purine base xanthine. This was a puzzle as xanthine is not found in proteins but is instead derived from guanine in nucleic acids. Only later did Albrecht Kossel show that the xanthine Miescher and Piccard had found in their preparations of protamine was due to contamination by nucleic acid that had co-purified. The situation was further complicated when Miescher described the isolation from ox-sperm of what he claimed to be a nuclein that contained sulphur, but which was clearly residual protein—most likely, protamine.[45]

And yet at the same time as Miescher was acquiring this reputation of 'the contaminator' and being eclipsed by his former student Altmann, he was nevertheless drawing support from an unlikely quarter.

7.5 'The Guild of Dyers'

Miescher's unlikely allies were cell biologists—which was somewhat ironic considering the scepticism that he had often expressed towards their experimental methods:

> *One of these [cellular] processes may completely hide another from the observation of a microscopist. One and the same analytical result can be interpreted differently... One waits willingly for a better survey of the field of microscopy.*[46]

When Hoppe-Seyler had set up his 'Journal of Physiological Chemistry' (*Zeitschrift für Physiologische Chemie)*, Miescher remarked with dry wit that 'I hope that you will not question that the chemical facts could be applied to histological questions; in the worst case this could lead even a partisan microscopist to take up your journal.'[47]

Miescher's disdain for cytology along with his conviction that a chemical approach was naturally superior, may have been rooted in his own brief and frustrating experience of using staining methods in microscopy. At the beginning of his 'sperm campaign' in the early 1870s, Miescher had studied the heads of sperm cells following the approach developed by Flemming which combined preparations of the heavy metal osmium with methylene green and

[45] (Miescher 1874) in (His 1897); pp. 101-102.
[46] Miescher, 20th Dec 1869, Letter V, in (His 1897); p. 39.
[47] Miescher, 20th Oct. 1870, Letter XII in (His 1897); p. 51.

Gentian violet dyes. While the outer shell of the sperm cell appeared not to take up the dye, the innermost part of the cell showed a rich staining. But on seeing that this sharp distinction rapidly disappeared when treated with hydrochloric acid, Miescher concluded the difference in staining was nothing more than an artefact and could offer no reliable information about the internal structure of the cell and its contents. Such results were, he wrote, 'a poor consolation for those histologists who are accustomed to regard the elective relationship to the aforementioned and other dyes without further question as showing direct and certain reactions to nuclein bodies.'[48]

Not that Miescher thought microscopy to be entirely without worth. But he believed that, for cytology and staining methods to be of any value, they must 'go daily hand in hand with chemical studies if anything is to be achieved.'[49] 'Histology…must first be further developed on a chemical basis, and not just with gold or osmium stains,' he argued, 'Chemistry must teach us to abandon or reorganise much of what has been dispersed or thrown together according to insignificant (optical) properties.'[50]

Yet although Miescher was dismissive of their microscopy-based approach, cell biologists such as the botanist Eduard Zacharias (1852–1911) were showing great interest in nuclein. Zacharias was sufficiently intrigued by Miescher's isolation of this substance from the nuclei of animal cells, to investigate whether it was also present in the cells of plants. Since Miescher had reported nuclein to be resistant to degradation by the proteases found in stomach juice, and insoluble in mild acids but dissolved by mild alkali, Zacharias subjected nuclei in cells from plants such as buttercup to these same tests from which he concluded that they too contained mostly nuclein. But he also went further, showing that the microscopic sub-cellular structures which at that time were simply known as 'threads' and had been observed by to line up equatorially during cell division, were made of nuclein.[51]

These 'threads' had first been observed by the eminent Polish-German botanist Eduard Strasburger and would soon become known by a new name thanks to German cell biologist Walther Flemming (1843–1905). When not giving away sizeable chunks of his salary to beggars in the city of Kiel where he was Professor of Anatomy, Flemming was busy cultivating a different kind of legacy thanks to a growing interest in nuclein. Using chemical fixatives such as osmic, chromic and acetic acid in combination with recently

[48] (Miescher and Schmiedeberg 1896) in (His 1897); p. 385.

[49] Miescher, 1870, Letter VIII, in (His 1897); p. 46.

[50] Miescher to Hoppe-Seyler 25th June 1876, Letter LXI, in (His 1897); pp. 83–86.

[51] (Zacharias 1881); p. 175.

discovered coloured aniline dyes that were becoming readily available due to the rise of the German chemical industry, Flemming pioneered methods to stain and observe structures within the cell.[52]

His main area of interest was in the thread-like structures that had been observed to appear in the nucleus when the cell underwent division. As these structures readily took up coloured stains, Flemming coined a new name for the material from which he supposed they must be made:

> *Since the term 'nuclear substance' would obviously be subject to so many misunderstandings, I will use the name chromatin for the time being. This is not intended to prejudge the fact that this substance must be a definitely constituted chemical body that remains constant in all nuclei; although this is certainly possible, we still do not know enough about nuclear materials to accept it. Chromatin should therefore only be designated for the little substance in the cell nucleus that absorbs the colour as treatments with dyes known in nuclear tinctures.*[53]

Chromatin derives its name from the Greek word for 'coloured substance', a reference to its capacity for taking up dyes. And although Flemming might well have intended the name 'chromatin' to be merely provisional and contingent upon new chemical discoveries, nearly 150 years later, his term is not only still in use but has become part of the lexicon of cell and molecular biology. Moreover, the staining of the nuclear scaffold with dyes suggested to Flemming that nuclein and chromatin were one and the same substance:

> *The scaffold owes its refractive power, the nature of its actions, and in particular its significant colorability to a substance which I have provisionally called chromatin with regard to the latter. It is possible that this substance is almost identical to nuclein body; in any case, it is clear from the experiments of Zacharias that it is the bearer of it, and if not of nuclein itself, at least of the compounds from which it is split. I retain the name chromatin until the decision is made by chemistry, and use it empirically to designate "the substance in the cell nucleus that absorbs colour during tinctions.*[54,55]

The scaffold of thread-like structures, to which Flemming referred, and which he proposed to be made of chromatin, had been called 'batonnets' by the Belgian cytologist Edouard van Beneden (1846–1910) on account of their rod-like appearance. But within only a few years of Flemming having coined

[52] (Williams 2019); p. 29.

[53] (Flemming 1880); p. 157–158.

[54] (Flemming 1882); p. 142.

[55] 'tinction' was a term used in the late nineteenth century to describe the act of staining or imbuing with colour.

'chromatin', his term had lent itself to a new name for the threadlike structures themselves thanks to German biologist Heinrich Wilhelm Gottried von Waldeyer-Hartz (1836–1921) who rechristened van Beneden's 'batonnets' as 'chromosomes.'[56]

Eager to know more about chromosomes, and the possible role they might play in fertilisation, the German anatomist and embryologist Oskar Hertwig (1849–1922) of the University of Berlin headed for the sunshine of the French Riviera where he became the first person to witness sexual reproduction—at the cellular level, at least. Around the same time as Miescher was comparing the unfertilised egg to an unwound clock that needs to be nudged back into action by the physical impact of sperm, Hertwig was extracting sperm and eggs from sea urchins and mixing them together under the microscope. What he observed was that, after the sperm had penetrated the egg, its nucleus fused with that of the egg to create a single nucleus of the newly formed embryo.

Using Flemming's staining reagents, Hertwig became convinced that chromatin and nuclein were indeed one and the same substance.[57] And Hertwig was confident that this substance played some crucial role in biology. Showing remarkable insight, he proposed that 'nuclein is a substance that transmits the characteristics of parents to children and is transferred from cell to cell during development.'[58] He even had an idea as to how this might be achieved—'as a hypothesis one can safely assume that the nuclein of both types of sex cells is not of the same composition and the nuclein of the male gametes is somehow different to that in the egg cells.'[59]

Würzburg histologist Albert von Kölliker (1817–1905) not only echoed Hertwig's views but extended them, by recognising that the role of nuclein might not necessarily be confined to fertilisation:

> *...the question arises as to whether perhaps the nuclein discovered by Fr. Miescher (more recently chromatin) is not only active in fertilization, as may be assumed on the basis of the new experience about the processes involved in fertilization, but also the substance which, as Sachs first said of plants, 'the fertilized embryos and the vegetation points that emerge from them owe their ability to shape, or, as we might also say, the substance that Nägeli calls idioplasm and which causes heredity.*[60]

[56] (Waldeyer 1888); p. 48.

[57] (Hertwig and Hertwig 1884); p. 11

[58] (Hertwig 1885); p. 310; (Hertwig and Hertwig 1884); p. 89.

[59] (Hertwig 1885); p. 310; (Hertwig and Hertwig 1884); p. 89.

[60] (Kölliker 1885); p. 40.

Considering the flak he had taken from fellow physiological chemists over nuclein, Miescher might have been expected to welcome this show of support from the cell biologists with open arms. But on the contrary, he complained instead that:

> *I must again defend my skin against the Guild of Dyers who insist that there is nothing but chromatin (nuclein), whereas the presence of various spatially separated substances in the sperm head can be easily demonstrated*[61]

Miescher's reference to 'the Guild of Dyers' was a perjorative term for cell biologists whom he felt to be over-reliant on staining techniques for microscopy. Miescher felt this approach was too prone to artefacts and misinterpretation as to be of much use when unaccompanied by the rigour of chemistry. It is rather ironic therefore that while physiological chemists traded blows on the pages of journals over the existence of nuclein, members of the 'Guild of Dyers' such as Zacharias, Flemming, Hertwig and von Kölliker had not only accepted its existence but were also showing remarkably clear insights into its function. Yet when the eminent German biologist August Weismann (1834–1914) published his own evidence that the nucleus was the carrier of heredity and postulated that this resided at the molecular level, Miescher's response was scathing—'The speculations of Weismann and others are afflicted with half-chemical concepts which are partly unclear, and partly correspond to an obsolete state of chemistry.'[62] As an example, he challenged Weismann's talk of 'Continuity' in the context of heredity as being vague in contrast to his own use of the term[63]:

> *Continuity does not only lie in the form, it also lies deeper than the chemical molecule. It lies in the constituent groups of atoms. In this sense I am to the utmost, an adherent of a chemical model of inheritance.*[64]

In his curt dismissal of the cell biologists, Miescher was evidently unaware of the adage 'not to bite the hand that feeds you.' And despite Miescher's confidence in the superiority of the chemical approach, it too was not without its pitfalls.

---

[61] Letter LXVII 10th Nov 1890 (His 1897); p. 107–108.

[62] Letter LXXVIII 13th Oct 1893 (His 1897); p. 122.

[63] See for example (Weismann 1883); p. 5—'Heredity in these single-celled organisms rests on the continuity of the individual, whose bodily substance is constantly multiplied by assimilation'; p. 6 'If it were only a question of the continuity of the substance of the germ cell from one generation to another… ' also (Weismann 1891).

[64] Letter LXXVIII 13th Oct 1893 (His 1897); p. 122.

7.6 Karyogen—Another Nuclein—Or a Chemical Mirage?

One reason that Miescher did not pay more attention to the work of the cell biologists was perhaps because he was too excited by his discovery of what he believed to be yet another novel and important cellular substance.[65] Calling this substance 'karyogen', Miescher believed that it differed from nuclein in several crucial ways. Unlike nuclein, karyogen appeared to contain iron but no phosphorus. Miescher was also convinced that it was karyogen, and not nuclein, which took up the coloured stains that cytologists such as Flemming had used to identify chromatin.[66,67] But along with nuclein, he believed that karyogen too played a central role in fertilisation:

> *Sexual reproduction results from the fact that the three fundamental substances of the male cell - nuclein, karyogen and protoplasmic protein…are brought together with the corresponding substances of the female cell and by some means are united through some karyokinetic pantomime dance.*[68]

Perhaps feeling somewhat eclipsed by his former student Altmann's coining of the term 'nucleic acid', Miescher may have hoped that karyogen presented him with an opportunity to reassert his authority in the field of physiological chemistry. 'If the majority of years were not behind one and there were a couple of hundred years in which to live,' Miescher enthused over his new discovery, 'it would be a joy to work on such things [as Karyogen].'[69] But sadly, Miescher had barely another 3 years in which to work on such things, let alone a hundred.

7.7 Death in Davos: The Not-So Magic Mountain

Throughout his life, Miescher had always been driven by a fearsome work ethic and perfectionism. Reflecting on what motivated him, Miescher likened it to 'the passion of the hunter, the soldier, the chess player, the charm of the

[65] Miescher, 'Ueber Bau und Mutmassliche Bildungsweise der Spermatozoen', 13th Oct 1892, Letter LXXIV, in (His 1897); p. 112–117.

[66] Miescher 22nd July 1893 to His, Letter LXXVI, in (His 1897); p. 117.

[67] Miescher 13th October 1893 to His, Letter LXXVIII, in (His 1897); p. 122.

[68] Miescher, 'Ueber Bau und Mutmassliche Bildungsweise der Spermatozoen', 13th Oct 1892, Letter LXXIV, in (His 1897); p. 116.

[69] Miescher, 28th July 1892, Letter LXXII, in (His 1897); p. 110.

struggle with the perils and caprices of natural phenomena, or in the mathematician with the laws of thought of one's own mind.'[70] But there was also another psychological force at work too—and one that was far less healthy. As the years rolled by, Miescher was plagued by an ever more painful sense of having failed to live up to his potential. An uncomfortable feeling of having squandered a golden opportunity gnawed away at him. Worse still, he feared that it was an opportunity upon which others might capitalise:

> *Only when, here and there, I read in the published work of another author, a half-formed fragment of that which I have put together, do I realize what could have been achieved with my material.*[71]

Miescher now slept less, hardly socialised, and insisted on working through holidays. All of which was taking a heavy toll on his health. He was becoming weak and pale, plagued by fits of sweating and fever with a dry cough. These were classic symptoms of tuberculosis—the onset of which might well have been exacerbated by all those long hours he had spent working in freezing cold conditions to prepare samples of Nuclein.

Treatment for tuberculosis at that time was severely limited. Often, the best that could be done was for patients to retire to a sanatorium in a remote location, in the hope that a regime of rest, fresh air and a special diet might aid their recovery (Fig. 7.3).

With fresh mountain air believed to be particularly beneficial, the Swiss resort of Davos was at the centre of the sanatorium movement and featured as the setting for Thomas Mann's famous 1924 novel 'The Magic Mountain.' But for Miescher there was to be nothing particularly magical about his visits to Davos, which began in 1890. Four years later his condition had become so severe that he had to take up permanent residence there.

Wary that nuclein was slipping from his grasp, and driven as ever by his ferocious work ethic, Miescher tried to use his period of recuperation in Davos to write up some unpublished work on nuclein.[72] In a letter of 1894, he outlined his extensive and ambitious plans for future research, including the isolation of a breakdown product from sperm nuclein that seemed to be analogous to the base that Kossel had identified as thymine.[73]

---

[70] (Suter 1944); p. 10.

[71] Miescher, 2nd Mar 1891,Letter LXIX, in (His 1897); p. 109.

[72] Miescher, 1st Dec 1894, Letter LXXXI, in (His 1897); pp. 124–126.

[73] Miescher, F. 'Publicationspläne', 1st Dec 1894, Letter LXXXI, in (His 1897); pp. 124–126; p. 126.

a

b

Nr. 70.40.171/4

Motiv:
Die Liegekur

Datum der Aufnahme:
um 1895

Herkunft/Besitzer der Aufnahme:

Erklärungen:

Liegekur.

No. 2139. Künzli-Tobler, Zürich.

Dr.Karl Turban gründete das erste geschlossene Tuberkulose-Sanatorium im Hochgebirge an der Oberwiesstrasse in Davos-Platz (Eröffnung 1889). Er führte für die Patienten eine straffe Disziplin mit genau geregeltem Tagesablauf ein. Das Fundament seiner Kuren war die Freiluft-Liegekur, die bei allem Wetter, im Sommer und im Winter, täglich auf den grossen Terrassen des Sanatoriums eingehalten werden mussten. Die Kuren Turbans hatten Erfolg, und sein Sanatorium wurde zum Vorbild vieler anderer Sanatorien und brachte dem Kurort Davos einen neuen Aufschwung.

Fig. 7.3 (**a**) In 1889 Dr. Karl Turban founded the first closed tuberculosis sanatorium in the high mountains on Oberwiesstrasse in the Swiss alpine resort of Davos. He introduced strict discipline for the patients with a precisely organised daily routine founded on the daily practice of patients lying on the large terraces of the sanatorium in the open air, in all weathers, summer and winter. In April 1894, Miescher became a patient at Dr. Turban's sanatorium (Meurolt-Landolt, 1969; p.19). (Photograph reproduced with kind permission of the Medizinmuseum Davos). (**b**) Photograph of patients at a sanatorium in Davos taken at roughly the same time as Miescher would have been there. Like these patients, much of his recovery would have involved resting whilst out on a balcony in the fresh open air. (Image reproduced with kind permission of the Municipality of Davos Document Library)

But sadly, these were destined to remain nothing more than just plans on paper. Sensing that time was against him, the burden of things left undone began to weigh ever more heavily upon him. The Sisyphean boulder that Miescher had spent so many years pushing up a slope, was now slipping out of his grasp to roll backwards and crush his spirits into the ground:

I will never know the happiness that belongs to the man who has lived up to their station in a harmonious way to the satisfaction of themselves and others, so my basic mood is the uncomfortable feeling of one who has lost a button from their braces. For years, I have had to get used to the demands of my work being four times that of my ability to carry it out, such that I go to bed each night feeling like a schoolboy who has failed to complete his tasks.[74]

In May the following year, Miescher wrote to the University of Basel, explaining that although his doctors believed he had a good chance of making a gradual recovery, he was unable to say when he would be able to return to his post. Following a visit by an official from the University, his pension was increased from 1840 to 2000 Swiss Francs and Miescher asked that the University relieve him of his teaching responsibilities from the 1st October of that year.[75]

But if Miescher hoped that his chances of recovery might improve with the burden of teaching lifted, he was to be sadly mistaken. Having officially ended his employment with the University of Basel on 15th June 1895, Miescher died on 26th August 1895, 2 weeks after his 51st birthday.

At a memorial service held in Miescher's honour, one of his academic associates at the University of Basel offered some reflections on his life and work that sounded more like an end-of-term appraisal given by a Professor to a B-grade undergraduate student, than a eulogy. His words were unlikely to offer much consolation to Miescher's grieving widow, Maria Anna Rüsch and his three children, two of whom died at a relatively young age whilst the third later went insane:

If he did not reach the highest peaks of achievement, that was due solely to certain weakening and obstructing factors in his organization. Thus, even if we did not lose in him a teacher and investigator whose words and works were pioneering and deci-

[74] (His 1897); p. 31.

[75] (Cohen and Portugal 1977); p. 28.

sive for the development of his science and science in general...still Friedrich Miescher, thanks to his strong interests and his relentless drive to do research, his competence and knowledge in his own field and his general knowledge, his sharp critical faculties, and his correct recognition of all that is involved in scientific research, was a well-known and able scholar.[76]

An obituary in *The British Medical Journal* took a less condescending tone. Miescher was described as having been 'much beloved of his pupils and to all of them he was the type of a noble, true and unselfish friend.'[77] As for his discovery of nuclein, very little was said other than that it was 'a new phosphorus-containing body..., which forms the chief part of the cell nuclei.' Far more attention was given to his research into the life history of the Rhine salmon which was described by the Swiss physician and historian of medicine Franz Merke (1893–1975) just over a century later as a piece of 'forgotten, groundbreaking physiological research.'[78]

Wilhelm His also recognised that his late nephew had laid the foundations of something great—and it was not the migratory habits of salmon. Looking ahead to the new century that was about to dawn, His made a confident prediction about the legacy of his nephew:

The appreciation of Miescher and his works will not diminish with time, instead it will grow, and the facts he has found and the ideas he has postulated are seeds which will bear fruit in the future.[79]

Wilhelm His was right about one thing—nuclein was a seed that would bear an abundance of fruit. But it would lie dormant in fallow ground for quite some time yet—and whether Miescher would indeed be appreciated for having sown it there as his Uncle believed he would, was by no means certain.

---

[76] Ibid., p. 29.

[77] Professor Friedrich Miescher, BMJ 28th Sept 1895, p. 812.

[78] (Merke 1973).

[79] (His 1897); p. 4.

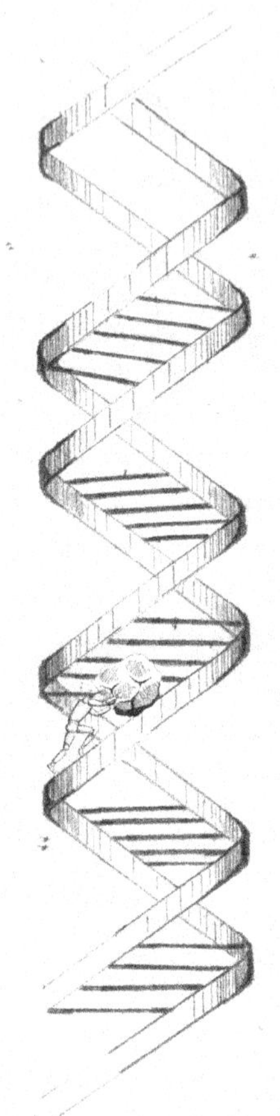

Drawing by Kersten Hall

References

Altmann, R. 1889. Ueber Nucleinsäuren. *Archiv für Physiologie* 1:524–536.

Callahan, M. P., K. E. Smith, H. J. Cleaves, J. Ruzicka, J. C. Stern, D. P. Glavin, C. H. House, and J. P. Dworkin. 2011. Carbonaceous meteorites contain a wide range of extraterrestrial nucleobases. *Proceedings of the National Academy of Sciences of the United States of America* 108 (34): 13995–13998. https://doi.org/10.1073/pnas.1106493108.

Cohen, J. S., and F. H. Portugal. 1977. *A century of DNA: A history of the discovery of the structure and function of the genetic substance*. Cambridge, MA: MIT Press.

Dahm, R. 2008. Discovering DNA: Friedrich Miescher and the early years of nucleic acid research. *Human Genetics* 122:565–581.

Flemming, W. 1880. Beiträge Zur Kenntniss Der Zelle Und Ihrer Lebenserscheinungen. *Archiv für Mikroskopische Anatomie* 18:151–259.
Flemming, W. 1882. *Zellsubstanz, Kern Und Zelltheilung*. Leipzig: F. C. W. Vogel.
Hertwig, O. 1885. Das Probleme Der Befruchtung Und Der Isotropie Des Eies, Eine Theorie Der Vererbung. *Jenaische Zeitschrift Für Medizin Und Naturwissenschaft* 18:276–319.
Hertwig, O., and R. Hertwig. Untersuchungen Zur Morphologie Und Physiologie Der Zelle, 1884.
His, W. 1897. *Die Histochemischen Und Physiologischen Arbeiten von Friedrich Miescher*. Leipzig: F. C. W. Vogel.
Kölliker, A. 1885. Die Bedeutung Der Zellenkerne Für Die Vorgänge Der Vererbung. *Zeitschrift Für Wissenschaftliche Zoologie* 42:1–46.
Kossel, A. 1879. Ueber Das Nuclein Der Hefe. *Zeitschrift Für Physiologische Chemie* 3:284–291.
Kossel, A. 1882. Zur Chemie Des Zellkerns. *Zeitschrift Für Physiologische Chemie* 7:7–22.
Kossel, A. 1883. Ueber Guanin. *Zeitschrift Für Physiologische Chemie* 8:404–410.
Kossel, A., and A. Neumann. 1894. Darstellung Und Spaltungsprodukte Der Nucleïnsäure (Adenylsäure). *Berichte der Deutschen Botanischen Gesellschaft* 27:2215–2222. https://doi.org/10.1002/cber.189402702206.
Macleod, J. J. R. 1899. Zur Kenntnis Des Phosphor Im Muskel. *Zeitschrift Für Physiologische Chemie* 28:535–558.
Mathews, P. 1927. Professor Albrecht Kossel. *Science* 66.
Merke, F. 1973. Bemerkungen Zu Vergessen, Grundlegenden Physiologischen Untersuchungen Am Wandernden Lachs Durch Den Basler Physiologen Friedrich Miescher. *Gesnerus* 30:47–52.
Miescher, F. 1874. Die Spermatozoen Einiger Wirbeltiere. Ein Beitrag Zur Histochemie. *Verhandlungen Der Naturforschenden Gesellschaft in Basel* 6:138–208.
Miescher, F. 1880. Statistische Und Biologische Beiträge Zur Kenntnis Vom Leben Des Rheinlachses Im Süsswasser. In *Abdruck Aus Dem Wissenschaftlichen Anhang Zum Katalog Der Schweizerischen Betheiligung an Der Internationalen Fischereiausstellung in Berlin*.
Miescher, F., and O. Schmiedeberg. 1896. Physiologische-Chemisch Untersuchungen Über Die Lachsmilch. *Naunyn-Schmiedeberg's Archives of Pharmacology* 37:100–155.
Osborne, T. B., and I. F. Harris. 1902. Die Nucleinsäure Des Weizenembryos. *Hoppe-Seylers Zeitschrift Für Physiologische Chemie* 36 (2–3): 85–133.
Suter, F. 1944. Prof. F. Miescher. Persönlichkeit Und Lehrer. *Helvetica Physiologica et Pharmacologia Acta Supplementa* 2:5–17.
Thess, A., I. Hoerr, B. Yazdan Panah, G. Jung, and R. Dahm. 2021. Historic nucleic acids isolated by Friedrich Miescher contain RNA besides DNA. *Biological Chemistry* 402:1179–1185.

Veigl, S. J., O. Harman, and E. Lamm. 2020. Friedrich Miescher's discovery in the historiography of genetics: From contamination to confusion, from Nuclein to DNA. *Journal of the History of Biology* 53:451–484. https://doi.org/10.1007/s10739-020-09608-3.

Waldeyer, W. 1888. Ueber Karyokinese Und Ihre Beziehungen Zu Den Befruchtungsvorgängen. *Archiv für Mikroskopische Anatomie* 32:1–122.

Weismann, A. 1883. *Über Die Vererbung*. Jena: Fischer.

Weismann, A. 1891. *Essays upon heredity and kindred biological problems*. Oxford: Clarendon Press.

Williams, G. 2019. *Unravelling the double helix: The lost heroes of DNA*. London: Weidenfeld and Nicholson.

Wilson, E. B. 1896. *The cell in development and heredity*. 1st ed. London: Macmillan.

Witze, A. 2025. Asteroid fragments upend theory of how life on earth bloomed. *Nature*. https://doi.org/10.1038/d41586-025-00264-3.

Zacharias, E. 1881. Ueber Die Chemische Beschaffenheit Des Zellkerns. *Botanische Zeitung* 39:170–177.

8

'The Wooden Stretcher Behind the Rembrandt'?

Contents

Abstract Although the discovery of nuclein (DNA) had been met with suspicion—and sometimes hostility by many in the field of physiological chemistry, it received support from an unlikely quarter. Cell biologists—whose methods had been dismissed by Miescher as 'the guild of dyers', showed that nuclein was the main component of chromosomes—the threadlike structures which many believed to mediate heredity. These ideas came tantalisingly close to unveiling the real function of DNA inside organisms. But by the start of the twentieth century, things had changed. This chapter explores why nuclein fell from favour as being the most likely candidate to be the material of heredity and why, for almost 50 years, it was relegated to being thought of as being little more than, in the words of the writer Horace Freeland Judson, it 'the wooden stretcher behind the Rembrandt.'

K. Hall, R. Dahm, *The Dawn Fisherman*, Copernicus Books,
https://doi.org/10.1007/978-3-032-14219-1_8

Keywords Levene • Tetranucleotide hypothesis • Nucleic acids • DNA • Structure • Genetics • Proteins • Molecules • Bases • Phosphate • Sugar • Biochemistry • History • Science • Theory • Experiment • Discovery • Critique • Revolution • Watson • Crick • Double helix • Genetic material • Molecular biology • John Mason Gulland • Presentism

Interest in Nuclein Starts to Wane

As Miescher lay dying in Davos, clinicians in the United States were making bold claims for nuclein. 'I believe, with others even more conservative and less sanguine than I, that it will mark an era in therapeutics' wrote one of them in the Journal of the American Medical Association, whilst another hailed nuclein treatment as having 'acted like a charm' against bacterial infections such as diptheria.[1,2] In a somewhat grim twist of irony, another condition against which nuclein appeared to be effective was tuberculosis—the very disease that was killing Miescher at the time.[3]

Within a year of Miescher's death, nuclein was being praised in the *Journal of Nervous and Mental Disorder* as having 'proved itself one of the most pronounced successes of modern therapy.'[4] But this optimism quickly evaporated. Several clinicians had observed that injections of nuclein were followed by elevated levels of leucocytes and inferred from this that its apparently miraculous effects might be due to the stimulation of leucocyte production.[5] But in what appears to have been a monumental example of confusing correlation with causation, the claims by American doctors that nuclein was an immune-system boosting anti-bacterial wonder-drug turned out to be false.

Except for a single article in the *Journal of the American Medical Association*[6] that mentioned Miescher's research into the metabolic changes during salmon migration, and another which included a lone footnoted reference to his work, the excited discussion around the alleged therapeutic benefits of nuclein made no mention at all of its recently deceased discoverer.[7]

Elsewhere too, interest in nuclein was starting to wane. In the immediate years following Miescher's death, cytologists had become ever more confident

[1] (Summers 1895); p. 966.

[2] (Jackson 1895); p. 872.

[3] (McClintock 1895).

[4] (Allbright 1896).

[5] (Ames and Huntley 1897).

[6] Ibid.

[7] (McCaskey 1906).

that, as a major component of chromosomes, nuclein must play a crucial role in the transmission of heredity. But within a few short decades, they were starting to have serious doubts about this.

Cytological staining methods had shown chromosomes to be made of chromatin and that this material was most likely synonymous with nuclein. But now they revealed a formidable problem. Staining of chromosomes showed them to become visible during only one phase of cell division. Speculating as to the reason for this, cytologist Eduard Strasburger suggested that 'In the stages preliminary to their division, the chromosomes become denser and take up a substance which increases their staining capacity; this is called chromatin.' But he then went on to explain why this made it unlikely that chromatin might be the carrier of hereditary traits: 'The chromatin cannot itself be the hereditary substance, as it afterwards leaves the chromosomes, and the amount of it is subject to considerable variation in the nucleus, according to its stage of development.'[8] The problem was simply this: that the presence of chromatin, as revealed by staining methods, appeared to vary depending on the physiological state of the cell. And if chromatin (and therefore nuclein) was indeed the carrier of hereditary traits, then surely its amount would be expected to remain constant.

The eminent American cytologist Edmund Beecher Wilson (1856–1939) echoed Strasburger's doubts. When Wilson's book, 'The Cell in Heredity and Development', first appeared in 1896, he had been confident that nuclein was the material of heredity: 'The chemical composition and staining-reactions of the nuclein-series,' he wrote, 'indicate that the substance known as nucleic acid plays a leading part in the constructive process [of the cell].'[9] In a 2nd edition of the book published 16 years later, his confidence had grown and he felt there to be 'considerable ground for the hypothesis that in a chemical sense this substance is the most essential nuclear element handed on from cell to cell, whether by cell-division or by fertilization; and that it may be a primary factor in the constructive processes of the nucleus.'[10]

But by the time his book had reached its 3rd edition in 1925, he was no longer quite so convinced. Staining with acidic and basic dyes had shown the nucleus to contain two different types of material which were termed basophilic and oxyphilic. The basophilic portion which reacted with basic dyes contained nucleic acid; while the oxyphilic portion which reacted with acidic dyes contained proteins. The problem was that the staining patterns obtained

[8] (Strasburger 1909); pp. 102–111.

[9] (Wilson 1896); p. 262.

[10] (Wilson 1911); pp. 358–359.

with these dyes appeared to change significantly depending on what stage of rest or division the cell was in. During some stages of division, the basophilic nucleic acid portion first increased, before showing a marked decrease, while the oxyphilic (protein) component appeared to remain constant.[11] Commenting on these observations, Wilson wrote:

> *So far as the staining-reactions show, therefore, it is not the basophilic component (nucleic acid) that persists, but the so-called "achromatic" or oxyphilic substance [protein]. The nucleic acid comes and goes in different phases of cell-activity, and it is the oxyphilic component that seems to form the essential structural basis of the nuclear organization.*[12]

And if the nucleic acid content showed such significant variation over the course of the cell cycle, could it really be the carrier of genetic traits? Surely any such material would have to be stable and constant in the cell. Proteins were starting to look like a much more likely candidate as hereditary material and, as Wilson pointed out, this was not without good reason:

> *The nucleic acids of the nucleus are on the whole remarkably uniform, showing with present methods of analysis no differences in any degree commensurate with those from the various species of cells from which they are derived. In this respect they show a remarkable contrast to the proteins, which, whether simple or compound, seem to be of inexhaustible variety*[13]

It was not until 1948 that, thanks to the work of French scientist Andre Boivin (1895–1949) with his colleagues Roger Vendrely (1910–1988) and Colette Vendrely (1924–2017) in Strasbourg, the DNA content of cells was finally shown to be constant in different tissues.[14] Had Miescher lived to see this discovery, he might well have felt that his criticisms and concerns about the limitations of microscopy and cell biology had been vindicated. While the apparent disappearance of nuclein under microscopical staining was starting to convince the cell biologists of the early twentieth century that it could not possibly be the hereditary material, Boivin and his colleagues had shown that, just because something doesn't show up under the microscope with a coloured stain, doesn't necessarily mean it's not there.

---

[11] (Wilson 1925): pp. 88–9; pp. 651–653.

[12] Ibid; p. 653.

[13] Ibid; p. 644.

[14] (Boivin, Vendrely and Vendrely 1948).

But even though cytology had now cast doubt on nucleic acid as the hereditary material, it was turning up powerful new evidence that the chromosomes themselves were indeed central to inheritance.

8.1 Peas, Flies, and Grasshoppers

We have already seen how, although Gregor Mendel is today hailed as the discoverer of the gene, his intention in crossing pea plants back in the mid to late 1850s was not to reveal the mechanism of heredity, but to uncover a law describing the formation and behaviour of hybrids. How then did Mendel the mid-nineteenth century hybridist, become the Mendel with whom we are all familiar today as the discoverer of the gene?

This transformation owes much to English biologist William Bateson (1861–1926). In 1901 Bateson introduced Mendel to the Anglophone scientific community thanks to having commissioned the first ever translation of Mendel's paper into English. In both an introduction to the translation and his book 'Mendel's Principles of Heredity: a Defence' which was published the following year, Bateson argued that Mendel's experiments offered vital experimental support for his own theory of heredity. This proposed that organisms are a complex of discrete traits determined by what he called 'unit-characters' that are transmitted through the reproductive cells.[15]

Historian Kostas Kampourakis has therefore suggested what we today call 'Mendelian genetics' should more accurately be described as 'Batesonian genetics.'[16] For it was from Bateson's interpretation and appropriation of Mendel's work that the idea of hereditary 'unit-characters' originated. But this in turn raised a new question—how were these 'unit-characters' transmitted physically from one generation to the next? Here too, plants—or more accurately a blight upon them—offered an answer, for while the grasshopper *Brachystola magna* may well have been the scourge of prairie farmers in the USA, research into this agricultural pest would prove to be a blessing for geneticists puzzling over the mechanism of heredity.

It was whilst working on a farm in his native Kansas during the summer of 1899, that Walter Sutton (1877–1916) had collected samples of the grasshopper for research into the formation of their sperm as part of his Master's degree

[15] For excellent accounts of how Bateson constructed the Mendel with which we are all familiar today, see 'Disputed Inheritance: The Battle Over Mendel and the Future of Biology' by Greg Radick, (2023, University of Chicago Press) and 'How We Get Mendel Wrong, and Why it Matters: Challenging the Narrative of Mendelian Genetics' by Kostas Kampourakis, (CRC Press Taylor & Francis Group, 2024).

[16] (Kampourakis 2024); p. 78.

at the University of Kansas.[17] Continuing this research at the University of Columbia with E. B. Wilson, he was able to follow precisely the movement of the insect's chromosomes. These were found in 11 pairs, one member of which was inherited paternally and the other maternally. By tracking the separation of these pairs during the formation of sex cells, Sutton realised that this explained the ratios that Mendel had found in his peas.[18]

This was of profound importance. Proposed independently by German biologist Theodor Boveri (1866–1915), what became known as the Sutton-Boveri theory was evidence that chromosomes were indeed the carriers of the hereditary units. And in 1909, the same year that Eduard Strasburger conceded chromatin was most likely just 'a nutritive material for the carriers of hereditary units', those very same hereditary units received the name by which we know them today thanks to Danish biologist Wilhelm Johannsen who called them 'genes.'[19]

Cytology had offered strong evidence that chromosomes were the carriers of hereditary traits. It is somewhat ironic then that, thanks to this very success, cytology now began to cede its grip on the study of chromosomes to an emerging new field.[20]

Three years before Johannsen coined the term 'gene', William Bateson had already suggested that the study of inheritance be known by a new name—'genetics.' But those working in this new field did not concern themselves too much with the question of what genes might be made of. One such figure was Thomas Hunt Morgan (1866–1945) who, working at the University of Columbia, had devoted himself to spending long hours in his lab peering at milk bottles full of mushed up rotten bananas over which swarmed hordes of the tiny fruit fly Drosophila melanogaster. Morgan was trying to induce variations in his flies in the hope of studying their evolution but with little success. Until that is, amidst all the red-eyed flies crawling over blackening bananas, he spotted one that stood out.

This fly had white eyes, and its effect on Morgan was profound. Until now he had been dismissive both of chromosomes and the new science of genetics, but his white-eyed fly gave him cause to reconsider. Of particular interest to him was that the white-eyed mutation seemed to be much more common in males. By 1912, he had shown why this was the case: of the four

[17] Like Miescher, Sutton's scientific career was cut tragically short but readers who would like to know more about him might find the following discussion of interest - https://soundcloud.com/user-287120959/dna-files-episode-3

[18] (Sutton 1902); (Sutton 1903).

[19] (Strasburger 1909); pp. 102–111.

[20] (Veigl et al. 2020); p. 473.

chromosomes possessed by Drosophila, it was the X sex chromosome that determined eye colour.

Morgan had confirmed the Sutton-Boveri theory that chromosomes carried the determinants of heredity. But he also went further. Studying other easily identifiable mutations, Morgan and his colleagues noted that some of them were often inherited together—a phenomenon known as 'linkage.' From this they inferred that linked mutations were found on the same chromosome and, the closer their degree of linkage, the closer were the physical locations of their genes on the chromosome. In this way they began to compile charts of the linear organisation of genes on the four chromosomes of Drosophila.

In 1933, Morgan was awarded the Nobel Prize in Physiology or Medicine, 'for his discoveries concerning the role played by the chromosome in heredity.'[21] This was somewhat ironic given that he had initially been sceptical about the role of chromosomes in heredity. But as was evident from his Nobel lecture given the following year, one issue on which he remained a sceptic was the question of whether the material nature of the gene should be of any concern to geneticists:

> *Now that we locate them in the chromosomes are we justified in regarding them as material units; as chemical bodies of a higher order than molecules? Frankly, these are questions with which the working geneticist has not much concerned himself, except now and then to speculate as to the nature of the postulated elements. There is no consensus of opinion among geneticists as to what the genes are—whether they are real or purely fictitious—because at the level at which the genetic experiments lie, it does not make the slightest difference whether the gene is a hypothetical unit, or whether the gene is a material particle....*[22]

To Morgan, it did not matter whether genes were solid material entities or hypothetical units—what was important was that they and the traits they determined could be mapped to precise locations on chromosomes. But not everyone was so sure about this. One dissenting voice was that of Hermann Muller (1890–1967) who had been a former student of Morgan, and a disgruntled one at that. Muller felt that during his time as a student in Morgan's lab at Columbia he had not been given sufficient recognition for his contributions. Perhaps it was this that drove him to distinguish himself by earning the

[21] https://www.nobelprize.org/prizes/medicine/1933/morgan/facts/. (Accessed 20 Aug 2024).

[22] (Morgan 1934); p. 315. MLA style: Thomas H. Morgan—Nobel Lecture. NobelPrize.org. Nobel Prize Outreach AB 2024. Wed. 3 Jan 2024. https://www.nobelprize.org/prizes/medicine/1933/morgan/lecture/

1946 Nobel Prize in Physiology or Medicine for his discovery made 20 years earlier that X-rays could induce genetic mutations.

Muller's discovery, made in 1926, suggested that, far from being a mere hypothetical unit, the gene was indeed a material substance. But if so, how big might it be? To answer this question, three workers, Nicolai Timofeef-Ressovsky; Carl Zimmer and Max Delbrück used X-ray induced mutagenesis to estimate the size of a gene. Their resulting publication titled 'On the nature of gene mutation and gene structure' or known more colloquially as 'The Three Man Paper', established the idea that genes were physical entities.[23] But if that were so, they must be made of something. The question was—what?

By this time, cytological approaches to the question of heredity had become a victim of their own success. Having enjoyed spectacular success in demonstrating that chromosomes were the carriers of hereditary units, cytology had now vacated the throne it once occupied making way for genetics to take its place. With the success of Morgan's approach of mapping mutations to loci on the chromosomes, cytology was relegated to being the handmaiden of genetics. And as the authority of these two disciplines underwent a tectonic realignment, Miescher and his work became somewhat obscured.[24]

Obscured perhaps,—but not completely hidden. Because as cytologists had turned their backs on nucleic acid, and geneticists showed more interest in constructing maps of the relative positions of genes than in what they were made of, someone else was taking a great deal of interest in Miescher's work. This was the organic chemist Phoebus Aaron Levene (1869–1940) who was so intrigued by the chemical structure of nuclein that he devoted his career to solving it. Many would later say that our understanding of DNA might well have advanced much faster had he not done so.

8.2 Phoebus Levene and the Tetranucleotide Theory[25]

Born in Sagor (now in Lithuania) in 1869, Levene had been one of the few Jewish graduates of the Imperial Military Medical Academy in St. Petersburg. But holding the rank of captain in the Russian Army sadly offered no

[23] (Timofeef-Ressovksy et al. 1935).

[24] (Veigl et al. 2020); p. 473.

[25] Readers who are interested in exploring Levene's work in greater depth—especially the debate amongst historians regarding to what extent the tetranucleotide theory cast nucleic into the shadows as the genetic material, might enjoy the discussion of these issues in Episode 5 of 'The DNA Papers' podcast—https://soundcloud.com/user-287120959/dna-papers-episode-5

protection against the tide of anti-Semitism that was rising in Russia towards the end of the nineteenth century. To escape persecution, he fled with his family to the USA in 1891 but returned to St. Petersburg a few months later to complete his medical studies. After completing his examinations, he returned to the USA where he qualified as a doctor and began to practice on New York's Lower East Side.

But his real passion was chemistry, a subject to which he had been introduced whilst in St. Petersburg by the scientist and composer, Alexander Borodin (1833–1887). Realising the importance of chemistry to medicine, Levene enrolled in the chemistry department of the School of Mines of Columbia University and, although he never actually received a chemistry degree, he was eventually given some space in a laboratory in the physiology department of the College of Physicians and Surgeons of Columbia University where he worked on the biochemistry of sugars. But after moving to the laboratory of the Pathological Institute at the New York State Hospital in 1896 to take up the post of Associate in Physiological Chemistry, his attention was turned towards nucleic acids by the American neurologist Ira Van Gieson (1866–1913).

No sooner had Levene begun his research in this field however, than he suffered a major blow. Like Miescher, Levene had a ferocious work ethic. He pushed himself hard and like Miescher, this punishing work schedule took a heavy toll on his health. Having contracted tuberculosis towards the end of 1896, he spent the next 2 years in recuperation, first at a sanatorium at Saranac Lake where he eventually met his future wife, and then in Davos, Switzerland where Miescher had died only a short while earlier.

After making a recovery, Levene returned to his work, but further misfortune soon followed when the laboratory of Physiological Chemistry was shut down as part of a reorganization. This cloud however was to have a shining silver lining for Levene and his future career. Having been forced by circumstance to become a scientific nomad, he moved amongst eminent European laboratories—one of which was that of Albrecht Kossel in Marburg, Germany. By now Kossel was a renowned authority on nucleic acids—a crown that he was soon to pass on to his student, Levene.

In his first paper, published in 1899 on nucleic acids extracted from the brain tissue of calves, Levene recognised that 'chromatin is the most important substance for the life of the cell and that most functions are connected with some change in that substance.'[26]

[26] (Levene 1899); p. 4.

As for what this change might be, Levene suspected that it had something to do with nucleic acids:

> *It is apparent that the distinction between different individual chromatins may be due to the nature of the different nucleic acids or to the other components entering in combination with the nucleic acid and thus forming the chromatin… distinction in function was, to a certain degree, caused by difference in chemical nature and there was a certain tendency to ascribe the individual function of a chromatin to the nucleic acid present in its molecule.*[27]

The task before him was therefore clear. The secret of nucleic acids and how they functioned, lay in discovering their chemical composition. And although the suggestion by Osborne and Harris that plant- and animal-derived nucleic acids could be distinguished by the presence or absence of uracil respectively, would eventually turn out to be wrong, it nevertheless spurred Levene to dig deeper into their chemical composition.

Levene wondered, for example, whether nucleic acids from different tissue types of an animal might also vary in their chemical composition? And if so, might such variation account for the formation of different tissues in an organism?

By this time, Levene was in a strong position to tackle this question. In 1905, he had taken up a post at the Rockefeller Institute for Medical Research (later to become Rockefeller University) which had been founded only a few years earlier as America's first biomedical research institute with the motto 'Science for the good of mankind.' In attempting to live up to this goal, its director Simon Flexner (1863–1946) sought to recruit the best and the brightest of scientific minds, but as he quickly found out with Levene, these traits did not always make for staff who were easy to manage. Nevertheless, despite bad-tempered clashes with Flexner over institutional administration and bureaucracy, Levene was quickly promoted to become director of the Division of Chemistry.

Levene's daily routine spoke of a man who had supreme confidence in his own abilities. He would rise at dawn to read a French novel before heading to his laboratory where he often summoned his lab staff to chauffeur him to the opera or an art gallery. Surprisingly, none of this activity appeared to detract from his productivity, which was truly formidable. Over the course of his career, Levene amassed over 700 scientific papers.

[27] (Levene 1903); p. 205.

Levene wondered whether biological differences might be rooted in variation of the chemical composition of nucleic acids.[28] This was a question that he believed to be at the heart of biology, for as he said in an address given in 1916 to the American Association for the Advancement of Science (AAAS):

Heredity, evolution, and all that is specifically biological about the living as contrasted with the non-living is pivoting on specificity. Can this characteristic of living organisms be shown to be a function of chemical structure? Is there among the many components of living tissue one or a group of several that may be regarded as carriers of the specificity of a tissue, an organ, of a species or of a genus?[29]

He was not, however, the first to ask this question. In 1883, August Weismann had already postulated that heredity might be due to 'Variations in the molecular structure of germ cells', but Miescher had gone even further.[30] In December 1892, just 3 years before his death, Miescher had speculated in a letter to his uncle about how a molecule might be able to transmit hereditary traits. And he had come up with a novel answer:

To me the key to the problem of sexual reproduction is to be found in the field of stereo-isomerism. The 'gemmules' of Darwin's Pangenesis are no other than the asymmetrical carbon atoms of organic substances....[31]

Gemmules were an agent postulated by Charles Darwin to explain a problem that was proving to be a major thorn in his side. His theory of evolution by natural selection required not only that organisms show variation within their populations, but that variations which conferred advantages in the struggle for survival, be passed on to successive generations. The problem for Darwin was—by what mechanism did this transmission of traits occur?

To address this, Darwin proposed 'Pangenesis'—a process by which organic entities that he called 'gemmules' are produced by the tissues of the adult organism and travel to the sex organs to be transmitted to the next generation.

Miescher however, went further. He had conceived a means by which variations in biological traits might be represented at the molecular level. This was through stereoisomerism—a concept that had its roots in Louis Pasteur's research into wine. Although Pasteur is better known for his work on germ theory, the early days of his career were spent studying tartaric acid, an organic

[28] (Cohen and Portugal 1977); p. 78.

[29] (Levene 1917); p. 829.

[30] (Weismann 1883); p. 18.

[31] Miescher, Letter LXXV, 17th December 1892 in (His 1897); pp. 116–117.

acid that occurs naturally in grapes and which often forms crystalline deposits at the bottom of a wine bottle. He observed that different forms of crystalline tartaric acid have differing optical effects—an explanation for which was given independently in 1874, by the Dutch chemist Jacobus Henricus van't Hoff (1852–1911) and the French chemist Joseph Achille le Bel (1847–1930). They suggested that while any two molecules of tartaric acid were composed of the same atoms, as given by their chemical formula, the physical arrangement of these atoms in space might be such that the two molecules could never be superimposed onto each other—in rather the same way that, although a left and right handed glove both contain four fingers and a thumb, they can never be superimposed. These two different forms of tartaric acid—both of which contain the same number and types of atoms but organised in a different spatial arrangement—came to be known as stereoisomers.

According to Miescher, stereoisomerism (Fig. 8.1) was the key to understanding how the genetic molecule might specify biological traits. Each different trait corresponded to a particular stereochemical configuration of the atoms in the molecule. He had also recognised that this meant that the genetic molecule must meet two key criteria—firstly, it must be large, and secondly, it must possess immense capacity for structural variation. The larger the

Fig. 8.1 Stereoisomerism can be exhibited by molecules which, although they share the same chemical formula, have a different spatial arrangement of their component atoms. The example shows two stereoisomers of the amino acid alanine. Both stereoisomers have the chemical formula $C_3H_7NO_2$ (Black spheres are atoms of carbon, C; blue = nitrogen, N; red = oxygen, O; white = hydrogen, H) but the spatial arrangement of these atoms is such that, despite being chemically identical, the two molecules could never be superimposed onto each other

molecule, and the more atoms of different types that it contained, the greater its potential number of stereoisomers—and therefore the greater its capacity to represent biological traits.

Miescher even had a candidate molecule in mind that met both these criteria—but it was not nuclein:

> *In the huge molecules of albumen compounds, or in the yet more complicated molecules of haemoglobin, etc. the many asymmetric carbon atoms provide a colossal amount of stereo-isomerism. In them all the wealth and variety of hereditary transmissions can be found just as all the words and concepts of all languages can be found in the expression of twenty-four to thirty alphabetic letters*[32]

Due to their huge molecular size and therefore their potentially vast number of stereoisomers, proteins - or 'albumens' as they were sometimes known in Miescher's day - seemed to be the most likely candidate for the genetic molecule. In a second letter written in the following year, Miescher even attempted to quantify the number of possible stereoisomers that might arise from a single type of protein molecule. He reasoned that, since there were two possible stereoisomers for each single asymmetrical carbon atom, a polymer made up of 40 such carbon atoms would give rise to 2^{40} possible permutations. This alone was a staggeringly large number, but as Miescher pointed out, since it was restricted only to carbon atoms and did not take account of the possible stereoisomers of other atoms such as nitrogen that were present in proteins, the real figure was likely to be vastly greater.[33]

Writing in response to what he considered to be the vague speculations of Weismann, he declared that 'In order to provide the incalculable diversity required by the theory of heredity, my theory is more suitable than any other.'[34] And although he turned out to be wrong about proteins being the genetic material, his idea that the genetic molecule could represent biological traits through variation in its structure or composition was a powerful insight. It would, in time, turn out to be crucial in understanding the secret of how DNA works.

Had the same methods of chemical analysis used by Kossel and Levene in their study of nucleic acids been available to Miescher, he might well have thought to see whether nucleic acids showed a similar potential for variation in their structure as proteins. But instead, it fell to Levene to do so. And his

[32] Ibid.

[33] Miescher 13th October 1893, Letter LXXVIII in (His 1897); pp. 122–123.

[34] Ibid.

conclusion was, in his own words, 'very discouraging'. After many years of analysing their chemical composition, Levene concluded that nucleic acids were:

> *Constant, invariable in their structure, present wherever life is present. They are indispensable for life, but carry no individuality, no specificity, and it may be just to accept the conclusion of the biologist that they do not determine species specificity, nor are they carriers of the Mendelian characters.*[35]

But despite being disappointed that nucleic acids did not appear to carry sufficient chemical variability to be the genetic molecule, Levene did not turn his back on them. Instead of asking grand questions about whether nucleic acids might be the carriers of biological specificity, Levene turned his attention to solving how the individual chemical components of bases, sugars and phosphates were linked together to form a nucleic acid chain. This change of direction turned out to becrucial for our understanding of the chemistry of nucleic acids - but it also resulted in what has been called 'A scientific catastrophe.

8.3 The Tetranucleotide Theory: 'A Scientific Catastrophe'?

Albrecht Kossel had shown that nucleic acids were composed of four bases, together with sugars and phosphorus. But what remained a mystery was how these different components were chemically linked together. Slowly, and by painstaking effort over many years, not without its blind alleys and mistakes, Levene now came up with an answer. In the process, he also coined a term that has become part of the vocabulary of modern biochemistry. Levene showed that each of the four bases identified by Kossel was linked by a chemical bond to the sugar that he had also detected to form a structural unit, which Levene called a 'nucleoside.' And in nucleic acids, each nucleoside was bonded chemically to a phosphate group, to become what Levene called a 'nucleotide' (Fig. 8.2).[36] It was these nucleotides that were the basic building blocks—or, to use Kossel's terminology, 'Bausteine' (building bricks)—of nucleic acids.

But he also went further. In 1909, Levene showed that the sugar to which the bases in yeast nucleic acid were bonded was D-ribose, which is composed

35 (Levene 1917); p. 835.

36 (Levene and Jacobs 1909); p. 2475; (Levene 1910); p. 238.

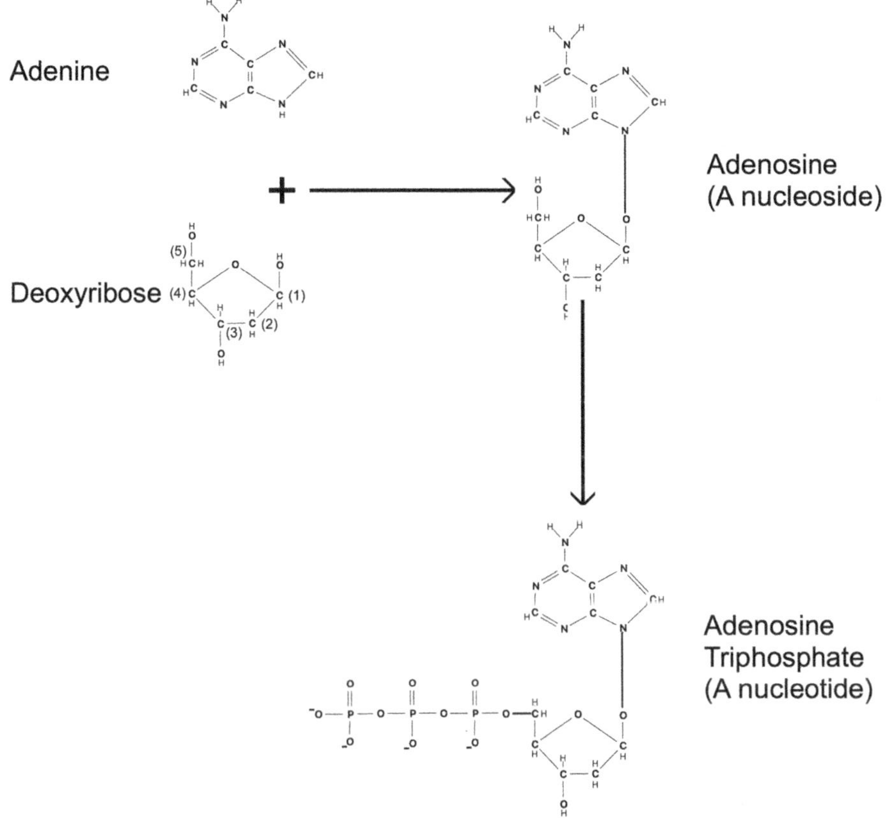

Fig. 8.2 (a) A **nucleoside** is a compound containing a nitrogenous base (such as adenine in the example shown here) chemically bonded to the first carbon atom (numbered clockwise) in a sugar ring. In RNA, this sugar is ribose which contains 5 carbon atoms, while in DNA, it is the five carbon sugar *deoxy*ribose in which the carbon atom at position 2 in the ring lacks the oxygen atom that is present in ribose. **(b)** A **nucleotide** is formed by the chemical attachment of one or more phosphate groups to the fifth carbon atom of the sugar in a nucleoside. In the example shown, the *nucleoside* deoxyadenosine is attached to three phosphate groups to become the *nucleotide*, deoxyadenosine triphosphate (ATP). Nucleotides are the fundamental units from which chains of nucleic acids are built

of a ring of five carbon atoms, before then turning his attention to solving the identity of the sugar in nucleic acids from thymus glands.[37] At the time, this was believed to be a hexose sugar (one which contains six carbon atoms) but proving that this was not the case was to take Levene over 20 more years of meticulous work. This was not least because the methods of extraction and

[37] (Levene 1909).

purification available at the time altered the sugars of thymus nucleic acid, causing them to undergo a chemical transformation during the process of isolation. Levene found a way around this problem—thanks in part to a New York mugger.

In 1923, Levene received a desperate request for help from another of his former mentors at the Academy in St. Petersburg. This was Ivan Pavlov (1849–1036), the scientist who is famous today for his experiment in which he used a bell to induce salivation in dogs, and who was visiting New York on his way to Paris to attend a celebration in honour of Pasteur. Having been mugged in Grand Central Station, the elderly Pavlov found himself with no money or passport and so turned to his former student for help. Paying no attention to the reservations of his director, Simon Flexner, Levene obtained financial support for Pavlov from Rockefeller funds and arranged a visa for him. Indebted to his former student, Pavlov invited him to visit his laboratory in St. Petersburg the following year.

Taking up the offer, Levene learned there of an alternative—and somewhat unusual—method of isolating the sugar component of nucleic acid. This involved introducing the nucleic acid sample into the stomach of a live dog via a fistula and then recovering digested material from a fistula in the animal's small intestine. Using this method, Levene was eventually able to show that the sugar in thymus nucleic acid was not a hexose sugar, but deoxyribose.

Levene did more than just showing how the three components—base, sugar and phosphate—are bonded together to form a nucleotide. He also explored how these individual nucleotides might form chemical bonds to link together with each other. In 1908, comparing the results of his own chemical analysis of nucleic acid composition with that of fellow researcher Hermann Steudel (1871–1967), he proposed that individual nucleotides might link together to form a basic structural unit of nucleic acids and, in so doing, coined a term that would cast a long shadow over his legacy:

> *If one ascribes the formula proposed by Stendel [sic] to the nucleic acids, then they appear to be composed of a tetranucleotide. The values obtained by one of us (Levene) over several years for Milznucleic acids favour the formula $C_{54}H_{71}N_{20}O_{37}P_5$ which can be represented as a pentanucleotide.*[38]

In the decades that followed, Levene explored the possible ways in which individual nucleotides might bond together. Did carbon atoms at specific positions on the sugar rings of two adjacent nucleotides form a direct

[38] (Levene and Mandel 1908); p. 1907.

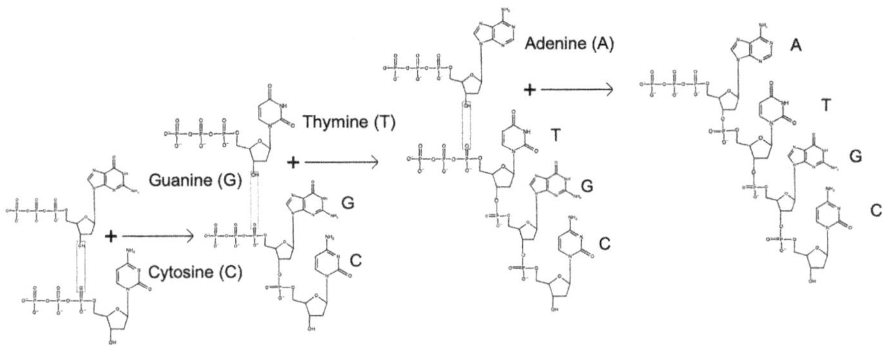

Fig. 8.3 Nucleotides are the building blocks of nucleic acids. In the diagram shown, a chemical bond is formed (indicated by the region within the dotted line) between the first phosphate group on the fifth carbon atom of the sugar ring in the cytosine nucleotide, and the hydroxyl (-OH) group on the third carbon atom of the sugar ring in the guanine nucleotide. Through the formation of this bond, the two nucleotides become linked together to create a dinucleotide. The first phosphate of the guanine nucleotide can then form a similar bond with the third carbon atom on the sugar ring of the thymine nucleotide, to form a trinucleotide. The phosphate of the thymine nucleotide can then form a bond with the adenine nucleotide to form a tetranucleotide. In this way, an entire nucleic acid chain can be assembled. (Diagram by K. Hall)

chemical bond with each other? Or was the base involved? Or the phosphate group? In 1935, in a paper that has been hailed as the pinnacle of his work, Levene and his co-worker R. Stuart Tipson gave an answer.[39] Nucleotides could link together with each other through the formation of chemical bonds (Fig. 8.3).[40,41]

This might seem to have little significance beyond organic chemistry, but its implications were huge. It showed not only that individual nucleotides could bond with each other to form a polymer, or polynucleotide as it became known, but that this bonding conferred a directionality upon the polynucleotide chain. This would later prove to be crucial in understanding not only how they can make copies of themselves, but also how the biological information that they carry can be expressed.

Levene also believed that he had made an important discovery about the way in which individual nucleotides were arranged to form a nucleic acid chain.

His analysis of nucleic acids found in yeast (later shown to be RNA), appeared to show that the four nucleotides were present in roughly equal amounts. From this, he inferred that yeast nucleic acid was built up of

[39] (Levene and Tipson 1932).
[40] (Levene and Tipson 1935).
[41] (Levene and Jacobs 1912).

repeating units which contained all four types of nucleotide. Initially he believed that each of these structural units contained five nucleotides, but he later revised this view in favour of a model in which nucleic acid chains were made of what he called tetranucleotides—repeating identical blocks of nucleotides containing all four bases.

Levene later extended this tetranucleotide model to thymonucleic acid (better known today as DNA) and singled out one researcher for having made his own work possible[42]:

> *it is noteworthy that when the analytical results of all workers are compared with the present-day theory, it is found that the figures nearest to theory were obtained by the man who discovered nucleic acid, namely, Miescher, and by those who undertook to complete his work, Schmiedeberg and his students.*[43]

In his 1931 monograph, 'Nucleic Acids', Levene had nothing but praise for what he called 'the great work of Miescher.'[44] Miescher had, in his eyes, made 'truly important discoveries in the domain of the chemistry of nucleic acids' and it was to him that 'the credit for the discovery of this group of substances belongs.'[45] Describing the early work on nucleic acids, Levene said that the progress made during this period had been 'due entirely and undisputedly to Miescher; personally and to his influence on other workers… Miescher not only initiated and carried out practically all the work of the first period, but to his influence may be attributed the work of the following period.'[46] Added to which was the fact that he had done all this with only very limited methods, making his achievement all the more impressive.[47]

Unfortunately, history would not be quite so kind to Levene. When he died in 1940, an obituary in the journal *Science* praised his 'insistent quest for the knowledge of life processes' and hailed one of his main achievements as having been to show 'that the nucleic acids are tetranucleotides.'[48,49] Yet within a short space of time, what Levene had called his tetranucleotide theory was

---

[42] (Levene 1919).
[43] (Levene and Bass 1931); p. 262.
[44] Ibid. p. 239.
[45] Ibid. p. 243.
[46] Ibid.
[47] (Levene and Bass 1931); p. 262.
[48] (Levene 1917); p. 591.
[49] (Bass 1940).

being condemned as 'a scientific catastrophe' that had severely delayed the discovery of the DNA structure.[50]

The problem was that Levene's tetranucleotide hypothesis made nucleic acids unlikely to be the carriers of genetic traits. Proteins seemed to be much more likely candidates. In a letter written to his uncle Wilhelm His in 1892, Miescher had already explained why this should be the case. He argued that biological traits might be represented through stereoisomeric variations in the structure of a macromolecule and that, because proteins contained up to 20 different types of amino acid, their capacity for immense stereochemical variation was immense. A few years later this same argument was made by Edmund Beecher Wilson and Albrecht Kossel.

If nucleic acid chains were composed of repeating identical tetranucleotide units, as Levene proposed, then they would have very little capacity for structural variation. This made them unlikely to be the genetic molecule. In his 1996 book, 'The Eighth Day of Creation', historian Horace Freeland Judson offered a memorable description.

> *And so the belief was held with dogmatic tenacity that the DNA could only be some sort of structural stiffening, the laundry cardboard in the shirt, the wooden stretcher behind the Rembrandt, since the genetic material would have to be protein.*[51]

Not everyone had accepted Levene's tetranucleotide hypothesis without question, however. The American geneticist Bentley Glass (1906–2005) despaired at how scientists had accepted Levene's tetranucleotide hypothesis with what he called 'dismal blindness', but there was at least one notable exception.[52] This was the British chemist John Mason Gulland (1898–1947) who, 4 years after Levene's death, reviewed the experimental evidence for the tetranucleotide hypothesis and found it wanting. If Levene had been correct and nucleic acid chains were repeats of identical tetranucleotide blocks then, regardless of the source from which nucleic acids were obtained, they should yield the four bases in equal proportions. But Gulland remarked that 'Several instances are recorded in which hydrolysis of nucleic acids from different sources yielded the nitrogenous bases in amounts differing considerably from those expected on the basis of the tetranucleotide ratio.'[53] Gulland's point was that if the tetranucleotide hypothesis as proposed by Levene were correct, then each of

---

[50] (Glass 1965); p. 229.

[51] (Judson 1996); p. 13.

[52] (Glass 1965); p. 230.

[53] (Gulland, Barker, and Jordan 1945); p. 190.

the four different bases should always be present in the same amounts. But even some of Levene's own work suggested that this was not the case![54] Gulland's conclusion was that at present there was 'no evidence to justify the recognition of the existence of the "structural tetranucleotide".[55,56]

He also pointed out that even if nucleic acids were made up of tetranucleotide blocks, this did not necessarily rule out significant variation in their composition. 'Even if the oligonucleotides were tetranucleotides,' he wrote, 'there is no reason why their nucleotides should be arranged uniformly in each molecule, and in the writer's opinion the experimental results could be explained in other ways which leave open the question of a regular sequence.'[57] Why, asked Gulland, should it be assumed that 'the four nucleotides are always combined in a fixed manner in an unchangeable sequence'?[58]

History has not been kind to Levene, or his tetranucleotide hypothesis. But the benefit of hindsight allows harsh judgements to be made all too easily. When viewed from the lofty vantage point of the early twenty-first century, the route to the present day seems all too obvious and inevitable. Commenting on the problems of taking this view, the historian Pedro Bernal has offered some insights that are particularly relevant when considering Levene:

> *The history of science is also useful for giving students a sense of place in the great enterprise that is science. It shows that scientists in the past were trying to make sense of the world with the experimental and conceptual tools at their disposal and that present scientists are, essentially, doing the same. True, our experimental tools are better and our theories may be more comprehensive, yet the nature of the engagement is the same. At times, we are tempted by the notion that our present scientific understanding is final and, in a sense, outside of history. We think that we have now arrived at the right answers and, that scientists in the past could have, in some cases, come to the same conclusions if they had carefully considered the available evidence. That is the problem that I have termed "presentism."*[59]

[54] (Levene 1921); p. 179.

[55] (Gulland, Barker, and Jordan 1945); p. 188.

[56] Ibid. p. 192.

[57] Ibid. p. 189.

[58] Ibid. p. 184.

[59] (Bernal 2006); p. 870.

Levene himself had admitted that, in the early stages of his work at least, he had considered his model of linkage between the nucleotides to be only 'of a provisional character and arbitrary in nature' and that his 'original graphic representation of the entire molecular of yeast nucleic acid had only an arbitrary schematic sense.'[60,61] Rather than holding the status of a formal theory such as the Theory of Relativity, Levene's tetranucleotide hypothesis seems to have been conceived initially as more of a heuristic, a 'rule-of-thumb.' But somewhere along the line, it appears to have been simply accepted as dogma.

The reasons for this may not have been entirely scientific. Historians of science have long recognised that factors external to science can play a powerful role in its development, and the case of the tetranucleotide theory is no exception. Levene's autocratic nature and the reverence and respect with which he was held have all been invoked to explain how what started out as a simple rule-of-thumb evolved into an accepted theory.

The historian Robert Olby described Levene's tetranucleotide hypothesis as 'a remarkable example of those paradigms which start off as working hypotheses in an ill-defined field: tools with which to open up a subject. They are successful and become entrenched so firmly that a revolution is required to dislodge them.'[62]

The revolution that would eventually bring the tetranucleotide hypothesis crashing down began very quietly. Its first gentle tremors were felt in the years that immediately followed Levene's death in 1940 and had their origins in two very different places. One was an experiment performed in 1944 at the Rockefeller Institute in New York City and which has since been described as a 'bombshell'[63]; the other was a discovery made in a former stable that, having been converted into an industrial textiles laboratory, was now billowing with solvent fumes.[64] Both would put nucleic acids firmly back at the heart of biology.

[60] (Levene 1917); p. 591.

[61] (Levene 1919); p. 415.

[62] (Olby 1994); p. 74.

[63] (Watson et al. 1987); p. 69.

[64] (Lederberg 2000); p. 194–195.

Drawing by Kersten Hall

References

Allbright, J. H. 1896. Protonuclein. *Journal of Nervous and Mental Disease* 21:431.

Ames, D. ', and A. A. Huntley. 'The nature of the leucocytosis produced by nucleinic acid'. Journal of the American Medical Association 29 (1897): 472–478.

Bass, L. W. 1940. Phoebus Aaron Theodor Levene 1869-1940. *Science* 92:392–395.

Bernal, P. J. 2006. From "greasy chemistry" to "macromolecule": Thoughts on the historical development of the concept of a macromolecule. *The Journal of Chemical Education* 83:870–879.

Boivin, A., R. Vendrely, and C. Vendrely. 1948. L'acide Désoxyribonucléique Du Noyau Cellulaire, Dépositaire Des Caractères Héréditaires; Arguments d'ordre Analytique. *Comptes Rendus De L'Académie Des Sciences* 226:1061.

Cohen, J. S., and F. H. Portugal. 1977. *A century of DNA: A history of the discovery of the structure and function of the genetic substance*. Cambridge, MA: MIT Press.

Glass, B. 1965. A century of biochemical genetics. *Proceedings of the American Philosophical Society* 109:227–236.

Gulland, J. M., G. R. Barker, and D. O. Jordan. 1945. The chemistry of the nucleic acids and nucleoproteins. *Annual Review of Biochemistry* 14:175–206.

His, W. 1897. *Die Histochemischen Und Physiologischen Arbeiten von Friedrich Miescher*. Leipzig: F. C. W. Vogel.

Jackson, A. C. 1895. The Protonucleins. *Journal of the American Medical Association* 24:872.

Judson, H. F. 1996. *The eighth day of creation*. Cold Spring Harbor, NY: Cold Spring Harbor Press.

Kampourakis, K. 2024. *How we get Mendel wrong, and why it matters: Challenging the narrative of Mendelian genetics*. Boca Raton, London, New York: CRC Press Taylor & Francis Group.

Lederberg, J. 2000. The dawning of molecular genetics. *Trends in Microbiology* 8:194–195.

Levene, P. A. 1899. On the nucleoprotid of the brain (cerebronucleoprotid). *Archives of Neurology and Psychopathology* 2:3–14.

Levene, P. A. 1903. On the chemistry of the chromatin substance of the nerve-cell. *Journal of Medical Research* 10:204–211.

Levene, P. A. 1909. Über Die Hefenucleinsäure. *Biochemische Zeitschrift* 17:120–131.

Levene, P. A. 1910. On the biochemistry of nucleic acids. *Journal of the American Chemical Society* 32:231–240.

Levene, P. A. 1917. The chemical individuality of tissue elements and its biological significance. *Journal of the American Chemical Society* 39:828–837.

Levene, P. A. 1919. The structure of yeast nucleic acid IV: Ammonia hydrolysis. *Journal of Biological Chemistry* 40:415–424.

Levene, P. A. 1921. Preparation and analysis of animal nucleic acid. *Journal of Biological Chemistry* 48:177–184.

Levene, P. A., and L. W. Bass. 1931. *Nucleic acids*. New York: Chemical Catalog Company.

Levene, P. A., and W. A. Jacobs. 1909. Ueber Die Hefe-Nucleinsäuren. *Berichte Der Deutschen Chemische Gesellschaft* 42:2474–2478.

Levene, P. A., and W. A. Jacobs. 1912. On the structure of thymus nucleic acid. *Journal of Biological Chemistry* 12:411–420.

Levene, P. A., and J. A. Mandel. 1908. Über Die Konstitution Der Thymonucleinsäure. *Berichte der Deutschen Chemischen Gesellschaft* 41:1905–1908.

Levene, P. A., and R. S. Tipson. 1932. The ring structure of adenosine. *Journal of Biological Chemistry* 94:809–819.

Levene, P. A., and R. S. Tipson. 1935. The ring structure of thymidine. *Journal of Biological Chemistry* 109:623–630.

McCaskey, G. W. 1906. The role of Nuclein in the animal economy. *Journal of the American Medical Association* 47:1800–1804.

McClintock, C. T. 1895. The disease resisting powers of the body: A review of the foundations of Nuclein and serum therapy. *Journal of the American Medical Association* 24:535–543.

Olby, R. 1994. *The path to the double helix: The discovery of DNA*. Garden City, NY: Dover Publications.

Radick, G. M. 2023. *Disputed inheritance: The battle over Mendel and the future of biology*. Chicago, IL: University of Chicago Press.

Strasburger, E. 1909. The minute structure of cells in relation to heredity. In *Darwin and modern science*, ed. A. C. Seward, 102–111. Cambridge: Cambridge University Press.

Summers, O. T. 1895. Leucocytes and Nucleins. *Journal of the American Medical Association* 24:963–966.

Sutton, W. 1902. On the morphology of the chromosome Group in Brachystola Magna. *Biological Bulletin* 4:24–39.

Sutton, W. 1903. The chromosomes in heredity. *Biological Bulletin* 4:231–251.

Timofeef-Ressovsky, N., K. G. Zimmer, and M. Delbruck. 1935. The nature of genetic mutations and the structure of the gene. *Nachrichten Aus Der Biologie* 1:189–245.

Veigl, S. J., O. Harman, and E. Lamm. 2020. Friedrich Miescher's discovery in the historiography of genetics: From contamination to confusion, from Nuclein to DNA. *Journal of the History of Biology* 53:451–484. https://doi.org/10.1007/s10739-020-09608-3.

Weismann, A. 1883. *Über Die Vererbung*. Jena: Fischer.

Wilson, E. B. 1896. *The cell in development and heredity*. 1st ed. London: Macmillan.

Wilson, E. B. 1911. *The cell in development and heredity*. 2nd ed. London: Macmillan.

Wilson, E. B. 1925. *The cell in development and heredity*. 3rd ed. London: Macmillan.

9

'The Grammar of Biology'

Contents

Abstract Almost 50 years after Miescher's death, an experiment was performed which began to make scientists think differently about nucleic acids. In 1944, the US-Canadian microbiologist Oswald Avery and his colleagues at the Rockefeller Institute Hospital, MacLyn McCarty and Colin Macleod showed that, in bacteria at least, nucleic acids could confer the ability to cause disease. Although their experiment did not usher in an overnight revolution, it did have a profound influence on a number of people. One of these was the Columbia biochemist Erwin Chargaff who said that Avery's experiment had hinted at a 'grammar of biology.' This chapter explores how, having been inspired by Avery's work to unlock the secrets of this grammar, Chargaff used a novel method of chemical analysis first developed in West Yorkshire, England, for the analysis of wool fibres in the textile industry to obtain the first hint that nucleic acids might carry biological information through the linear sequence of their bases.

K. Hall, R. Dahm, *The Dawn Fisherman*, Copernicus Books,
https://doi.org/10.1007/978-3-032-14219-1_9

Keywords DNA • RNA • Genetics • Molecular biology • Nucleic acids • Proteins • Amino acids • Chromatography • Partition chromatography • Erwin Chargaff • Friedrich Miescher • James Watson • Francis Crick • Hershey-Chase experiments • Bacteriophages • Gene expression • Transcription • Translation • Double helix • Base composition • Adenine • Thymine • Guanine • Cytosine • Biochemistry • Archer Martin • Richard Synge • Wool Industries Research Association

Nuclein returns to the centre stage and, thanks to another white stringy material, begins to reveal its secrets…

John Mason Gulland was one of the first scientists to recognise the seismic importance of what was happening. Having shown that the tetranucleotide theory did not necessarily rule out a role for nucleic acids in heredity, he drew attention to an experiment carried out in 1944 which he believed showed they 'must be regarded not merely as structurally important but as functionally active in determining the biochemical and specific activities…and its implications are of the greatest importance in the fields of genetics, virology, and cancer research.'[1]

9.1 'Avery's Bombshell'

The experiment to which Gulland referred had been carried out by the Canadian-American microbiologist Oswald T. Avery (1877–1955), and has since been hailed in the pages of one undergraduate textbook as 'Avery's Bombshell.'[2] Working with the American geneticist Maclyn McCarty (1911–2005) and the Canadian-American geneticist Colin MacLeod (1909–1972) at the Rockefeller Institute Hospital, Avery found that avirulent (i.e. non-disease causing) forms of the pneumococcus bacterium could become virulent (be able to cause disease) by undergoing a process known as 'transformation'. And the crucial ingredient in this transformation was the addition of nucleic acid (in this case, DNA) that had been isolated from forms of the bacterium that were able to cause disease. Put simply, Avery's experiment appeared to show that DNA alone—without the presence of

[1] (Gulland, Barker, and Jordan 1945); p. 199–200.

[2] (Watson et al.1987); p. 69.

protein—was sufficient to confer the ability to cause disease upon the non-virulent version of the pneumococcus bacterium.[3]

Avery's experiment was the first solid piece of evidence that DNA could pass on hereditary properties but it by no means ushered in a dramatic revolution overnight. While a handful of other researchers such as microbiologist Hattie Alexander (1901–1968) at Columbia University and Andre Boivin of the Institut Pasteur in Paris (1895–1949) attempted to replicate the phenomenon of transformation in other bacterial species, Avery's critics such as his fellow colleague at the Rockefeller, Alfred Mirsky (1900–1974), argued that the ability to become virulent might be due not to DNA, but to traces of residual protein in the preparation.[4]

In fairness to Mirsky, it was a justifiable criticism at the time but one that began to look much less likely thanks to a set of experiments published in 1952 by American microbiologist Alfred Hershey (1908–1997) and American geneticist Martha Chase (1927–2003). Hershey and Chase were studying bacteriophages (also known as 'phages')—a type of virus that infects bacterial cells and consists simply of a nucleic acid carried inside an outer shell of protein. To reproduce itself, the bacteriophage must first bind and then enter the cell of a host bacterium. But what Hershey and Chase wanted to know was, which of the two components of the bacteriophage—protein, DNA, or both—enters the host cell?

To find out, they grew bacteriophages in which the protein component had been labelled with radioactive sulphur, while the nucleic acid was labelled with radioactive phosphorus. If protein was indeed the genetic material, then after being infected with the radioactively labelled bacteriophage, the host bacterium would be expected to contain radioactive sulphur. But this is not what they found. Instead, the radioactive sulphur in the protein component of the bacteriophage remained bound to the outside of the bacterial cells, while the nucleic acids labelled with radioactive phosphorus were found to have entered them. It appeared that the protein component of the phage served only as a delivery vehicle, to land on the surface of the bacterium and discharge the nucleic acid into the host cell.

Like that of Avery, McCarty and MacLeod, the Hershey and Chase experiment is today regarded as a classic. But at the time, Hershey and Chase were restrained in their conclusion: 'The protein probably has no function in the growth of intracellular phage. The DNA has some function. Further inferences should not be drawn from the experiments presented.'[5]

---

[3] (Avery, MacLeod, and McCarty 1944).

[4] (Alexander and Leidy 1951a; Alexander and Leidy 1951b; Leidy, Hahn, and Alexander 1953; Alexander and Leidy 1953; Boivin 1947).

[5] (Hershey and Chase 1952); p. 56.

Hershey and Chase's conclusion, although modest, was an important one, for it strengthened the case that nucleic acids, and not proteins, were indeed the genetic material. And only a few years later, scientists working in Tübingen where Miescher had first discovered DNA, found that it was not the only nucleic acid able to carry genetic information. Studies of tobacco mosaic virus, a pathogen which damages tobacco plants, had shown that it was composed only of protein and RNA. Scientists at the Max Planck Institute for Virus Research in Tübingen found that, even when all the protein was removed, the RNA remained infectious leading them to conclude that the ability of the virus to replicate itself resided in the RNA component.[6] A year later, a team of researchers at Berkeley, California working with TMV independently obtained a similar result and many other viruses including influenza and SARS-CoV 2 have been since been found to use RNA as their genetic material.[7]

The discovery that certain viruses used RNA instead of DNA as their genetic material led one of Avery's colleagues, Rollin Hotchkiss to propose what became known as the 'RNA world' hypothesis. Speaking at a symposium in 1957, Hotchkiss argued that far back in Earth's distant geological past, RNA was the molecular ancestor of DNA:

> Perhaps the confusing relations between RNA and DNA may be illuminated by the speculation that, as a genetic determinant, RNA was replaced during biochemical evolution by the more molecularly and metabolically stable DNA. Cell lines have preserved the RNA entities which, evolutionwise, were primary to DNA ... Viruses, as products of retrograde evolution by loss of function, may have had a choice of either RNA or DNA when specializing to get themselves made in the ample environment of the host cell.[8]

And although what later became known as the 'RNA world' hypothesis has not been universally accepted, one thing is certain—without RNA, DNA couldn't do very much—as will become very clear later in the book.[9,10]

---

[6] (Gierer & Schramm 1956).

[7] (Fraenkel-Conrat, Singer & Williams 1957).

[8] (Hotchkiss 1995; p. 70—citing Hotchkiss (1957) Special Publications of the New York Academy of Sciences, 5; pp. 226–227.)

[9] For a review of the RNA world hypothesis see (Robertson and Joyce 2012).

[10] See for example, (Caetano-Anolles and Seufferheld 2013).

9.2 The Nobel Prize That Never Was

Unlike Hotchkiss, Avery was not given to making grand pronouncements. Like Miescher, he was reclusive and withdrawn, preferring simply to remain absorbed at the lab bench rather than be drawn into arguments with his critics, or blow his own trumpet about his work. In the case of Miescher, it has been suggested that this reclusive nature may well be why his name faded from the DNA story for, having never strived to attract a sizeable entourage of research students, he left no intellectual heirs to continue his legacy.[11]

Avery was similarly shy and reclusive. Although his experiment was soon being discussed at international scientific meetings, he attended none of them. Neither an honorary degree from the University of Cambridge, the Pasteur Gold Medal from the Swedish Medical Society, or the prestigious Copley Medal from the Royal Society of London were sufficient to entice Avery away from the safety and comfort of his laboratory bench.

It has been said that Avery's work was worth not one, but two Nobel Prizes.[12] Yet despite being nominated 44 times—(mostly for other studies in microbiology besides his work on nucleic acids)—Avery never received this prestigious honour. His reclusive nature was well known and perhaps the Nobel Committee opted simply not to risk the embarrassment of conferring this accolade upon Avery, only for him then not to show up at the award ceremony. But whatever the reasons, the decision not to grant Avery a Nobel Prize was later described by Swedish biochemist and Nobel Laureate Arne Tiselius (1902–1971) who went on to become President of the Nobel Foundation from 1960–1964, as 'the most conspicuous omission' in the history of the prize.[13]

9.3 Erwin Chargaff and 'The Grammar of Biology'

Tiselius was not alone in holding this view. Columbia University biochemist Erwin Chargaff (1905–2002) (Fig. 9.1) was one of those who had nominated Avery for the Nobel Prize and was adamant that, in being denied the award, he had been short-changed.[14]

[11] (Dahm and Banerjee 2019).

[12] (Judson 1996); p. 72.

[13] (Judson 2003).

[14] (Portugal 2010); p. 563.

Fig. 9.1 Inspired by Avery's work, Erwin Chargaff (1905–2002) applied the method of partition chromatography that had originally been developed by Archer Martin and Richard Synge for the chemical analysis of wool, to quantifying the chemical composition of DNA. His work offered the first hint at how it carried biological information and was crucial to James Watson and Francis Crick's discovery of its double-helical structure. (Image reproduced with kind permission of the American Philosophical Society)

Chargaff had good reason to feel a debt of gratitude to Avery. Born in 1905 in what is now Ukraine, Chargaff had studied chemistry at the University of Vienna but spent most of his career in the USA, despite once confessing that he felt 'afraid of going to a country that was younger than most of Vienna's toilets.'[15] In 1935, he had taken up a post in the Department of Biochemistry at Columbia University to work on blood coagulation but 9 years later, Chargaff had his own Damascene moment when he learned of Avery's experiment:

> *I saw before me in dark contours the beginning of a grammar of biology.... Avery gave us the first text of a new language, or rather he showed us where to look for it. I resolved to search for this text.*[16]

Chargaff's plan to find the 'grammar of biology' was essentially the same as that with which Levene had begun: to search for variation in the chemical composition of different nucleic acids. But where Levene had studied nucleic acids from different tissues—and found no difference, Chargaff had grander ambitions. He planned to isolate nucleic acid from different species and then

[15] (Chargaff 1978); p. 37.

[16] (Chargaff 1971); p. 639.

quantify the amounts of the four bases to see whether there was any significant variation in their composition.

If so, this would be strong evidence that variation between different types of organism was determined at the level of the chemical composition of the nucleic acids.

But he faced a formidable technical problem. At that time, there was no method of chemical analysis available with the level of precision and accuracy that Chargaff needed to separate and quantify the four bases. Levene's own attempts to quantify the base composition of nucleic acids had all been hamstrung by this same problem. Despite this, Levene had remained optimistic and, in 1931, when he was presented with the Willard Gibbs medal, he looked to the future with hope:

> *Step by step, one mystery of life after another is being revealed. Whether the human mind will ever attain complete and absolute knowledge of and complete mastery of life is not essential. It is certain, however, that the revolt of the biochemist against the idea of a restriction to human curiosity will continue. New discoveries in physics, in mathematics, in theoretical chemistry furnish new tools to biochemistry, new tools for the solution of old problems and for the creation of new ones. So long as life continues, the mind will create mysteries and biochemistry will play a part in their solution.*[17]

In 1941, just a year after his death, one of Levene's predictions came true. A development in chemistry did indeed furnish a powerful new tool and it was one with which Erwin Chargaff would finally unlock 'the grammar of biology.'

The origin of this new development, however, was far removed from the quest to solve grand questions in biology such as the mechanism of heredity. Rather, it emerged from the more pragmatic, down-to-earth task of trying to improve raw materials for the textiles industry.

9.4 'In Praise of Wool'

Thanks to the work of Viktor von Bruns, textile fibres had already played a crucial role in the story of DNA. The absorbent cotton bandages developed by von Bruns had provided Miescher with an abundant source of leukocytes from which he had first isolated nuclein. Now, another textile fibre would prove to be crucial in unlocking its secrets. This was wool, a material which

[17] Levene cited in (Bass 1940); p. 395.

the eminent British textile scientist William Astbury (1898–1961) recalled had once been dismissed as being 'thoroughly dead, unbelievably dull and unprofitable scientifically, and altogether the kind of protein ... that no respectable, aspiring biochemist would touch with a barge pole.'[18] Astbury himself could hardly have disagreed more. Through his X-ray studies of the keratin proteins in wool fibres, Astbury had established himself as a pioneer in the emerging new science of molecular biology—an approach that sought to understand living systems in terms of molecular structure. For him, wool was nothing less than 'the most exciting protein in the world.'[19]

His fellow textile scientists Archer Martin (1910–2002) and Richard Synge (1914–1994) with whom he collaborated most likely shared this view—and with good reason. Martin and Synge worked about a mile down the road from Astbury's laboratory at the University of Leeds in Headingley, a suburb of the city that is renowned to cricket fans around the world for legendary acts of sporting prowess, such as Sir Ian Botham's stunning performance in the 1981 Ashes series against Australia. Less well known however is that Headingley has also been home to some acts of spectacular scientific prowess. One of which was Martin and Synge's research into wool fibres which brought about a revolution in our understanding of life at the molecular level that would have left Miescher utterly delighted.

Martin and Synge had first met in Cambridge where Synge's PhD research was focussed on trying to develop a new method of separating the amino acids found in wool proteins. This was an area of applied research that was of great economic importance at the time. Although wool was a major raw material for the textile industries of several countries in the British Commonwealth, it faced growing competition from synthetic fibres. With funding from the International Wool Secretariat (IWS), an organisation composed of wool growers from Australia, New Zealand and South Africa, Synge's research sought to deliver insights into the chemical structure of wool that might give it a new competitive edge.

Initially Synge began attempting to separate the individual amino acid building blocks of the wool proteins by hand. But when this quickly proved to be too onerous a task, he was relieved to be introduced to Martin whose expertise in chemical separation had been honed thanks to an unusual teenage hobby. Like any other adolescent, Martin had spent long hours behind closed doors away from his parents, but the activity that preoccupied him there was a somewhat unconventional pursuit. Having collected old tin cans, Martin

[18] (Astbury 1955); p. 220.

[19] Ibid.

would happily spend his time welding them together and filling them with coke powder to make distillation columns. It was a skill that served him well when, whilst carrying out research at Cambridge into pellagra, a nutritional disease in pigs, he used his expertise to construct a machine that used the flow of two different solvents in opposite directions for the separation and purification of Vitamin E.

The device could easily be adapted for Synge's analysis of wool, but it was felt that the best place to continue this research was not Cambridge. Rather it was in Leeds, West Yorkshire where, ever since the Middle Ages when Cistercian monks at Kirkstall Abbey near today's city centre had raised sheep and sold their fleeces to foreign wool merchants, wool and textiles had become the economic lifeblood of the city. It was for this reason that, just after the First World War, the British Government had set up the technical headquarters of the Wool Industries Research Association (WIRA) in Headingley, Leeds. WIRA was just one of several technical associations established by the UK government at that time as it sought to emulate the enviable success shown by Germany in applying basic science to industrial research.

The operation of Martin's machine was far from pleasant. It involved sitting for long hours at a time in a former stable that had been converted into a laboratory, and which was now billowing with chloroform fumes leaking from the device (Fig. 9.2). When one half of the duo came to relieve the other of his shift, he would invariably find himself on the receiving end of a string of colourful expletives all thanks to his colleague having become intoxicated from chloroform inhalation.

It came as a relief to discover that the entire separation process could be carried out just as effectively using chromatography with two solvents on filter paper. And in so doing, Martin and Synge did much more than just develop a new lab technique for the analysis of amino acids in proteins; they also proposed that these amino acids might be arranged in a specific order, or 'sequence.' Known as partition chromatography, this method was soon used by the Cambridge scientist Fred Sanger (1918–2013) to determine the precise order of amino acids in the protein hormone insulin. This marked a milestone in biochemistry, for it was the first time that the sequence of amino acids in any protein had been determined, and in 1958, this achievement earned Sanger the Nobel Prize in Chemistry.[20]

[20] This would turn out to be the first of two Nobel Prizes for Sanger. For having won this accolade in 1958 for having developed a method of determining the sequence of amino acids in proteins, he then shared the 1980 Nobel Prize in Chemistry for the invention of a method to determine the sequence of nucleotides in DNA.

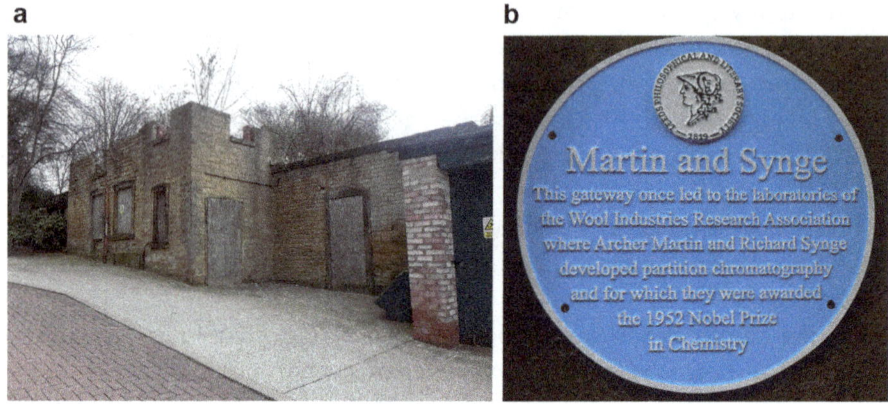

Fig. 9.2 (**a**) The ruin of the former stable that was converted into a textiles research laboratory for the Wool Industries Research Association (WIRA) on Headingley Lane, Leeds, UK. It was here in 1941 that Archer Martin and Richard Synge developed the chemical separation method of partition chromatography for which they were awarded the 1952 Nobel Prize in Chemistry—an achievement that was commemorated with this plaque (**b**) unveiled by Leeds Philosophical and Literary Society in 2019. Although initially developed for the analysis of wool, this transformed our understanding of the structure of proteins as well as being crucial in allowing Erwin Chargaff to analyse the base composition of DNA from different species. (Photographs by K. Hall)

Six years earlier, Martin and Synge had themselves been awarded the Nobel Prize in Chemistry for their development of partition chromatography. In a lecture entitled 'In Praise of Wool', William Astbury praised Martin and Synge's achievement, describing it as an 'episode to the glory of wool that catches the imagination.'[21] Erwin Chargaff was similarly impressed and wrote to Martin and Synge offering his congratulations to them on their award of the Nobel Prize. Much more so than Astbury, Chargaff had good reason to feel a particular debt of gratitude for the invention of partition chromatography.

In one of the early papers on partition chromatography that he had written with Synge, Martin had speculated that partition chromatography might well be 'by no means limited to the separation of amino acids.'[22] Chargaff would prove him to be spectacularly correct. When Martin visited Columbia to give a seminar on partition chromatography, one of Chargaff's post-doctoral researchers, Ernst Vischer was sitting in the audience.

[21] Ibid; p. 235.

[22] (Consden, Gordon, and Martin 1944); p. 231.

Grabbing a paper chromatogram that Martin had brought with him, Vischer dashed back to show it Chargaff who realised that this method was the answer to his prayers.

Using partition chromatography to separate and quantify the four bases from several different species of micro-organism, as well as tissues including bovine thymus and spleen, Chargaff made a crucial discovery. The quantities of the four bases varied considerably between species[23] and were 'not in accord' with the values that would be expected if nucleic acids were simply a monotonous repeat of the same four nucleotides.[24,25] 'I think that there will be no objection to the statement that, as far as chemical possibilities go,' declared Chargaff, 'they [nucleic acids] could very well serve as one of the agents, or possibly as the agent, concerned with the transmission of inherited properties.'[26]

Through his application of partition chromatography to nucleic acids, Chargaff had shown that they did indeed possess the key quality needed for them to be the genetic molecule—that of rich structural variation.

It was tragic enough that Miescher went to his grave plagued by an undeserved sense of never having lived up to his potential. But to have glimpsed how a molecule might represent biological traits without ever having realised that nuclein was the very substance that did this adds an extra pathos to his story. Not that Miescher can be blamed for failing to make this connection. He was constrained by the technical and conceptual limitations of his time. So his groundbreaking idea about how biological variation might reside at the level of molecular structure was destined to lie buried in an obscure letter written to his uncle. And when this idea did eventually become part of the scientific vocabulary, it was not Miescher who would be remembered for it.

---

[23] It is worth noting here that the G:C ratio (and therefore, by implication, the A:T ratio) is not the same in every species. The genomes of monocotyledonous plants (these are one of the two main groups of flowering plants and are characterised by their seed containing a single embryonic leaf. Their members include cereals and agricultural crops such as rice, wheat, corn, sugar cane, bamboo, onion, and garlic) have on average around 46% G:C, whereas those of dicotyledonous plants are around 34%. In bacteria, this figure can be as a high as 50%, whereas in fungi it is around 47% and in protists (defined as single-celled eukaryotic organisms that are not animals, plants or fungi, it can be as low as 26% (Li and Du 2014).

[24] (Chargaff et al., 1949); p. 415.

[25] (Vischer, Zamenhof, and Chargaff 1949); p. 429.

[26] (Chargaff 1950); p. 209.

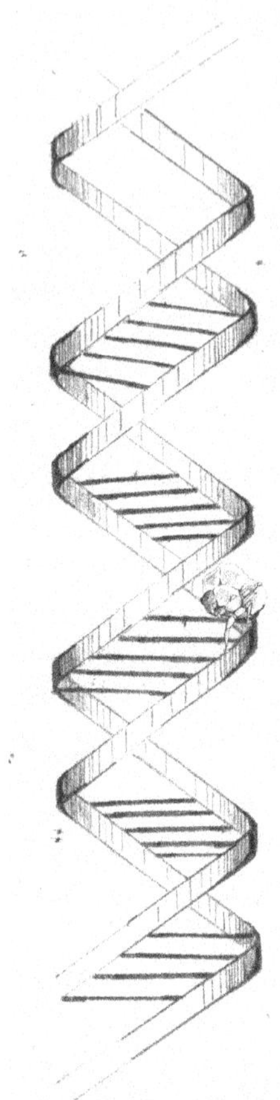

Drawing by Kersten Hall

References

Alexander, H. E., and G. Leidy. 1951a. Determination of inherited traits of H. influenzae by desoxyribonucleic acid fractions isolated from type-specific cells. *Journal of Experimental Medicine* 93:345–359.

Alexander, H. E., and G. Leidy. 1951b. Induction of heritable new type in type-specific strains of Hemophilus influenzae (19161). *Proceedings of the Society for Experimental Biology and Medicine* 78:625–626.

Alexander, H. E., and G. Leidy. 1953. Induction of streptomycin resistance in sensitive Hemophilus influenzae by extracts containing desoxyribosenucleic acid from resistant Hemophilus influenzae. *Journal of Experimental Medicine* 97:17–31.

Astbury, W. T. 1955. In praise of wool. In *Proceedings of the International Wool Textiles Research Conference, B*, 220–243.

Avery, O. T., C. MacLeod, and M. McCarty. 1944. Studies on the chemical nature of the substance inducing transformation of pneumococcal types: Induction of transformation by a desoxyribonucleic acid fraction isolated from pneumococcus type III. *Journal of Experimental Medicine* 79:137–158.

Bass, L. W. 1940. Phoebus Aaron Theodor Levene 1869-1940. *Science* 92:392–395.

Caetano-Anollés, G., and M. J. Seufferheld. 2013. The coevolutionary roots of biochemistry and cellular organization challenge the RNA world paradigm. *Microbial Physiology* 23 (1–2): 152–177. https://doi.org/10.1159/000346551.

Chargaff, E. 1950. Chemical specificity of nucleic acids and mechanism of their enzymatic degradation. *Experientia* 6:201–240.

Chargaff, E. 1971. Preface to a grammar of biology. *Science* 172:637–642.

Chargaff, E. 1978. *Heraclitean Fire: Sketches from a Life Before Nature*. New York, NY: The Rockefeller University Press.

Consden, R., A. H. Gordon, and A. J. P. Martin. 1944. Qualitative analysis of proteins: A partition chromatographic method using paper. *The Biochemical Journal* 38:224–232.

Dahm, R., and M. Banerjee. 2019. How we forgot who discovered DNA: Why it matters how you communicate your results. *BioEssays* 41:1–3. https://doi.org/10.1002/bies.201900029.

Fraenkel-Conrat, H., B. Singer, and R. C. Williams. 1957. Infectivity of viral nucleic acid. *Biochimica et Biophysica Acta* 25:87–96. https://doi.org/10.1016/0006-3002(57)90422-5.

Gierer, A., and G. Schramm. 1956. Infectivity of ribonucleic acid from tobacco mosaic virus. *Nature* 177 (4511): 702–703. https://doi.org/10.1038/177702a0.

Gulland, J. M., G. R. Barker, and D. O. Jordan. 1945. The chemistry of the nucleic acids and nucleoproteins. *Annual Review of Biochemistry* 14:175–206.

Hershey, A. D., and M. Chase. 1952. Independent functions of viral protein and nucleic acid in growth of bacteriophage. *Journal of General Physiology* 36:39–56.

Hotchkiss, R. D. 1995. DNA in the decade before the double helix. *Annals of the New York Academy of Sciences* 758 (1): 55–73. https://doi.org/10.1111/j.1749-6632.1995.tb24809.x.

Judson, H. F. 1996. *The eighth day of creation*. Cold Spring Harbor, NY: Cold Spring Harbor Press.

Judson, H. F. 2003. No Nobel prize for whining. *The New York Times*. Accessed 20 October 2003.

Portugal, F. H. 2010. Oswald T. Avery: Nobel laureate or Noble luminary? *Perspectives in Biology and Medicine* 53:558–570.

Robertson, M. P., and G. F. Joyce. 2012. The origins of the RNA world. *Cold Spring Harbor Perspectives in Biology* 4 (5): a003608. https://doi.org/10.1101/cshperspect.a003608.

Vischer, E., S. Zamenhof, and E. Chargaff. 1949. Microbial nucleic acids: The Desoxypentose nucleic acids of avian tubercle bacilli and yeast. *Journal of Biological Chemistry* 177:429–438.

10

'What Is Life?'

Contents

Abstract 1944 was a landmark year in the story of DNA not only thanks to Avery's experiment, but also for the publication of 'What is Life?', a short book written by physicist Erwin Schrödinger. Although Schrödinger is more famous for being one of the founders of quantum mechanics, in this book he turned his attention to how a physicist might approach biology—and in particular the molecular nature of heredity. Schrödinger proposed that in its chemical nature, the genetic molecule must possess two key properties: it must be big, and it must be capable of structural variation which, through stereoisomerism would allow biological traits to be represented at the molecular level. But although Schrödinger's book has been hailed as landmark in this history of molecular biology, these key ideas about the genetic molecule had already been proposed by Miescher in a letter written three years before his

K. Hall, R. Dahm, *The Dawn Fisherman*, Copernicus Books,
https://doi.org/10.1007/978-3-032-14219-1_10

death. Alongside the discovery of DNA, this is Miescher's other great contribution to biology and this chapter explores why he is not better known for it, along with tracing another little known story—how the use of X-rays to study wool fibres helped lay the foundations for the later work of James Watson, Francis Crick, Rosalind Franklin and Maurice Wilkins to solve the double-helical structure of DNA.

Keywords DNA • Double helix • Rosalind Franklin • James Watson • Francis Crick • Molecular biology • Genetics • X-ray Crystallography • Base pairing • Adenine • Thymine • Guanine • Cytosine • Replication • Genetic material • Friedrich Miescher • Nuclein • Pus-soaked bandages • Physiological chemistry • Biochemistry • Morphology • Architecture of Life • Code of life • Genetic code • Protein synthesis • Amino acid sequence • The Double Helix (book) • DNA structure • Scientific discovery • History of Science • William Astbury • Florence Bell • Elwyn Beighton • William Bragg • Lawrence Bragg

Along with his discovery of nucleic acids, Miescher left another monumental scientific legacy – but someone else is remembered for it…

During 1895, as Miescher's life steadily ebbed away, his former mentor from his days in Leipzig, Carl Ludwig, wrote to him in the hope of offering some consolation:

> *As grievous as it may be for you to be sick, you have the comfort of having achieved everlasting accomplishments; you have made the centre, the core of all organic life accessible to chemical analysis; and however often the cell will be studied and examined during the centuries to come, the grateful descendants will remember you as the ground-breaking researcher.*[1]

Ludwig proved to be correct about one thing at least. Miescher had indeed made two monumental accomplishments. The first of these had been the discovery of nucleic acids; the second was to have glimpsed how a molecule might carry hereditary traits through structural variation.

But Ludwig's confidence that Miescher would be remembered 'as the ground-breaking researcher' behind these discoveries was to prove less well-founded. Certainly, there were researchers, such as Jack Schultz (1904–1971)

[1] Ludwig to Miescher, cited in (His 1897); p. 12.

who, having started his career washing bottles in Thomas Hunt Morgan's 'Fly Room' had gone on to become an eminent biochemist and geneticist with a particular interest in the chemical constitution of the genetic material. In a paper of 1941, Schultz pointed out that not only had Miescher discovered nucleic acids, but he had also realised that whatever the nature of the genetic molecule—whether it be protein or nucleic acid—it must have one key property:

> *The problem [of the nature of the gene] was first attacked by Friedrich Miescher and you will remember that he came to the conclusion that the substances he discovered in sperm nuclei, relatively simple substances that we now know as thymonucleic acid and protamines, might be the material basis of heredity, if the possibility of isomerism were kept in mind.*[2]

But within only a couple of years of Schultz having praised Miescher for conceiving of this idea, someone else was having very similar thoughts about the genetic molecule.

10.1 Cats and Code Scripts: Erwin Schrödinger

Having dreamt up a famous thought experiment involving a quantum cat that has been referenced on numerous occasions in the hit TV science comedy 'The Big Bang Theory', Austrian physicist Erwin Schrödinger (1887–1961) (Fig. 10.1) has achieved a level of recognition in popular culture way beyond that of most scientists. Schrödinger proposed his famous thought experiment to highlight the apparent paradoxes thrown up by quantum mechanics, a field of physics which he had helped to found. But his influence on the emerging field of molecular biology is much less well known.

In 1944, whilst living as an exile in Dublin having fled the annexation of his native Austria by the Nazis, Schrödinger published a small book which posed a huge question. Called 'What is Life?', it was based on a series of lectures that he had given a year earlier and despite its grand title, Schrödinger declared that its aim was modest. This was, he said, 'to convey one small comment' on a much larger question, which was 'How can the events *in space and time* which take place within the spatial boundaries of a living organism be accounted for by physics and chemistry?'[3]

---

[2] (Schultz 1941); p. 55.

[3] (Schrödinger 1944); p. 1.

Fig. 10.1 Erwin Schrödinger (1887–1961) is better known for his role in the founding of quantum mechanics, but his book 'What is Life?', published in 1944 while he was living as a political exile in Dublin, proved to be influential in steering several physicists towards biology. (Photograph reproduced with kind permission of the School of Theoretical Physics, Dublin Institute for Advanced Studies)

Schrödinger wondered how it could be that all living systems appeared to operate in defiance of a fundamental principle in physics. This is the second law of thermodynamics, which states that all physical processes move in a direction of increasing disorder or, to use the technical term, maximum entropy. Schrödinger observed that living organisms however appear to be the exact opposite of a disordered system: not only do they maintain stability and order in the face of increasing entropy, but they are also able to pass on stable biological traits to successive generations.

To account for how living organisms could do this, Schrödinger proposed that they possessed a property that he called 'negative entropy' but he also recognised that if heritable traits really were somehow represented at the molecular level then there was a paradox. Molecules, by virtue of their small size, were subject to random thermal motion and disorder, and early experiments based on the generation of mutations in flies by X-rays, had estimated that the size of the gene was of the order of only a few hundred atoms. This was far too small to overcome the disordering effect of random atomic motion. So, if the gene really was a molecule, how then could it possibly overcome these effects to retain sufficient order in its structure to represent and transmit hereditary features in a fly or, to use Schrödinger's own favourite example, the characteristic lip shape of the Habsburg dynasty?[4]

[4] ibid, p. 46.

But Schrödinger had a solution in mind. He postulated that the genetic molecule, whatever its chemical nature, could not be a small molecule composed of only a few atoms in the same way as water or carbon dioxide, but instead must be a huge chain molecule of atoms bonded together. Only in this way could biological traits be represented at the molecular level, without collapsing into disorder due to random thermal motion. As for the question of how a long chain molecule might represent biological traits, Schrödinger had a simple answer:

> *We shall assume the structure of a gene to be that of a huge molecule, capable only of discontinuous change, which consists of a rearrangement of the atoms, and leads to an isomeric molecule.*[5]

Schrödinger's idea bears strikingly similarity to Miescher's own suggestion that isomerism in a giant molecule might be the molecular basis of heredity. And the similarities did not end there. Due to the vast number of stereoisomers arising from its structure, the genetic molecule was described by Schrödinger as 'a code-script' representing heritable biological traits and determining the development of the organism.[6,7] He also proposed that mutations, or variations in a single trait such as eye colour (known as 'alleles') arose from transitions between isomeric forms of the genetic molecule.

Once again, Schrödinger was echoing Miescher's earlier proposal that what today we might call mutations could arise 'As a result of minute changes and external conditions' which caused their carbon atoms to 'suffer positional changes and thus produce structural defects.'[8]

Thanks to Oswald Avery's work discussed in the previous chapter, we now know that it is nucleic acids, and not proteins that are the genetic material. Moreover, thanks to Erwin Chargaff's application of chromatography to the analysis of DNA also discussed in the previous chapter, we now know that genetic information is not represented at the molecular level through stereoisomerism as Miescher believed, but rather through variation in the sequence of the four bases. But even though they were wrong on this technical aspect, both Miescher and Schrödinger had grasped that biological variation could reside at the level of molecular structure.

---

[5] (Schrödinger 1944; 1967 ed); p. 56.

[6] Ibid; p.19.

[7] Ibid.

[8] Miescher Letter LXXV 17th December 1892 (His 1897); p. 116.

Yet today it is Schrödinger, and not Miescher who is largely remembered for this insight. This should perhaps come as no surprise. 'What is Life?' was translated into seven languages and sold over 100,000 copies at the time of its publication. Even today the book is still regarded as a classic - such as when British celebrity physicist and TV presenter Professor Brian Cox cited it in the opening of his 2013 BBC TV documentary series 'Wonders of Life.'[9]

Miescher's ideas by contrast, powerful though they were, lay buried and largely unnoticed by the world, with the odd exception such as Jack Schultz. In William Shakespeare's play, Henry V declares that 'In peace, there's nothing so becomes a man as modest stillness and humility': Miescher had both these qualities in abundance—and they did not serve his memory particularly well. He might have been better remembered for his ideas about the genetic molecule had he followed King Henry's further advice—'But when the blast of war blows in our ears, then – imitate the tiger. Stiffen the sinews, summon up the blood. Disguise fair nature with hard-tempered rage. Lend the eye a terrible aspect.'

10.2 'A Delightful Little Book'

Shortly after the publication of 'What is Life?', the British evolutionary botanist and cytologist Irene Manton (1904–1988) sang its praises in the journal *Nature* saying that 'When a great physicist takes the trouble to explain in simple language some of his matured thoughts on topics of general interest outside his own subject, it is an event for which one cannot be too grateful'.[10] Manton called it a 'delightful little book' which had inspired her own research into chromosome structure and its implications for understanding plant phylogeny.[11,12]

But not everyone was quite so impressed. 'When I first read this book, over 40 years ago,' wrote US chemist and twice Nobel laureate Linus Pauling (1901–1994), 'I was disappointed. It was, and still is, my opinion that Schrödinger made no contribution to our understanding of life.'[13] Moreover, Pauling went on to suggest that by invoking such concepts as 'negative entropy', Schrödinger had obscured, rather than addressed, the question of

[9] 'Wonders of Life' BBC (2013) Episode 1; 'What is Life?'

[10] (Manton 1945); p. 471.

[11] ibid; pp. 471–473.

[12] (Williams 2015).

[13] (Pauling 1987); p. 229.

'What is Life?'. Austrian-British biologist and Nobel laureate Max Perutz (1914–2002) was equally unimpressed and pointed out that Schrödinger's much vaunted concept of the 'aperiodic crystal' was little more than a reformulation of an idea already articulated by the German-American biophysicist Max Delbrück (1906–1981) who, with his co-workers, the Soviet biologist Nikolaj V. Timofeef-Ressovsky (1900–1981) and the German chemist Karl G. Zimmer (1911–1988) had already attempted to estimate the size of a gene using X-radiation.[14] 'What was true in his book was not original,' Perutz complained, 'and most of what was original was known not to be true even when the book was written.'[15]

Miescher might well have had some sympathy with Perutz's claim that few of Schrödinger's ideas were truly novel. Two years before his death, Miescher had reflected on problem of heredity and concluded that 'Continuity does not only lie in the form, it also lies deeper than the chemical molecule. It lies in the constituent groups of atoms. In this sense, I am to the utmost, an adherent of a chemical model of inheritance.'[16]

Yet even though 'What is Life?' cast Miescher's idea about the genetic molecule work into the shadows, it is a safe bet that he would most likely have been grateful for the astounding success of Schrödinger's book had he only been alive at the time to see it.

10.3 'Uncle Tom's Cabin'

The growing a sense of disappointment that Miescher felt with passing years at having failed to live up to his potential was confined not just to himself but to others too. Having championed an approach that sought to understand the cell in terms of physics and chemistry, he felt more than a little let down that physicists and chemists did not take up this cause with the same degree of enthusiasm:

> *I believe that the physiologist does science a great service when he arrives at precise, pure chemical problems in order to produce fully proven experts, as opposed to when he struggles with incomplete knowledge of the field. Unfortunately, few physicists and chemists are so inclined to sacrifice themselves for the accomplishment of tasks, the significance of which is to be found in applied, rather than pure disciplines.*[17]

---

[14] (Perutz 1987).

[15] (Perutz 1987); p. 243.

[16] 13th Oct 1893, Letter LXXVIII, in (His 1897); p. 122.

[17] Miescher to Hoppe-Seyler, 1870, Letter XV in His 1897, pp. 55–57; p. 56.

Schrödinger's book would change this situation dramatically. Each of the three scientists who were awarded the 1962 Nobel Prize in Medicine for their discovery of the structure of DNA, cited 'What is Life?' as having had a crucial influence in directing their work towards the question of the genetic molecule. In his 1968 memoir, 'The Double Helix', James Watson praised 'What is Life?' for having 'very elegantly propounded the belief that genes were the key components of living cells and that to understand what life is, we must know how genes act.'[18]

Watson had started out studying zoology as an undergraduate, with vague ideas of becoming an ornithologist, but Francis Crick and Maurice Wilkins (1916–2004) with whom he shared the Nobel Prize had come from even further afield. Having trained as a physicist, Crick had spent the Second World War working on the development of magnetic mines for the British admiralty, but after reading 'What is Life?' became captivated by its suggestion 'that biological problems could be thought about in physical terms.'[19] In the summer of 1953, after he and Watson had first unveiled their double-helical model of DNA, Crick even wrote to Schrödinger to say how influential his book had been:

> *Dear Professor Schrödinger,*
>
> *Watson and I were once discussing how we came to enter the field of molecular biology, and we discovered that we had both been influenced by your little book, "What is Life?". We thought you might be interested in the enclosed reprints - you will see that it looks as though your term "aperiodic crystal" is going to be a very apt one.*[20]

Like Crick, Maurice Wilkins was also a physicist who had found a military application for his skills during the War. This was not something of which he was proud. Having worked on the preparation of uranium for the Manhattan Project, Wilkins was sickened with guilt when he learned of the horrors unleased upon Hiroshima and Nagasaki. Perhaps as a means of seeking absolution, he now turned to the study of living systems—rather than developing yet more means for its mass destruction—and found inspiration in the pages of 'What is Life?':

---

[18] Watson JD (1968) The double-helix. Weidenfeld & Nicolson. Phoenix 2010 edition, p. 11.

[19] (Crick 1965); pp. 183–186.

[20] Letter from Francis Crick to Erwin Schrödinger 12th Aug 1953, Archives of the Dublin Institute for Advanced Studies, SCH/M/44.

> *the main impact of Schrödinger's book was that it set me in motion...Schrödinger used the language of physicists and that stimulated me, as a physicist, to persevere with his book and its introduction to genetics, and to decide that this was the general area that I wanted to explore, as a 'biophysicist.'*[21]

A fellow traveller with Wilkins on this road from the physical sciences into biology was the German-American scientist Gunther Stent (1924–2008). And once again, 'What is Life' had pointed the way. Born in Berlin, Stent had fled the Nazi regime when he was 14 years old and made his way to the United States where he had eventually earned a doctorate in physical chemistry. But after reading 'What is Life', he had become convinced that molecular biology was the future and headed to Caltech to join Max Delbrück who was blazing the trail in the new field with his work on bacterial viruses.

Reflecting on this change of direction, Stent once compared 'What is Life?' with the novel 'Uncle Tom's Cabin.' Written in 1852 by American author Harriett Beecher Stowe (1811–1896), this novel promoted the abolitionist cause and has been interpreted as a rallying cry for the anti-slavery movement in the years preceding the American Civil War of 1861–1865. According to Stent, 'What is Life?' was similarly a call to arms for physicists and chemists to turn to the science of life and usher in what he described as a 'revolution in biology that, when the dust had cleared, left molecular biology as its legacy.'[22] Stent was confident that, after reading the book, physical scientists who might otherwise have dismissed biology as being 'confined to stale botanical and zoological lore' would no longer consider it to be such a 'sissy subject.'[23,24]

'What is Life?' took a sledgehammer to such prejudices. Scientist and historian Jan Witkowski has written that it 'alerted physical scientists to the problems awaiting solution in biology'.[25,26] It was exactly such a rallying cry for physical scientists to turn their attention to physiological chemistry that Miescher, just over half a century earlier, had lacked. But even had Miescher published and disseminated his ideas as widely as Schrödinger, he still lacked a vital factor—one that was recognised by the French biochemist and Nobel laureate Jacques Monod (1910–1976) as having been crucial to the success of 'What is Life?' and its powerful effect upon the physics community. 'Just to hear one of the leaders in quantum mechanics asking 'What is Life?", wrote

[21] (Wilkins 2003); p. 84.

[22] (Stent 1968); p. 392.

[23] (Stent 1968); p. 3.

[24] (Symonds 1986).

[25] (Witkowski 1986).

[26] Monod, The Logic of Living Systems: A History of Heredity. Translated by B.E. Stillman.; p. 259.

Monod, 'was enough to fire the enthusiasm of certain young physicists and to bestow some sort of legitimacy on biology.'[27] According to the historian Robert Olby, the combined kudos of being both a Nobel laureate, together with his role as a founder of quantum physics conferred an authority upon Schrödinger and his pronouncements that Miescher never had:

> *even though Miescher had already come up with a concept rooted in stereoisomerism, it took a physicist of Schrödinger's calibre to stir the interest of other physicists….*[28]

Olby went on to point out that, following Miescher, there were others who had all suggested that hereditary traits might be represented through isomeric arrangements within a large molecule. In 1917 German chemist and Nobel laureate Emil Fischer (1852–1919) published calculations showing that if proteins were chains of different types of amino acids linked together, then the number of potential structural permutations was vast. He also noted that this might be 'of some interest' to biological phenomena such as heredity.[29] Five years earlier, Albrecht Kossel had echoed a similar view in his Harvey Lecture[30]:

> *We must remember that the proteins are composed of Bausteine united in very different ways…The number of Bausteine which may take part in the formation of the proteins is about as large as the number of letters in the alphabet. When we consider that through the combination of letters an indefinitely large number of thoughts may be expressed we can understand how vast a number of the properties of the organism may be recorded in the small s vbpace which is occupied by the protein molecules. It enables us to understand how it is possible for the proteins of the sex-cells to contain, to a certain extent, a complete description of the species and even of the individual.*[31]

According to the maths at least, isomerism of proteins was looking to be the most likely candidate by which heredity traits were passed on. Addressing the British Association for the Advancement of Science in 1926, British physiologist John Beresford Leathes (1864–1956) calculated that, even with certain structural restrictions in place, a protein made up of 50 amino acids could

[27] Ibid.

[28] (Olby 1994); p. 246.

[29] (Fischer 1917); p. 54.

[30] For a further exploration of Kossel's work, including a discussion about the impact of the alphabet analogy that both he and Miescher proposed, readers might like to listen to Episode 2 of 'The DNA Papers'—https://soundcloud.com/user-287120959/dna-papers-episode-2-albrecht-kossel

[31] (Kossel 1911; p. 45).

give rise to as many as 10^{48} different conformational variants.[32] Mathematician-turned theoretical biologist, Dorothy Wrinch (1894–1976) took this idea even further. In 1934, she proposed that it was this variation in the amino acid sequence of chromosomal proteins which conferred genetic specificity.[33]

Yet in time, Miescher's insight—along with that of Kossel, Fischer, Leathes and Wrinch—that the key property of the genetic molecule was structural variability would all be eclipsed by Schrödinger. Because, as Robert Olby concluded 'it required a physicist, and an eminent one at that, to make physicists aware of the concept, and perhaps most important of all, to relate the code-script to crystallography in the conception of an aperiodic crystal.[34]

Perhaps so. But there were at least two physicists who did not need the kudos of Schrödinger and his book to convince them that biology was much more than just 'stale botanical and zoological lore'. And their inspiration came not from asking 'What is Life?' but rather—what is wool?

10.4 Wool Weaves a Path to the Double Helix

Thanks to Archer Martin and Richard Synge's development of partition chromatography for the chemical analysis of proteins in wool, the textiles industry had played a crucial role in unravelling how nucleic acids could carry genetic information. But how could a molecule pass that information on down the generations? Here too research into the humble wool fibre helped to offer an answer.

In 1938, the same year as Archer Martin arrived in Leeds to begin the chemical analysis of wool by sitting in an old stable that was billowing with chloroform fumes, physicist Florence Bell (1913–2000) (Fig. 10.2) was working in equally unpleasant conditions in the Department of Textiles, about a mile down the road at the University of Leeds. These involved working for long hours in a darkened room with dangerously high voltages and red-hot glass tubes from which X-rays were emitted. Yet through her willingness to endure these conditions, Bell helped to lay the foundations for the discovery of the structure of DNA.

The origins of Bell's work can be traced to a letter written in December 1914 by William Bragg (1862–1942), who was then Cavendish Professor of Physics at the University of Leeds to the then Vice-Chancellor Michael Sadler.

---

[32] (Leathes 1926); p. 388.

[33] (Wrinch 1934).

[34] (Olby 1994); p. 246.

Fig. 10.2 Physicist Florence Bell (1913–2000) who came to Leeds in 1937 to work William Astbury on his X-ray studies of wool fibres and a year later showed that this same method could be used to reveal the regular, ordered structure of DNA. (Portrait reproduced with kind permission of Mr. Chris Sawyer)

In the closing paragraph of the letter, Bragg suggested that a new X-ray method discovered by himself and his son Lawrence (1890–1971), might well be useful in the study of the textile fibres that were of such importance to the economy of Leeds.

William and Lawrence Bragg had discovered that X-rays could reveal much more than just broken bones. During the summer of 1912 when he was Cavendish Professor of Physics at the University of Leeds, Bragg and his son were on holiday near Scarborough on the East Yorkshire coast when he learned through a letter from an old colleague about an experiment that intrigued him. This was the discovery by German physicist Max von Laue (1879–1960) that when X-rays passed through a crystal, they were scattered (or to use the technical term, 'diffracted') giving rise to a distinct pattern of black spots on a photographic film.

Dashing back to Leeds in excitement, the Braggs spent the following year working into the small hours of the morning in a freezing cold laboratory at the University of Leeds to show that, by applying certain mathematical techniques to the patterns of spots made by the scattered X-rays, they could be used to deduce the regular, ordered arrangement of atoms or molecules in the crystal.

Fig. 10.3 Plaque on the entrance foyer to the Brotherton Library at the University of Leeds, UK which commemorates the award of the 1913 Nobel Prize in Physics to father and son duo, William and Lawrence Bragg for their development of X-ray crystallography which uses the scattering of X-rays to reveal the arrangement of atoms and molecules in crystals. To date, 28 Nobel Prizes have been won using this method—including the 1962 Nobel Prize in Physiology or Medicine for the discovery of the structure of DNA. (Photograph by K. Hall)

Their method, known as X-ray crystallography, has since transformed our understanding of matter. In 1915 it earned them the honour of becoming the only parent and child team ever to be awarded a Nobel Prize (Fig. 10.3) and to date, 28 Nobel Prizes have been won using the Braggs' method It has allowed us to solve the molecular structure of biomedically important materials such as the hormone insulin, Vitamin B12 and the blood protein haemoglobin and has even been used by the Curiosity rover to make studies of Martian rock.

Of all the Nobel prizes awarded which used this method, the most famous must surely be the award of the 1962 Nobel Prize in Physiology or Medicine to James Watson, Francis Crick and Maurice Wilkins for solving the double-helical structure of DNA. And this owed much both to Florence Bell, and the study of wool fibres.

William Bragg had initially applied his method to the study of simple crystalline structures such as ice, diamond and common salt, but by the time he left Leeds for London in 1915, he was starting to wonder whether it might be able to reveal the structures of larger, fibrous molecules found in living systems. Thanks to his time in Leeds, and the importance of textiles to the city, there was one material that Bragg felt would be an ideal candidate for this work.

Shortly before leaving for London, Bragg warned the University of Leeds that its Textiles Department 'does not know enough physics.'[35] He advised that this could be remedied by finding 'a keen young man' who might be able

[35] Letter from W.H. Bragg to A. Smithells, 26th Mar 1915. Cited in *William Henry Bragg 1862–1942: Man and Scientist.* Caroe, G. (1978); Cambridge University Press. p. 138.

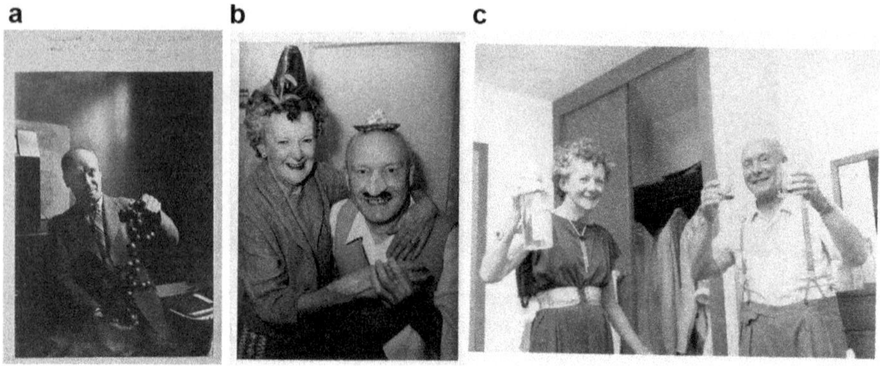

Fig. 10.4 (**a**) Leeds-based physicist William Astbury (1898–1961) who, through his application of the method of X-ray crystallography developed by William and Lawrence Bragg to the study of biological fibres such as wool was an early pioneer of the emerging new science of molecular biology that sought to understand life in terms of molecular structure. Although, as is quite evident from photographs (**b**) and (**c**), he and his wife Frances also knew how to throw a good party. (Photographs reproduced with kind permission of Special Collections, Brotherton Library, University of Leeds, and the late Mr. Bill Astbury (grandson))

to apply X-ray crystallography to the study of wool fibres.[36] Being the product of a particular time and culture, he appears not have considered the possibility that 'a keen young woman' might do just as good a job.

On his arrival in London, Bragg found the ideal candidate. After graduating from Cambridge University, William Astbury (1898–1961) (Fig. 10.4) had joined William Bragg's team in London where he was given the task of applying X-ray crystallography to the study of wool fibres. But, just as Archer Martin and Richard Synge would leave Cambridge and move to Leeds to continue their work on wool, in 1928 Astbury was encouraged by his mentor to leave London and head North to take up the newly created post of Lecturer in Textile Physics at the University of Leeds.

Astbury did not share Bragg's enthusiasm for this suggestion—well, not initially at least. To a young scientist eager to build a career in the new field of X-ray crystallography, the prospect of leaving Bragg's laboratory in London for industrial research in the provinces seemed like a massive backward step. Writing to his friend and fellow Bragg protégé, J. D. Bernal, Astbury lamented that he felt as if he was 'heading into the wilderness.'[37] He could not have been more wrong.

[36] Ibid.;p. 139.

[37] Letter from W. T. Astbury to J. D. Bernal, 13th Sept 1928. John Desmond Bernal. Personal and Scientific Papers. Cambridge University Library.Department of Manuscripts and University Archives.

Astbury's X-ray analyses of the molecular structure of wool fibres helped to transform our basic understanding of proteins and offered a new approach to understanding living systems at the molecular level. His X-ray studies of keratin, the main protein in wool fibres, showed that its molecules were long chains of smaller compounds called amino acids connected like beads strung together on a necklace. Moreover, he showed that these protein chains could change shape to take on compacted, or elongated forms. This ability of proteins to change shape not only explained the elasticity of wool that made it so attractive as a raw material for the textile industry but also how muscles contract, how bacteria can swim and, more recently, how the now infamous spike protein on the surface of SARS-CoV2 enables the virus to bind and enter human cells.

For Astbury, this approach came to define a whole new science that he popularised as 'molecular biology.' This sought to understand living systems through the structure of the giant molecules from which they were made and how these change shape. Through this work he soon became recognised as an international authority in the study of biological fibres using X-rays and in the late 1930s was sent some samples of a very different kind of biological fibre. This was not a protein but what at the time was known at the time as thymonucleic acid and is today known by the more familiar name of DNA.

Astbury once confessed to believing that, when it came to science, although women lacked what he called a certain 'creative spark', they nevertheless 'do so often such marvellously conscientious and thorough work after the spark has been struck.'[38] The arrival of Florence Bell in his laboratory in 1937 should certainly have made him think twice about holding such views. Quickly recognising Bell's formidable talent and intellect, he gave her the task of applying X-ray crystallography to the study of the DNA fibres that he had been sent.

In 1938, she obtained the very first X-ray diffraction images of DNA fibres. At first sight, Bell's images hardly looked impressive. They appeared to be little more than a messy smear of black spots that would not have looked out of place in a gallery of contemporary art (Fig. 10.5). But they revealed a crucial piece of information.

From their interpretation of these photographs, Bell and Astbury proposed an early model of the DNA molecule—much of which is today known to be incorrect. But Astbury and Bell's work on DNA is nevertheless important for several reasons. Firstly, despite other errors, their model showed that the bases,

GBR/0012/MS Add.8287J.2.

[38] Letter from W.T. Astbury to W.L. Bragg, 5th November 1943. Private papers of Professor A.C.T. North. Cited with kind permission of Professor A.C.T. North.

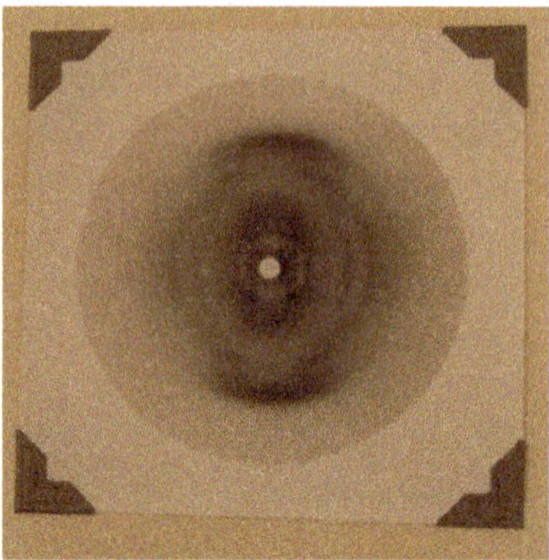

Fig. 10.5 The first successful X-ray diffraction image of DNA fibres, taken by Florence Bell in 1938 and published in her PhD thesis. With this photograph, Bell showed that the method of X-ray crystallography first developed by William and Lawrence Bragg could be used to reveal the structure of DNA. (Reproduced with permission of Special Collections, Brotherton Library, University of Leeds)

which form the rungs of the DNA ladder, stack up on top of each other, as they described it, 'like a pile of plates.'[39,40] They were also able to measure the distance between adjacent bases—which James Watson later acknowledged, had given him and Crick a vital foothold when they began building their own model of the DNA molecule.[41]

Secondly, when Astbury presented this work at a meeting in Cold Spring Harbor in 1938 he suggested that nucleic acids might not simply be a repeating tetranucleotide but instead contain regions of structural variation that might well be of some significance:

> *You know that the thymonucleic acid is said to be a tetranucleotide. We have not worked this out completely, but it is clear from the photographs that the true period along the nucleotide column is at least 17 times the thickness of a nucleotide. So the nucleotides do not follow each other always in the same order, and this gives you another chance of great variation.*[42]

But most importantly of all, Bell and Astbury's work on DNA showed that it had a regular, ordered structure and crucially one that could be revealed using X-ray crystallography. Fourteen years later, armed with this insight, another female crystallographer would take an X-ray image of DNA that has since

[39] (Astbury and Bell 1938a).

[40] (Astbury and Bell 1938b).

[41] (Watson 1968; 2010 edition); p. 40.

[42] (Astbury & Bell 1938a); p. 119.

been proclaimed as being 'one of the most important photographs in the world.'[43]

10.5 One of the Most Important Photographs in the World?

At its centre, the image showed a striking pattern of black spots arranged into the form of a cross. This had been made by X-rays scattered after having passed through fibres of B-form DNA—a conformational variant of the molecule which forms at higher humidity than the more crystalline A form. The photo had been taken in the spring of 1952 by Rosalind Franklin (1920–1958) (Fig. 10.6) and her PhD student Raymond Gosling (1926–2015) at King's College, London who labelled it simply as 'Photo 51' in their lab notes (Fig. 10.7).

Yet despite this modest name, the pattern at the heart of 'Photo 51' has become something of an icon. When the UK Royal Mint released a commemorative 50 pence coin in 2022 to mark the centenary of Franklin's birthday, one of its faces was inscribed with the pattern of 'Photo 51' (Fig. 10.7). And in his memoir, 'The Double Helix' published in 1968, James Watson famously recalled the dramatic effect that this pattern had upon him when he was first shown 'Photo 51':

> *The instant I saw the picture my mouth fell open and my pulse began to race…the black cross of reflections which dominated the picture could arise only from a helical structure.*[44]

The cross-shaped pattern of the spots told Watson that DNA was coiled into a helical shape. But it is worth bearing in mind that Watson's memoir was written 15 years after the events it describes and needs to be read with a critical eye.[45] His mouth may well have fallen open, and perhaps his pulse did

[43] Inscription on a commemorative plaque outside King's College, London where Franklin and Gosling worked.

[44] (Watson 1968; 2010 edition); p. 120.

[45] Some commentators have gone even further. Franklin's sister Jenifer Glynn has described The Double Helix as 'a novel. It is not history.' (cited in Markel, 2021); p.308. It was a view shared by Francis Crick who, after reading an early draft of the book, wrote a scathing letter to Watson: 'The book is not a history of the discovery of DNA as you claim in the preface. Instead, it is a fragment of your autobiography which covers the period when you worked on DNA…It is thus absolutely clear that your book is not history as it is normally understood…Should you persist in regarding your book as history, I should add that it shows such a naïve and egotistical view of the subject as to be scarcely credible.' (Letter from F.H.C.Crick to J.D. Watson, 13th April 1967. London, Wellcome Library, PP/CRI/I/3/8/4).

Fig. 10.6 Rosalind Franklin (1920–1958) enjoying a holiday in Tuscany around 1950 before moving to King's College, London where, with Raymond Gosling, she took the now famous 'Photo 51' X-ray image of DNA. This photograph was taken by her colleague Vittorio Luzzati. (Reproduced with kind permission of Churchill Archives Centre, Cambridge and the Franklin Family Estate)

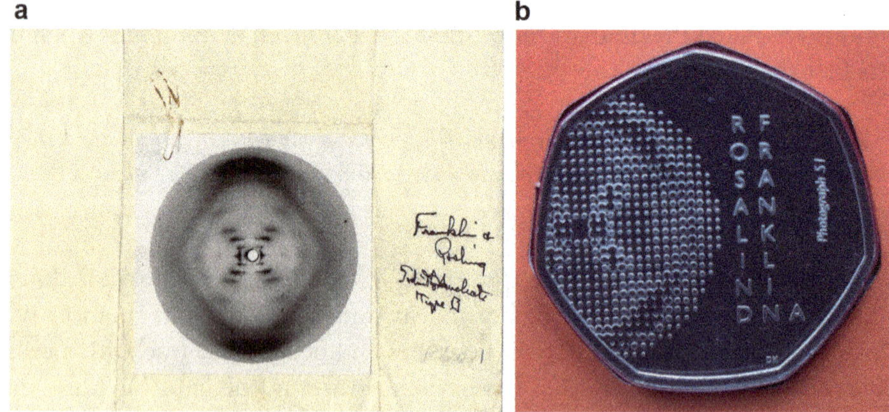

Fig. 10.7 (**a**) X-ray diffraction photograph of B-form DNA taken by Rosalind Franklin and Raymond Gosling in 1952. Labelled in their lab notes as 'Photo 51', it shows a striking cross-pattern of black spots that is characteristic of a helical molecule. Ava Helen and Linus Pauling Papers (MSS Pauling), Oregon State University Special Collections and Archives Research Center, Corvallis, Oregon. (Reproduced with kind permission of Oregon State University Special Collections and Archive Research Center) (**b**) Commemorative 50p coin released in 2020 by the UK Royal Mint to mark the centenary of Rosalind Franklin's birth. Designed by artists Jody Clark and David Knapton, one face is engraved with the distinctive cross shaped pattern at the centre of 'Photo 51'. (Photograph by Ed Hall)

indeed race, but this was not quite the Eureka moment that it might appear to be. Photo 51 was certainly a vital clue but it alone was not enough to allow Watson and Crick to solve the structure of DNA. For this, two other crucial pieces of information about the molecule were needed - both of which also came from work done by Franklin. The first of these was that the DNA molecule consisted of two polynucleotide chains, not one as in Astbury and Bell's model; and the other was that these two chains ran in opposite directions to each other.

Like Photo 51, these two additional clues had been provided by Rosalind Franklin's X-ray studies of DNA and were contained in a report that she had written to the Medical Research Council (MRC). But the circumstances by which Watson and Crick came to acquire both the information contained in Photo 51 as well as that in Franklin's MRC report have proven to be controversial.

Watson first saw Photo 51 when he was shown it by Maurice Wilkins on a visit to King's College in January 1953. Sitting in a freezing cold railway carriage on the train journey back to Cambridge, he raced to scribble down the details of the photograph on his newspaper. The crucial information contained in Franklin's report to the MRC meanwhile had been shown to Watson and Crick by the director of the Cambridge Medical Research Council Biophysics Unit, Max Perutz.

The problem was that in both cases, this data was shared without Franklin's knowledge or permission. Unsurprisingly, much has since been written about the possible motivations of all those involved and an exploration of the rights and wrongs of their actions with the depth and attention that they deserve easily requires a book itself, so we refer readers who are interested to know more about this episode to several other titles in which it is explored in greater detail.[46]

The author of one of these books, historian Howard Markel, has described the sharing of Photo 51 by Maurice Wilkins with James Watson as 'one of the most egregious rip-offs in the history of science' but it was nevertheless one which Wilkins himself came to regret.[47]

---

[46] See for example 'Rosalind Franklin: Dark Lady of DNA' by Brenda Maddox (Harper Collins 2002); 'Unravelling the Double-Helix: The Lost Heroes of DNA' by Gareth Williams (Weidenfeld and Nicholson, 2019); 'The Secret of Life: Rosalind Franklin, James Watson, Francis Crick and the Discovery of DNA's Double-Helix' by Howard Markel (W.W. Norton 2021) and 'My Sister Rosalind Franklin' by Jenifer Glynn (Oxford University Press, 2012).

[47] (Markel 2021); p. 315.

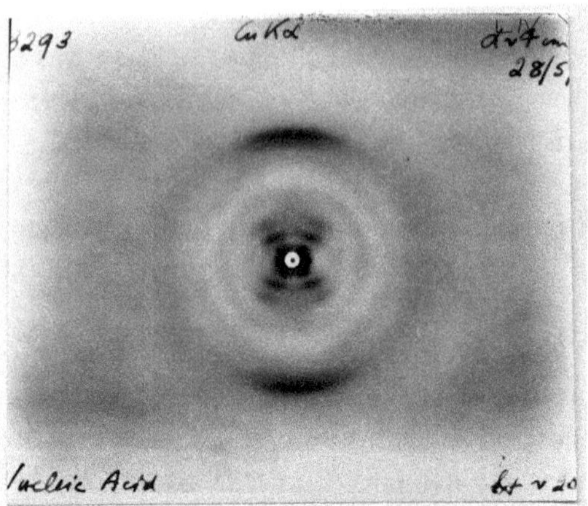

Fig. 10.8 X-ray diffraction image of the B-form of DNA taken by William Astbury's research assistant Elwyn Beighton at the University of Leeds, UK, in late May/early June 1951. The arrangement of the black spots made by the scattered X-rays clearly shows the same striking cross-shaped pattern that was so distinctive in the more famous 'Photo 51' taken by Rosalind Franklin and Raymond Gosling at King's College, London a year later. (Reproduced with kind permission of Special Collections, Brotherton Library, University of Leeds)

In his 2003 autobiography, 'The Third Man of the Double-Helix', Wilkins described 'Photo 51' as 'possibly the most famous X-ray diffraction pattern ever photographed' before lamenting that:

I had been rather foolish to show it to Jim [Watson] during our hurried conversation in the corridor…Jim later wrote that seeing the pattern had spurred him on tremendously to build helical models of DNA. If I had known that, I might well not have shown him the pattern.[48]

Few people are aware however that neither Wilkins, Franklin, or Watson were the first to see this pattern. Just a year before Franklin and Gosling took 'Photo 51', Elwyn Beighton, research assistant to William Astbury at Leeds had taken a set of X-ray images of DNA that were almost identical (Fig. 10.8). But Astbury's response to this image could not have been more different from that of Watson. Astbury's jaw did not drop, nor did his pulse race—instead he just filed the images away never to be published or even presented at a conference.

This was not because Astbury was ignorant of the importance of DNA. Quite the opposite in fact. In fact, he had been one of the very first scientists to sit

[48] (Wilkins 2003); pp. 218–19.

up and take notice of the importance of Oswald Avery's work. But, believing structure to be at the heart of biology, it is more probable that rather like Miescher, he believed that the genetic molecule carried hereditary traits through variation in its three-dimensional structure.

Beighton's photos would have given no indication whatsoever of such structural variation. Thus, rather than setting Astbury's pulse racing, they would most likely have been a disappointment to him.

If only Astbury had been as good a comedian as he was a physicist however, Beighton's images might well have come to the attention of the world. Because, at the same time as Beighton was taking them, Astbury was attending a conference at the Zoological Station in Naples where James Watson was also present. At the time Watson was working in a lab in Copenhagen on bacterial viruses, but having read Schrödinger's book and knowing about Avery's experiment he was becoming very interested in DNA—particularly in the idea that its role as the genetic material might reside in its physical structure. He also knew that X-ray crystallography held the key to unravelling the physical structure of the DNA molecule but by his own admission knew nothing about this technique.

Maurice Wilkins was also present at the conference where he had shown X-ray diffraction patterns that he and Raymond Gosling had obtained from crystalline samples of DNA. Eager to learn more about this X-ray method, Watson now faced a choice. Should he approach Astbury, the elder statesman—or Wilkins, the new kid on the block?

The decision did not turn out to be a very difficult one. Maurice Wilkins had been delighted to meet Astbury for the first time at the Naples meeting, 'because he seemed a real human being, he liked making little jokes.'[49] But Watson felt very differently. His first impression of Astbury was of a man who, with his best years far behind him, was reduced to telling what were described as 'off colour jokes.'[50] Disappointed, Watson decided not to waste his time with Astbury and went to speak instead with Wilkins.

Had Astbury's jokes been a little funnier, history might have unfolded rather differently. Watson might have been sufficiently intrigued to take the train up to Leeds and pay Astbury a visit when he arrived in Cambridge later that year. Had this been the case, it is likely that it would have been Beighton's X-ray images of DNA, not 'Photo 51' that made his jaw drop and his pulse race—and had events unfolded in this way, Rosalind Franklin's family might well have been spared a great deal of pain.

---

[49] (Wilkins 2003); p. 137.

[50] (Olby 2009); p. 129.

10.6 'The Dark Lady of DNA'[51]

When Watson and Crick's first paper proposing the double-helical structure of DNA appeared in the journal *Nature* in April 1953, its closing paragraph gave a muted nod of recognition to Franklin: 'We have also been stimulated by a knowledge of the general nature of the unpublished results and ideas of Dr. M.H.F. Wilkins and Dr. R.E. Franklin'.[52] A year later, in a much more technical paper delivered to the Royal Society of London, Watson and Crick seemed to be a bit more candid in acknowledging Franklin's contribution:

> *The information reported in this section was very kindly reported to us prior to its publication by Drs. Wilkins and Franklin. We are most heavily indebted in this respect to the King's College group and we wish to point out that without this data the formulation of our structure would have been most unlikely, if not impossible.*[53]

But if this was intended to give Franklin some credit, it was somewhat diminished by the following curious sentence:

> *We should at the same time mention that the details of their X-ray photographs were not known to us, and that the formulation of the structure was largely the result of extensive model building in which the main effort was to find any structure which was stereochemically feasible.*[54]

But while Franklin may have received only limited acknowledgement for her contribution in these papers, a recent discovery by historians Matthew Cobb and Nathaniel Comfort of two previously unknown archival sources has added a new and intriguing twist to this story.[55]

The first of these documents is a copy of Notes and Records of the Royal Society of London from January 1954 which lists the names of scientists who had contributed to a Conversazione event that had been held in June of the previous year. These included:

---

[51] A reference to Franklin which was coined by Maurice Wilkins and is an allusion to a character in what are known as William Shakespeare's 'Dark Lady' sonnets (127–152). In his book 'The Secret of Life', Howard Markel gives an interesting analysis of why Wilkins chose this particular reference from literature to describe Franklin. (Markel 2021); p.365.

[52] (Watson and Crick 1953); p. 738.

[53] (Crick and Watson 1954); p. 82.

[54] Ibid.

[55] Nor, when Watson and Crick gave their addresses at the award ceremony when they received their Nobel Prizes in 1962 did she receive any mention.

A group from King's College, London, the Medical Research Council Unit of King's College and the Cavendish Laboratory, Cambridge (Dr R. E. Franklin, Mr R. G. Gosling, Dr A. R. Stokes, Dr H. R. Wilson, Dr M. H. F. Wilkins, Mr F. H. C. Crick and Dr J. D. Watson) showed how a structure for DNA (desoxyribonucleic acid) had been derived from X-ray data and from stereochemical considerations.[56]

The source suggests that, in some circles at least, Franklin was not only being recognised at the time for her role in the discovery of the DNA structure, but also that this achievement was perceived as having been the result of a collaborative effort between the Cavendish Laboratory in Cambridge, and King's College, London. As does the second source—an article written by Joan Bruce, a London-based journalist working for '*Time*' magazine who interviewed Franklin and wrote of the discovery of the double helix as being the collaborative effort of two groups, one based in King's College to gather the experimental data, and the other in Cambridge working on the theory. In her piece, Bruce wrote about how these two teams 'linked up, confirming each other's work from time to time, or wrestling over a common problem.'[57]

Yet despite a photographer being dispatched to take pictures of Watson and Crick, the article was never actually published in '*Time*' magazine, possibly because Franklin expressed reservations about the scientific content of the draft. And, as Matthew Cobb and Nathaniel Comfort explain, had Bruce's article been published, our perception of the discovery of the double-helix might have been very different indeed:

From the outset, Franklin would have been represented as an equal member of a quartet who solved the double helix, one half of the team that articulated the scientific question, took important early steps towards a solution, provided crucial data and verified the result.[58]

Although Joan Bruce's article was never published, Franklin's vital contribution to the discovery of the double helix is well recognised today. Alongside the Royal Mint commemorative coin, her name is now included on plaques both inside and outside 'The Eagle' pub in Cambridge. Other accolades have

[56] Royal Society programme—Notes and Records of the Royal Society of London, Volume 11 (1954): pp.1–5; p. 4.

[57] Bruce, J. Draft article on discovery of double helix, May 1953. Franklin Papers FRKN 6/4, Churchill College Cambridge, UK. Cited in (Cobb and Comfort 2023); p. 660.

[58] (Cobb and Comfort 2023).

included a Mars rover being named after her, and an award-winning portrayal by the actor Nicole Kidman in a hit play in London's West End.

Whether any of these accolades brought any consolation to Franklin's friends and family who were hurt both by the way in which her data had been used without her knowledge or permission, and the far from gracious portrayal of her in James Watson's 1968 memoir, 'The Double Helix.' In the book, Watson admitted that Franklin 'of course, did not directly give us her data. For that matter, no one at King's realized they were in our hands.'[59] Franklin's sister, Jenifer Glynn, has since said that, had Franklin known the extent to which Watson and Crick had used her data to confirm their model, 'there would have been an almighty explosion.'[60] She might well have reacted similarly had she lived long enough to see her portrayal in Watson's book. But sadly, by the time that the book appeared, Franklin was dead, having succumbed to ovarian cancer 10 years earlier at the tragically young age of 37.

To Franklin's friends and family, the book was an insult to her memory. Watson treated Franklin with a mixture of contempt and condescension. At times he cast her as a pantomime villainess who was so hell-bent on thwarting his attempts to solve the structure of DNA that on one occasion he claimed to have feared she might strike him.[61] Elsewhere he dismissed her as having a 'good brain', but one that was hindered by an inability to 'keep her emotions under control.'[62] To add insult to injury, he insisted on referring to her as 'Rosy'—a name which her friends and family would never have used. When used by Watson it was not to convey familiarity, but rather to patronise. Reflecting on the difficult working relationship between Franklin and Maurice Wilkins (a situation that had arisen in no small part due to poor communications by their boss John Randall), Watson observed:

> *I suspect that in the beginning Maurice hoped that Rosy would calm down. Yet mere inspection suggested that she would not easily bend. By choice she did not emphasize her feminine qualities. Though her features were strong, she was not unattractive and might have been quite stunning had she taken even a mild interest in clothes. This she did not. There was never lipstick to contrast with her straight black hair, while at the age of thirty-one her dresses showed all the imagination of English blue-stocking adolescents... Clearly Rosy had to go, or be put in her place.*[63]

[59] (Watson 1968; 2010 edition); p. 131.

[60] (Devlin 2013).

[61] Ibid.; p. 166.

[62] (Watson 1968; 2010 edition); p. 18.

[63] (Watson 1968; 2010 edition); p. 14.

One means by which Watson took it upon himself to put Franklin 'in her place' was through snide remarks about her appearance. Sitting in a lecture given by her in late 1951 in which she reported on her latest X-ray studies of DNA, Watson remarked: 'There was not a trace of warmth or frivolity in her words. And yet, I could not regard her as totally uninteresting. Momentarily I wondered how she would look if she took off her glasses and did something novel with her hair.'[64] This preoccupation with her appearance would prove to be more than the musings of an immature young man—it would also be a stumbling block to Watson and Crick's progress.

After the publication of Watson's book, Franklin's family were understandably hurt by her portrayal. Hoping to offer some well-intended consolation, Franklin's friend and colleague Aaron Klug suggested to her mother that Franklin's vital role in the discovery of the double helix would at least now be remembered for posterity. Her reply was blunt and painful - 'I would rather she be forgotten than remembered in this way.'[65] Franklin, however, was not to be forgotten—instead, in the wake of Watson's book, a new image of her began to emerge- this time as a victim who struggled for recognition amidst a culture of misogyny and male chauvinism. But for some of her family, this portrayal too was somewhat at odds with their memories of Franklin. In her book, 'My Sister Rosalind Franklin', Jenifer Glynn observes that 'it suited the feminism of the 60s and 70s to show her as a victim of male dominance, and to trumpet her achievements as the triumph of a woman in a man's world.'[66] But had Franklin lived long enough to see herself adopted as a martyr, she would, according to Glynn, have been left feeling 'astonished and embarassed' as well as most likely unimpressed.[67] 'She was,' writes Glynn, 'never a feminist—she would have thought of herself simply as a scientist whose achievements should be judged on their own terms, not as a 'woman scientist' striking a blow for the rights of women.'[68] Glynn's conclusion is that, rather than being cast as a victim, Franklin would have wanted to be remembered for her science:

> *Rosalind became a symbol, first of an argumentative swot, then of a downtrodden woman scientist, and finally of a triumphant heroine in a man's world. She was none of these things, and would have hated all of them. She was simply a very good scientist with an ambition....*[69]

---

[64] (Watson 1968; 2010 edition); p. 49.

[65] (Maddox 2002); p. 312.

[66] (Glynn 2012); p. 157.

[67] Ibid.

[68] Ibid.; p. 158.

[69] (Glynn 2012); p. 160.

And the real tragedy of Franklin's story according to her sister 'was not that she was a woman, but that she did not have enough time' to fulfil this formidable scientific ambition.[70]

Along with the extensive discussions about Watson's treatment of Franklin, much has also been written about why she herself did not discover the double-helical structure of DNA. Watson maintained that she was 'anti-helical' and doubted that DNA had a double-helical structure but a careful study of her notebooks by Aaron Klug showed that this was not the case.[71] Franklin had found that the DNA molecule could take on two possible structural conformations (designated as A and B) depending on its water content. The striking cross pattern seen in Photo 51, which indicated a helix, had been obtained from studies of the B-form, which had a higher water content. The more crystalline A-form meanwhile did not give this pattern, but it did show many more spots from the scattered X-rays. To an X-ray crystallographer's trained eye, it would have been the A-form that appeared to offer more structural information and would therefore have been the natural first choice for study.

But as Klug showed, far from being 'anti-helical', by 1952 Franklin was turning her attention to the B-form and the possibility that it might be helical. Francis Crick meanwhile has since acknowledged that Franklin was only two steps away from solving the structure of DNA herself: she needed only to realise that the two chains of the molecule ran in opposite directions and the complementary relationship between bases on opposite chains.[72]

As to why these two final insights eluded Franklin, it is worth remembering that even though she was working on a biological molecule, Franklin did not really think of herself as a biological scientist. Her doctoral research had been funded by the British Coal Utilisation Research Association (BCURA) and her thesis title was 'The physical chemistry of solid organic colloids with special relation to coal and related materials.' After completing her PhD, she took up a research post in Paris where she used X-rays to study the physical properties of coal, graphite and charcoal. She even once described herself as 'a physical chemist who knows very little about physical chemistry but a lot about the holes in coal.'[73] And this knowledge of 'the holes in coal' served her well, for the first and last papers she ever published in the journal *Nature* both concerned the physical structure of carbon, marking the start and end of a distinguished career.

---

[70] Ibid.; p.xiii.

[71] (Klug 1968); (Klug 1974).

[72] (Olby 1996); p. vi.

[73] (Harris and Suarez-Martinez 2021); p. 290.

Little wonder then, that when first considering the move from Paris to King's College, London, she confessed, 'I am, of course, most ignorant about all things biological, but I imagine most X-ray people start that way.'[74] There is an intriguing parallel to be made here between Franklin and William Astbury. Both had converged upon the study of the DNA structure thanks to X-ray studies made for the purposes of industrial research. For Astbury, this had been the study of wool fibres; for Franklin it was coal and graphite. And perhaps this needs to be taken into consideration when seeking to explain why Franklin never solved the double-helical structure of DNA herself. She was first and foremost a materials scientist with a background in the study of graphite—which makes her achievements in the field of biological sciences even more impressive.[75]

As for why Watson felt so compelled to cast her as the villainess of his story, it has been suggested that the reasons may lie in his own immaturity towards women at that time—particularly those he saw as a threat due to their intellect.[76] For the historian Howard Markel, the most revealing example of this is an incident, which occurred in late November 1951. Crick had invited Wilkins to visit Cambridge for the triumphant unveiling of the model of DNA that he and Watson had constructed with cardboard and wire. Franklin came along too and although there are many accounts of what unfolded next, they all agree that it was humiliating for Watson and Crick.

Franklin pointed out to them that their model was hopelessly wrong for several reasons, one of which was that it contained the wrong number of water molecules. This oversight was no doubt a result of James Watson having been too preoccupied with Franklin's lack of lipstick and her choice in clothes to pay attention to her discussion of the water content in DNA when he had attended her seminar. As a result of this slip-up, the backbone of sugar and phosphate in Watson and Crick's model was on the inside of the molecule which left the bases jutting outwards. Franklin's data however showed that the molecule contained far too much water for this conformation to be possible. Watson and Crick had made a basic error in chemistry—one which left their boss, Lawrence Bragg, Director of the Cavendish laboratory, utterly livid. Furious at the embarrassment that Watson and Crick had brought upon the Cavendish, he imposed a moratorium on the construction of any more speculative models of DNA. Bragg issued a strict edict that Crick return to his PhD research; Watson meanwhile was ordered to focus solely on the work that he

[74] (Garman 2020); p. 699.

[75] (Franklin 1950); (Watt & Franklin 1957).

[76] (Markel 2021); p. 308.

was being officially funded to do—research into the structure of tobacco mosaic virus. Solving the structure of DNA was to be left to the group at Kings College.

Had Franklin not pointed out this error, Watson and Crick might have gone on stumbling in the dark for much longer. But rather than feeling grateful to her for this insight, the incident appears to have left Watson with a festering resentment towards her that eventually found its expression in his mean-spirited portrayal of her in his book. Yet even after this humiliation and Bragg's moratorium, it was hard to let DNA go.

10.7 A Change in the Rhythm of the Heartbeats of Biology

Watson and Crick's first model of DNA in which the sugar and phosphate groups were arranged on the inside now lay in ruins. Forced to go back to square one, they began to consider alternative possibilities. One of which was that the backbone of sugar and phosphate was on the outside of the molecule, while the bases were on the inside. But if this were so, how could the four different bases pack together internally to give a regular structure.

Whilst pondering this question over a pint of beer in 'The Bun Shop', a pub near the Cavendish Laboratory, Crick thought he might have glimpsed an answer. Having just attended a lecture entitled 'The Perfect Cosmological Principle', he wondered whether there might not also be a perfect biological principle based on the ability of DNA to make copies of itself. Crick had a hunch that both strands of DNA might somehow be held together by forces of attraction between bases, and that this was in some way crucial to the replication of the molecule.

The four different bases (A, C, G, and T) found in DNA belong to two kinds of chemical group called the purines and the pyrimidines: cytosine (C) and thymine (T) are pyrimidines, while adenine (A) and guanine (G) are purines. Over a pint of beer in a Cambridge pub, Crick put it to John Griffiths, a young mathematician and theoretical chemist, that bases of a specific type on one chain, such as adenine, interleaved with those on the opposite chain in such a way that attractive forces between them held the two strands of DNA together.

When Griffiths met Crick in a lunch queue a short while later, he had an answer. His calculations had shown that pairing between bases was indeed possible but not in the way Crick had envisaged. Through a type of

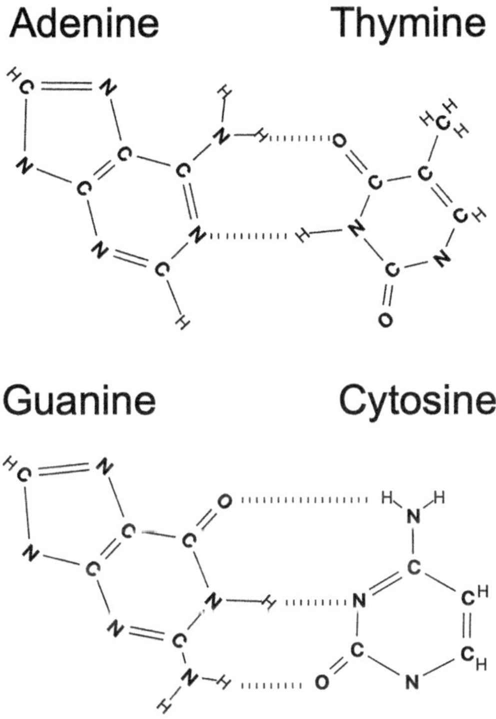

Fig. 10.9 Through electrostatic forces of attraction (shown as dashed vertical lines) between specific hydrogen atoms on one base and those of nitrogen or oxygen on another, adenine can pair with thymine, and guanine with cytosine. This specific pairing holds the two strands of the double-helix together and allows DNA to replicate. (Diagram by K. Hall)

intermolecular attraction known as hydrogen bonding Griffith showed that the pairing would occur between adenine and thymine, and guanine and cytosine. In other words, purines always paired with pyrimidines (Fig. 10.9).

The significance of this finding struck Crick like a bolt of lightning when he was introduced to Erwin Chargaff during a visit he made to Cambridge in 1952. Chargaff's demonstration that the base composition of DNA varied between different species had dealt a deadly blow to the tetranucleotide hypothesis, which he had once dismissed as 'absurd.'[77] It had also offered the first hint at how DNA might carry the genetic message - through variation in the bases. But there was another reason why it was so important. As well as finding that the amounts of the four bases varied between species, Chargaff had observed what he described as 'certain striking, but perhaps meaningless,

[77] (Hunter 1999); p. 98.

regularities' in their relative proportions. In the DNA of all the organisms he had studied, which also included human sperm, he noted that the ratios of 'adenine to thymine and of guanine to cytosine, were not far from 1.'[78]

This meant that the amounts of adenine and thymine must be equal, as must also be those of guanine and cytosine—but Chargaff remained cautious about reading too much into this observation, saying only 'whether this is more than accidental, cannot yet be said.'[79] These regularities in the proportions of the bases were, he wrote at the time, merely 'noteworthy.'[80]

This would prove to be a monumental understatement. Because when Crick met Chargaff in person for the first time, it was starting to become clear to him that the ratios Chargaff had found were no mere accident.

Chargaff was not impressed by Watson and Crick. He had a caustic wit and did not spare it on the Cambridge duo. He described Crick as having 'the looks of a faded racing tout' who spoke 'in an incessant falsetto, with occasional nuggets glittering in the turbid stream of prattle' while Watson was 'quite undeveloped at twenty-three, a grin, more sly than sheepish; saying little, nothing of consequence.'[81]

But as Chargaff spoke, Crick recalled hanging on his every word:

> *The effect was electric. That is why I remember it. I suddenly thought: "Why, my God, if you have complementary pairing, you are bound to get a one to one ratio."*[82]

Suddenly it all seemed clear to Crick how the two chains of DNA were held together—through hydrogen bonding, adenine on one chain was paired with thymine on the opposite chain, while guanine one one chain paired with cytosine on the other (Fig. 10.10). The 1:1 ratios between A:T and C:G that Chargaff had found were experimental confirmation of the theoretical calculations made by John Griffith. Unfortunately, the significance of this only became apparent to Chargaff with the benefit of hindsight. In his autobiography, published 26 years later, he reflected: 'I believe that the double-stranded model of DNA came about as a consequence of our conversation.'[83] But this reflection was not without a hint of bitterness, regret and, possibly self-recrimination at having missed the monumental significance of his findings at

[78] (Chargaff 1950); p. 206.
[79] Ibid.
[80] Ibid.
[81] (Chargaff 1978); p. 101.
[82] (Crick cited in Olby 1994; p. 388).
[83] Ibid.

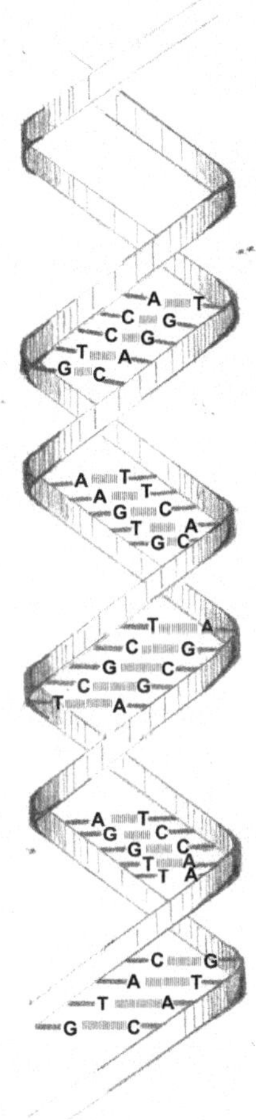

Fig. 10.10 Base-pairing through hydrogen-bonding (indicated by dashed vertical lines) holds opposite strands of the double-helix together. (Diagram by K. Hall)

the time: 'I seem to have missed a shiver of recognition of a historical moment…a change in the rhythm of the heartbeats of biology.'[84]

But there remained another puzzle to solve. And it was a formidable one. Two of the four bases, guanine and thymine, could adopt one of two possible structural conformations known as the 'enol' and 'keto' forms, depending on the position of a particular hydrogen atom that could jump between various chemical groups within the molecular structure of the base. Based on their knowledge of organic chemistry derived from the textbooks of the day, Watson

[84] (Chargaff 1978); p. 101.

and Crick believed that these two alternate structural forms (known as 'tautomers') should be found in equal proportions with the DNA molecule. The problem was that guanine, and thymine would be able to flit between 'keto' and 'enol' forms making it impossible for them to form the stable hydrogen bonds between opposite chains needed to hold the DNA molecule together (Fig. 10.11).

Watson was sitting in his office with cardboard cut-outs of the bases laid out in front of him like some kind of molecular jigsaw puzzle, desperately trying to work out a way in which they could pair up when his colleague Jerry Donohue (1920–1985) came to the rescue. A chemist and crystallographer by training, Donohue pointed out that, quantum mechanical considerations and the latest research literature suggested that, for each base, the 'keto' form of the molecule was favoured over the 'enol' form. And as Watson slid his cardboard cut-out shapes for the 'keto' form into pairs with each other, he saw what this meant.

In the 'keto' form, stable hydrogen bonds could form between thymine and adenine, or cytosine and guanine on opposite strands. When Watson and Crick finally published their model in the journal *Nature* in April that year, this insight led to what must surely be one of the greatest understatements in the history of science:

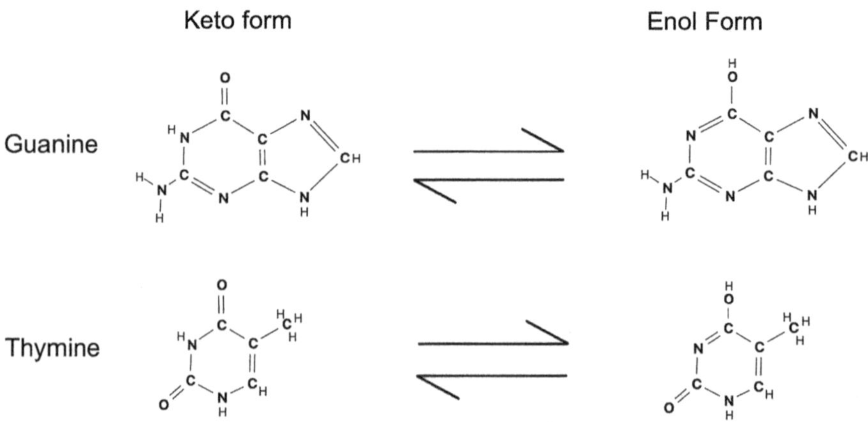

Fig. 10.11 Guanine and thymine can exist in two interchangeable isomeric forms depending on whether a hydrogen atom is attached to the first nitrogen in the ring (known as the keto form), or on the oxygen bonded to its neighbouring carbon atom (known as the enol form). It is only when in the keto form that guanine and thymine can form hydrogen bonds which allow pairing with cytosine and adenine, respectively. (Diagram by K. Hall)

It has not escaped our notice that the specific pairing we have postulated immediately suggests a possible copying mechanism for the genetic material.[85]

In a letter to his young son, Crick explained the full significance of this finding:

. . . the exciting thing is that while these are 4 different bases, we find that we can only put certain pairs of them together . . . we find that the pairs we can make—which have one base from one chain joined with one base from another are only A with T and G with C. Now on one chain, as far as we can see, one can have the bases in any order, but if their order is fixed, then the order on the other chain is also fixed . . . It is like a code. If you are given one set of letters you can write down the others. You can now see how Nature makes copies of the genes. Because if the two chains unwind into two separate chains, and if each chain makes another come together on it, then because A always goes with T, and G with C, we shall get two copies where we had one before . . . In other words, we think we have found the basic mechanism by which life comes from life.[86]

The specific pairing between bases on opposite strands explained how the DNA molecule could be copied (Fig. 10.12). It was this revelation that supposedly sent Watson and Crick sprinting through the streets of Cambridge and bursting into 'The Eagle' pub to declare that they had found the secret of life. Or at least, so it says on the plaque that hangs on the wall of 'The Eagle' today. But as well as commemorating two events that never happened—Watson and Crick's dramatic announcement, together with their apparent 'discovery' of DNA the wording on the plaque is misleading for another reason.

It gives the impression that Watson and Crick's discovery of the double-helix was the climax of the story of DNA. In truth, it was nothing of the sort, for it raised an entire host of new questions. How, for example, did the double helix unwind and make copies of itself? How did the information contained in the base sequence of DNA become physically manifest in the amino acid sequence of proteins? How did the genetic code work?

Miescher had already sensed the scale and grandeur of the questions that physiological chemistry would face in the coming century. In a letter written on sixth August 1895, only a few weeks before his death, he outlined the size of the task ahead:

---

[85] Watson and Crick 1953.

[86] F. Crick to M. Crick, March 19, 1953. Wellcome archive, PP/CRI/D/4/3: Box 243.

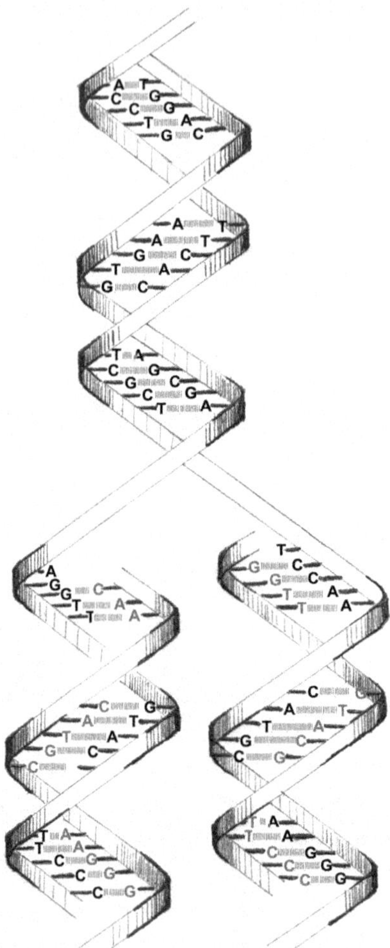

Fig. 10.12 The specific pairing (shown as a series of vertical lines to indicate hydrogen bonding) between purine and pyrimidine bases on opposite strands of the double-helix allows DNA to make copies of itself. In the diagram, the two strands of the parental molecule (with bases shown as single letters in black) unwind, and each is copied into a daughter strand with a complementary sequence (bases shown as single letters in grey). (Diagram by K. Hall)

> *There is no morphological continuity of the nuclein granule, only a chemical one... These are enormous questions of principle that will have to be fought out between morphologists and biochemists in the twentieth century. Is it just the substance, or is it the form as such that is inherited?*[87]

And although the plaque inside 'The Eagle' pub incorrectly ascribes the discovery of DNA to Watson and Crick, Miescher might well have taken some consolation had he been able to see it hanging there. Along with bemoaning the lack of interest shown by chemists and physicists in the life sciences, Miescher had once complained 'that experimental physiology suffers from a drought and needs to be irrigated with new ideas...physiology needs architects, who are searching for the building blocks of organisms and who want to understand how these building blocks work together.'[88]

The plaque in 'The Eagle' is a tribute to the fact that nearly a century later, those very architects for whom Miescher had called, had at last come forth. And their ranks were comprised of the very people whom Miescher had once despaired of showing no interest in the study of life in his own time. Physical scientists such Crick, Wilkins, Schrödinger, Astbury, Bell and Chargaff had at last turned their attention to nuclein and begun to unlock its secrets. But solving the structure of DNA was by no means the end of the story that had first begun with Miescher's pus-soaked bandages. In fact, it was only the beginning.

---

[87] Miescher 6th August 1895, Letter LXXXIII, in (His 1897); pp. 127–128.

[88] (Jaquet 1944); pp. 413–414.

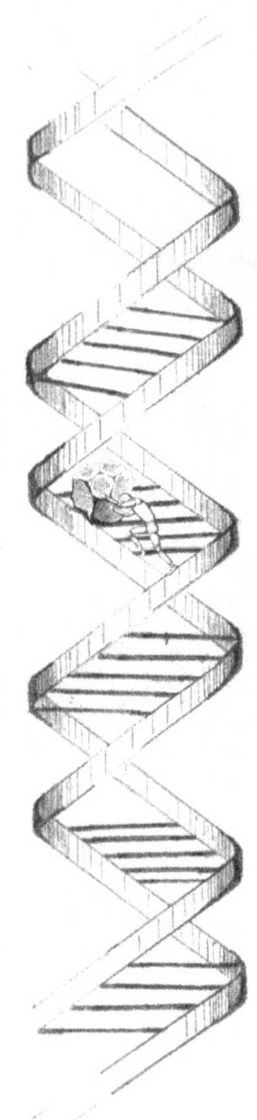

Drawing by Kersten Hall

References

Astbury, W. T., and F. O. Bell. 1938a. Some recent developments in the X-ray study of proteins and related structures. *Cold-Spring Harbor Symposia on Quantitative Biology* 6:109–121.

Astbury, W. T., and F. O. Bell. 1938b. X-ray studies of thymonucleic acid. *Nature* 141:747–748.

Chargaff, E. 1950. Chemical specificity of nucleic acids and mechanism of their enzymatic degradation. *Experientia* 6:201–240.

Chargaff, E. 1978. *Heraclitean Fire: Sketches from a Life Before Nature*. New York, NY: The Rockefeller University Press.

Cobb, M., and N. Comfort. 2023. What Watson and Crick really took from Franklin. *Nature* 616:657–660.

Crick, F. H. C. 1965. Recent research in molecular Biology: Introduction. *British Medical Bulletin* 21:183–186.

Crick, F. H. C., and J. D. Watson. 1954. The complementary structure of deoxyribonucleic acid. *Proceedings of the Royal Society of London* 223:80–96.

Devlin, H. 2013. My sister, her DNA breakthrough, and the Nobel prize that never. *The Times (London)*.

Fischer, E. 1917. Isomerie Der Polypeptide. *Zeitschrift für Physiologische Chemie* 99:54–66.

Franklin, R. E. 1950. The interpretation of diffuse X-ray diagrams of carbon. *Acta Crystallographica* 3 (2): 107–121. https://doi.org/10.1107/S0365110X50000264.

Garman, E. F. 2020. Rosalind Franklin 1920–1958. *Acta Crystallographica Section D, Structural Biology* 76 (7): 698–701. https://doi.org/10.1107/S2059798320008827.

Glynn, J. 2012. *My Sister Rosalind Franklin*. Oxford: Oxford University Press.

Harris, P. J. F., and I. Suarez-Martinez. 2021. Rosalind Franklin, Carbon Scientist. *Carbon* 171:289–293. https://doi.org/10.1016/j.carbon.2020.09.022.

His, W. 1897. *Die Histochemischen Und Physiologischen Arbeiten von Friedrich Miescher*. Leipzig: F. C. W. Vogel.

Hunter, G. K. 1999. Phoebus Levene and the Tetranucleotide structure of nucleic acids. *Ambix* 46:73–103.

Klug, A. 1968. Rosalind Franklin and the discovery of the structure of DNA. *Nature* 219 (5156): 808–810. https://doi.org/10.1038/219808a0.

Klug, A. 1974. Rosalind Franklin and the double helix. *Nature* 248:787–788.

Kossel, A. 1911. The chemical composition of the cell. *Harvey Lectures Series* 7:33–51.

Leathes, J. B. 1926. Function and design. *Science* 64:387–394.

Maddox, B. 2002. *Rosalind Franklin: The dark lady of DNA*. New York: Harper Collins.

Manton, I. 1945. Comments on chromosome structure. *Nature* 155:471–473.

Markel, H. 2021. *The secret of life: Rosalind Franklin, James Watson, Francis Crick and the Discovery of the Double Helix*. New York: W. W. Norton.

Olby, R. 1994. *The path to the double helix: The discovery of DNA*. Garden City, NY: Dover Publications.

Olby, R. 2009. *Francis Crick: Hunter of life's secrets*. Cold Spring Harbor Press, NY: Cold Spring Harbor Press.

Pauling, L. 1987. Schrödinger's contribution to chemistry and Biology. In *Schrodinger: Centenary Celebration of a Polymath*. Cambridge: Cambridge University Press.

Perutz, M. 1987. Erwin Schrödinger's what is life? And molecular biology. In *Schrodinger: Centenary Celebration of a Polymath*. Cambridge: Cambridge University Press.

Schrödinger, E. 1944. *What is life?* Cambridge: Cambridge University Press.

Schultz, J. 1941. The evidence for the nucleoprotein nature of the gene. *Cold Spring Harbor Symposia on Quantitative Biology* 9:55–65.

Stent, G. S. 1968. That was the molecular biology that was. *Science* 160:390–395.

Symonds, N. 1986. What is life? Schrödinger's influence on Biology. *The Quarterly Review of Biology* 61:221–226.
Watson, J. D. 1968. *The double-helix*. London: Weidenfeld & Nicolson.
Watson, J. D., and F. H. C. Crick. 1953. Genetical implications of the structure of deoxyribonucleic Acid. *Nature* 171:964–967.
Watt, J. D., and R. E. Franklin. 1957. Changes in the structure of carbon during oxidation. *Nature* 180 (4596): 1190–1191. https://doi.org/10.1038/1801190a0.
Wilkins, M. 2003. *The third man of the double helix*. Oxford: Oxford University Press.
Witkowski, J. A. 1986. Schrödinger's "what is life?": Entropy, order and hereditary code-scripts. *Trends in Biochemical Sciences* 11:266–268.
Wrinch, D. M. 1934. Chromosome behaviour in terms of Protein pattern. *Nature* 134:978–979.

11

Opening the Book of Life

Contents

Abstract The discovery of the structure of DNA in 1953 is often presented as if it were the triumphant climax of the story of the genetic material, but this was far from being the case. The discovery of the double-helix raised new and important questions—how, for example, did the molecule make copies of itself to pass on hereditary information? How did the order of bases in DNA specify biological traits—did it somehow determine the arrangement of amino acids in a protein? And how if so, how was this information encoded

K. Hall, R. Dahm, *The Dawn Fisherman*, Copernicus Books,
https://doi.org/10.1007/978-3-032-14219-1_11

in the DNA sequence transmitted from the nucleus to the sites of protein synthesis in the cytoplasm? By the late 1960s, one of the key figures in this work, Marshall Nirenberg had already predicted that the answers to these questions might well allow us not only to understand how DNA worked, but to deliberately manipulate the genetic material. With the development of recombinant DNA tools in the 1970s, Nirenberg's prediction became a reality. In this chapter we explore this important aspect of Miescher's legacy by tracing how our understanding of DNA has allowed the development of powerful technologies such as PCR and DNA fingerprinting as well as in achievements such as the Human Genome Project (HGP). When the first draft of the HGP was completed in 2000, it was described in the media as 'the book of life' and heralded by some politicians as promising to usher in a new age of revolutionary medical treatments such as gene therapy. But this would not come without the potential for considerable ethical, social and political challenges.

Keywords Genetics • Genomics • RNA • DNA • Gene regulation • MicroRNAs (miRNAs) • Small interfering RNAs (siRNAs) • Exosomes • ExRNA • Human Genome Project • Non-coding DNA • Gene expression • Molecular biology • Cell communication • RNA interference (RNAi) • Gene silencing • Diagnostic biomarkers • Therapeutic agents • Cancer • Alzheimer's disease • Genetic information • Medical therapy • Biotechnology • Gene control • Cellular differentiation • Developmental biology • RNA molecules • Messenger RNA (mRNA) • Ribozymes • Epigenetics

Miescher's discovery opens a Pandora's Box of possibilities…and problems

Whereas Miescher first isolated nucleic acids by toiling in what had once been the freezing cold kitchen of Tübingen castle, this entire process can today be performed in the comfort of your own kitchen without suffering the inconvenience of first having to convert it into a biochemical research laboratory. Moreover, this can be achieved using a far less unpleasant procedure than Miescher's original method of preparation. Instead of discarded surgical bandages oozing with pus, all that is needed is some washing-up liquid, table salt, a dash of ice-cold vodka and a handful of strawberries.

First, the strawberries are mashed up in washing-up liquid to burst open their cell membranes—an ideal activity for prising children away from tablets, phones and other devices (for a short while at least), with the bonus that any remaining strawberries can be eaten. Once, the strawberries have been mashed to a pulp, some salt is added which reduces the solubility of cellular proteins, causing them to precipitate out of solution. The precipitated protein is then removed by passing the mixture through a coffee filter and in the final step, ice-cold vodka is added to the filtered liquid. As the drops of freezing cold vodka hit the filtered liquid, a floating mass of white, stringy fibres start to appear in the solution. These fibres are made of nucleic acid, most of which will be DNA containing the genetic code of the strawberry plant. But whatever their source, whether from strawberries or leukocytes extracted from pus, our growing understanding of these fibres and how they work raises profound questions about what we want from science, and the kind of society in which we wish to live.

Almost a century and a half after Miescher first discovered it, DNA is rarely out of the news. Understanding how it works has fundamentally changed biology and medicine, but the impact of this knowledge is also felt way beyond hospitals and academic research labs. Its use in forensic science and paternity disputes has made it an indispensable plot device in countless TV dramas, and a recent report that DNA testing is being used to deter irresponsible dog owners from letting their pets foul the pavement would suggest that the future will see many such inventive applications of Miescher's discovery.[1]

Fortunes have been made from DNA, legal battles have been fought over it, and it has even sent people to prison. Entire books have been written about these topics and in what follows we cannot hope to do full justice to either the history or complexity of these different areas, but what we can do is to convey some sense of the monumental legacy that Miescher left, and the questions—scientific, political and ethical, that it poses for us today.[2]

To begin with, Watson and Crick's discovery of the double helix is often presented as the climactic finale to the story of DNA. Witness for example Watson and Crick's dramatic slow-motion sprint through the streets of Cambridge to The Eagle pub in the 1986 BBC drama 'Life Story.' But although it was without doubt one of the milestones in the history of science, the discovery of the double helix was by no means the final word in the story of DNA, for it raised many new questions.

[1] 'Need to track down a dog poo-pertrator? There's a DNA test for that.' The Guardian, 27th Jan 2019; https://www.theguardian.com/business/2019/jan/27/dog-dna-test-pooprints.

[2] In particular for an excellent and in depth explorations of how the genetic code was cracked, and the history of genetic engineering, see 'Life's Greatest Secret' (Profile Books 2015), and 'The Genetic Age' (Profile Books 2022), by Matthew Cobb.

11.1 'The Most Beautiful Experiment'

One of which was how exactly did the molecule make copies of itself? In their paper of April 1953, Watson and Crick had made what must surely be one of the biggest understatements in the history of science. This was their observation that 'it has not escaped our notice that the specific pairing we have postulated immediately suggests a possible copying mechanism for the genetic material'.[3] The double-helical structure made it possible for each strand of the parent molecule to act as a template for the synthesis of a daughter strand, giving rise to two new duplexes. Each of these would contain one strand derived from the original parent, and the other from the newly synthesized daughter molecule.

This model was known as 'semi-conservative' replication and one formidable problem that it faced was that the DNA molecule was twisted into a helix. This meant that the two parental strands would first have to be unwound before the molecule could copy itself. But how did this unwinding occur—and did it require that the strands of the parental molecule be broken in the process?

In a paper presented in 1956 at a symposium on the chemical basis of heredity, Max Delbrück and Gunther Stent considered some alternatives to the model of semi-conservative replication.[4] One of these was that the parent molecule might not unwind at all but somehow be copied directly into a daughter duplex. The other model, which they called 'dispersive replication', proposed that the parent DNA molecule was unwound by being broken down into fragments. These acted as templates for the synthesis of new daughter fragments which were then reassembled with sections of the parent molecule, giving rise to a duplex that contained adjacent regions of both original and newly synthesized DNA.

It was thanks to an initially brusque encounter between Delbrück and graduate student Matthew Meselson, that an answer eventually began to emerge. Meselson was working at the time as a doctoral student with Linus Pauling and although his PhD research was in the field of physical chemistry, he had expressed an interest in meeting Delbrück who, through his studies of bacteriophage, (a type of virus that infects bacteria), had established himself as one of the pioneers in the burgeoning field of molecular biology.

In the eyes of a young student such as Meselson this made Delbrück the perfect mentor, but their first meeting did not get off to a good start. When

[3] (Watson and Crick 1953a); p. 738.

[4] (Delbrück and Stent 1956).

Delbrück asked Meselson what he thought of Watson and Crick's paper on the double-helical structure, he confessed that he hadn't heard of either of them. Delbrück's response was to pick up a pile of reprints of the paper that Watson had sent him and hurl them at the young graduate student with the instruction to 'Read these and don't come back until you have!'[5]

With Delbrück's words ringing in his ears, Meselson did just that. Although not a biologist by training, Meselson's background in physical chemistry gave him an important insight into designing an experiment to test how DNA makes copies of itself. The idea relied on the same principle that allows swimmers to float in the Dead Sea or the Great Salt Lake. This occurs because the salt concentration of the water is so high that its density exceeds that of a swimmer, thus allowing them to float. Meselson wondered whether the same principle might be applied to DNA. He reasoned that if a DNA molecule were to be labelled with a heavy isotope of hydrogen, such as deuterium, its density might become sufficiently high that it would no longer float but instead sink in solution. Unlabelled DNA, with a lower density, on the other hand would continue to float in solution. In this way, Meselson reasoned that the labelled and unlabelled DNA molecules could easily be separated and distinguished by differences in their density.

It was whilst working as James Watson's teaching assistant at the Marine Biological Station in Woods Hole, Massachusetts, that Meselson met post-doctoral researcher Frank Stahl, who was sipping gin under a tree having taken a break from a course in physiology. When Meselson explained the outline of his idea, Stahl was interested and the two teamed up to combine Meselson's knowledge of physical chemistry with Stahl's expertise in biology.

Four years later they published a paper that has since been described as 'the most beautiful experiment in biology'.[6] In it they showed that when bacteria are grown on medium containing the heavy isotope of nitrogen ^{15}N, their DNA molecules incorporate the isotope and are therefore denser than if they had contained regular ^{14}N. When the bacteria were then switched to being grown on medium containing the regular ^{14}N however, something very interesting happened. The DNA molecules isolated from bacteria that had undergone one generation of division were found to be neither as dense as those containing only ^{15}N, nor as light as those containing only ^{14}N. Their density was of an intermediate value between the two extremes, suggesting that they contained half the amount of each of the two isotopes—which is exactly what the semi-conservative model of replication would have predicted.

[5] 'Interview with Matthew Meselson' Bioessays 2003 Vol. 25; p. 1238.

[6] (Judson 1996) p. 163.

Meselson and Stahl were modest in their claims. In the conclusion to their paper, they said only that 'The results of the present experiment are in exact accord with the expectations of the Watson-Crick model for DNA duplication.'[7] But today it is hailed as a milestone. Yet its presentation in science textbooks is a classic example of philosopher Thomas Kuhn's warning about the pitfalls of learning the history of science from such sources. Meselson and Stahl's work is usually presented as if it had been carried out as a single, polished experiment, performed in response to a simple question. In reality it took 4 years, most of which was spent working with bacteriophage before realising that this was a poor choice for the experiment in question and opting instead to use bacteria.[8] Much effort was also spent in trying to use the chemical reagent 5-bromouracil (5-BU) as the label, with the decision to use heavy nitrogen instead having been made only relatively late in the day. Technical optimisation of experimental methods and instrumentation also took up a great deal of time and energy. Having toyed with the idea of separating DNA molecules labelled with 5-BU according to their different charge in an electrical field, Meselson and Stahl eventually opted to separate them according to differences in density by spinning them through dense solutions of caesium chloride at speeds exceeding 20,000 rpm using an ultracentrifuge. But this in turn required exhaustive work in such areas as developing the mathematical theory behind centrifugation of macromolecules through a dense solution.

The Meselson-Stahl experiment is a good example of how what are presented in science textbooks as classic experiments often turn out to be gross oversimplifications. Textbooks often give the impression that, what are today hailed as landmark experiments, sprang into existence out of an intellectual vacuum, fully formulated and forged in a moment of genius. Such accounts strip them of any historical context. On closer inspection they are revealed to be complex historical processes shaped by a multitude of different factors, most of which have been trimmed away for the sake of convenience when teaching a particular scientific principle to students.

None of which is intended to diminish the achievement of Meselson and Stahl's work. It certainly left most of the scientific community convinced that DNA molecules were replicated by the semi-conservative method. But having shown how genetic information was copied from one generation to the next, this raised yet another question—how did the DNA molecule carry this information in the first place?

[7] (Meselson and Stahl 1958); p. 678.

[8] For an excellent and exhaustive historical reconstruction of Meselson and Stahl's work, see Meselson, Stahl, and the Replication of DNA: A History of 'the Most Beautiful Experiment in Biology' by F.L. Holmes (Yale University Press, 2001).

Six weeks after proposing the structure of DNA, Watson and Crick suggested an answer:

> *It follows that in a long molecule many different permutations are possible, and it therefore seems likely that the precise sequence of bases is the code which carries the genetic information*[9]

At first sight this might not seem to be such a novel proposal. In 1967, a paper appeared in the journal *Nature* arguing that with his suggestion of how biological traits might be represented through molecular isomerism, Miescher had in fact made 'An Early Reference to Genetic Coding.'[10] But tempting as it might be to ascribe some kind of prescience to Miescher and portray him as a man ahead of his time, this would be a mistake. The popular idea that certain scientists were somehow way 'ahead of their time' is a mirage resulting from distortions of the historical landscape when it is viewed from the vantage point of the present-day. In Miescher's case, the concept of a genetic code and biological information with which we are familiar today were not part of the intellectual landscape in which he worked and would have been unrecognisable to him.

Erwin Schrödinger's book 'What is Life?' is often hailed as having first suggested that the genetic molecule functioned as a 'code-script' to determine the development of the organism. This was true enough, but what Schrödinger did not talk about in his book was the idea that the code might carry 'genetic information.' This is because, by the time that Watson and Crick were writing, the concept of 'information' had acquired a very different meaning. This had its origins not in biology, but in a branch of mathematics and systems engineering that had emerged during the Second World War.[11] One of these was the field of 'information theory' pioneered by American mathematician Claude Shannon (1916–2001) who, while working at Bell Labs, had developed a precise mathematical concept of information. The other was the field of cybernetics which sought to understand the flow and control of information and was developed by American mathematician Norbert Wiener (1894–1964) with the aim of improving anti-aircraft defences. Along with the American-Hungarian mathematician John von Neumann (1903–1957), Wiener began to think about how these concepts might be applied to the study of living systems and how they replicate. And within only 2 months of

[9] (Watson and Crick 1953b); p. 965.

[10] (Olby 1967).

[11] (Cobb 2015a); Lily Kay 'Who Wrote the Book of Life: A History of the Genetic Code (Writing Life)' (Stanford University Press 2000).

publishing their double-helical structure for DNA, Watson and Crick received fan mail from someone else who was thinking along these very same lines.

11.2 George Gamow: Whisky, Card Tricks, and Codes

Former Soviet cosmologist-turned defector, George Gamow (1904–1968) had spent much of his career thinking about the Big Bang. But on learning of Watson and Crick's model of the double-helix he became excited about biology, and particularly by the idea that DNA might act as template directing the synthesis of proteins. The idea that the genetic material might in some way act as a template was not itself a novel one,[12] but what Gamow did was to outline a hypothetical means by which DNA might do this.

His other interests included magic card tricks, as well as drinking copious amounts of whisky and his letter to Watson on eighth July 1953 in which he introduced himself certainly looked as if it had been written whilst indulging in his favourite tipple. Gamow's words written in his own hand oozed all over the page with no fixed style or form, and in utter defiance of lines and margins. Describing Gamow's letter as 'a zany communication', Watson admitted that it 'had so many whimsical qualities that we did not know how serious he [Gamow] might be.'[13] But they did right to take Gamow seriously.

The letter outlined his idea that every different organism might be represented by a different number made up of the four bases arranged in a specific sequence. As an example, Gamow suggested that 'the animal will be a cat if Adenine is always followed by cytosine in the DNA chain [sic].'[14] This turned out not quite to be the case, but writing a year later, Gamow took his idea in an important new direction.

After 10 years of gruelling and often tedious work, the Cambridge biochemist Fred Sanger (1918–2013) had just shown that the individual amino acids making up the protein hormone insulin were arranged in a precise linear order, or 'sequence.' In a paper published in the journal *Nature* in 1954, written in response to Watson and Crick's papers on DNA, Gamow introduced the idea of a code by which amino acids at specific positions in a protein chain were determined by the order of bases in DNA.

---

[12] One such scheme in which genes acted as templates to copy themselves had been outlined in 1937 by British biologist J.B.S. Haldane (cited in Pollock 1970; p. 13).

[13] (Watson 2001); p. 24.

[14] G. Gamow to J. D. Watson and F. H. C. Crick, July 8, 1953. Wellcome Library, PP/CRI/D/4/1:Box 29.

Francis Crick liked this idea but, as the Russo-French geneticist Boris Ephrussi (1901–1979) pointed out to him, it was still mere speculation: no-one had yet shown experimentally that the bases in DNA determined the order of amino acids in proteins. To test the hypothesis Crick needed to find a clear example in which a genetic mutation resulted in an alteration in the amino acid sequence of a protein.

It was a challenge that drove him to tears—quite literally. To obtain the enzyme lysozyme, which is found in human tears, in sufficient quantities he resorted to chopping onions as well as thinking sad thoughts to make himself cry.[15] This came to nothing, but his colleague the German-American biologist, Vernon Ingram (1924–2006) had far more success. Ingram found a single genetic mutation in the DNA sequence encoding the blood protein haemoglobin that results in the substitution of the negatively charged amino acid glutamic acid by the electrically neutral valine. As a result of this single substitution, the overall charge of the protein changes and causes it to misfold into the wrong three-dimensional structure. This in turn lowers the solubility of the haemoglobin leading to its crystallisation and a deformation of red blood cells which gives rise to the debilitating condition known as sickle-cell anaemia.

The Times newspaper of London praised Ingram's discovery as 'a landmark in genetics.'[16] It was a view very much shared by Crick. Inspired by Ingram's discovery, Crick gave a lecture entitled 'On Protein Synthesis' at University College, London, to the Society for Experimental Biology later that year in which he is said to have 'permanently altered the logic of biology' with what he called 'The Sequence Hypothesis':[17]

> *. . . the specificity of a piece of nucleic acid is expressed solely by the sequence of its bases, and that this sequence is a (simple) code for the amino acid sequence of a particular protein.*[18]

But Crick's 'Sequence Hypothesis' still faced a formidable problem. Protein chains were made up of twenty different types of amino acids, whilst DNA was made of only four different bases. How then could the order of only four different bases determined the linear arrangement of twenty different amino acids in a protein?

---

[15] (Judson 1996); p. 301.

[16] The Times, London 23rd Aug 1957.

[17] (Judson 1996); p. 330.

[18] (Crick 1958); p. 152.

11.3 Cracking the Code

George Gamow offered a solution. He proposed that the assembly of new proteins took place by joining individual amino acids together directly on the surface of the DNA molecule. And each amino acid added to the growing chain was determined not by a single base in the DNA sequence, but by a group of bases.[19] In this way, all twenty of the naturally occurring amino acids could be represented through different permutations of the four bases in DNA. But this in turn raised two very important questions that needed to be answered.

11.3.1 Question 1: The Mysterious Messenger

Miescher had shown that DNA resided in the cell nucleus, but in late 1930s and early 1940s, Torbjörn Caspersson (1910–1997) at the Karolinska Institute in Stockholm, and Jean Brachet (1909–1988) based in Belgium had independently shown that cells which were dividing rapidly and thus synthesizing large amounts of protein also contained large amounts of RNA in their cytoplasm.[20] Both Caspersson and Brachet had also noticed that most of this RNA was associated with small spherical protein-rich structures in the cytoplasm. These structures had first been observed by the Belgian-American cell biologist Albert Claude (1899–1983) whilst he had been fractionating the components of cells in an attempt to identify the causative agent of tumours in chickens. Pondering the function of these tiny cytoplasmic particles, Claude speculated that they 'should lead to interesting developments'—and he was not wrong.[21]

At the time, these structures became known by several different names such as secretory granules, zymogen granules, and the highly imaginative, 'small particles.' But on account of their being made of RNA and protein, Brachet called them ribonucleoproteins and proposed that they were responsible for protein synthesis in the cell.[22] Today they go by the name of ribosomes and are composed of two structural components—one of which is protein, and the other a type of RNA known as ribosomal RNA (rRNA).

This raised a problem for the model Gamow had proposed. If protein synthesis took place on ribosomes in the cytoplasm, and not in the nucleus where

[19] (Gamow 1954).

[20] (Caspersson and Schultz 1939; Brachet 1942).

[21] (Claude 1940); p. 78.

[22] Ibid., p. 245.

DNA resided, how did the information in the DNA sequence that told a cell which amino acids to use and the order in which to join them together, get out of the nucleus and to the ribosomes?

Reflecting on this problem, Gamow likened the nucleus to the office of a factory manager in which the designs and blueprints for products are stored, while the ribosomes in the cytoplasm were the factory floor on which those products were made. Gamow proposed that, rather like a foreman bringing the necessary instructions from the manager's office down onto the factory floor, information carried in the DNA sequence might be brought from the nucleus to the ribosomes in the cytoplasm by a chemical intermediate.

But what might this chemical intermediate be? In 1947, the French scientists André Boivin (1895–1949) and Roger Vendrely (1910–1988) had already proposed that RNA might well fulfil this role. But the problem was—which type of RNA? For by the mid- to late 1950s it was becoming evident that there were different kinds of RNA in the cytoplasm. Ribosomal RNA (rRNA) was a poor candidate for the hypothetical messenger molecule because it was found in only two different sizes or lengths, but more importantly, its base composition showed very little variation when compared with that of DNA.

The second type of RNA had originally been called soluble, or sRNA, because when cell extracts were spun in a centrifuge, it was not found to be associated with the ribosomes but instead remained in the soluble fraction.[23] And Paul Zamecnik at the Massachusetts General Hospital found that sRNA had a very intriguing property.

Zamecnik had long been passionate about protein synthesis. There were several theories at the time about how the cell might make different proteins from individual amino acids. One idea proposed that new proteins were made by exchanging amino acids in and out of their structure; another was that they were assembled from precursors. Zamecnik however found that proteins were assembled at the ribosomes by the joining together of amino acids to form a growing molecular chain. And soluble RNA seemed to play a vital role in this process.[24] By attaching a radioactive label to specific amino acids, Zamecnik and his co-workers found that the soluble RNA molecules became radioactive.[25] This was because they had formed a chemical bond with the radiolabelled amino acid. Moreover, the amino acid was always attached at one end

[23] (Judson 1996; p. 313).

[24] See for example, (Zamecnik and Keller,1954); (Hoagland, Zamecnik & Stephenson, 1957).

[25] (Hoagland et al. 1957).

of the RNA molecule, which carried the sequence CCA—cytosine, cytosine, adenine.

These soluble RNA molecules later became renamed as transfer, or tRNA, and they turned out to play a vital role in protein synthesis by bringing amino acids to the ribosomes. But unfortunately, they too did not fit the criteria for the enigmatic messenger molecule. As for what these criteria might be, a suggestion came thanks to a feat of dietary acrobatics by the common gut bacterium *Escherichia coli*, and the work of French scientists Francois Jacob (1920–2013), Jacques Monod (1910–1976) along with their American collaborator, Arthur Pardee (1921–2019).

E. coli normally metabolises the sugar glucose to obtain energy, but if glucose is not available then certain mutant forms of the bacterium, known as lac+, can instead use the sugar lactose as a source of fuel. In the absence of glucose, but the presence of lactose, these lac+ mutants can rapidly synthesise an enzyme called beta-galactosidase which allows them to metabolise lactose. What Jacob, Monod and Pardee found was that something very intriguing happened when genetic material from lac+ mutants was introduced into lac-strains of *E. coli*, which were unable to synthesize this enzyme. Suddenly, the lac-bacteria began to rapidly synthesize beta-galactosidase, allowing them to metabolise lactose.

From these experiments, Jacob, Monod and Pardee proposed that the production of the beta-galactosidase enzyme was not controlled at the level of protein synthesis. Rather it was regulated through direct control of the activity of the gene itself, which they proposed acted to produce a chemical intermediate which directed synthesis of the beta-galactosidase enzyme. Extending these experiments with Monica Riley at Berkeley, they found that the lac+ DNA needed to be continually active to maintain production of the enzyme. This ruled out that the hypothetical chemical intermediate was stable—rather, it must be short-lived and with a high turnover. And as it happened, a candidate which fitted these criteria perfectly had just been identified.

In 1956, a year before Jacob, Monod and Pardee began their work, Eliot Volkin and Lazarus Astrachan at Oak Ridge National Laboratory in Tennessee had made a curious observation in phage-infected bacteria. They contained a type of RNA that was present in low amounts and was distinct from ribosomal and transfer RNA. Most curious of all however was that the four bases making up this RNA were found to occur in strikingly similar proportions to those found in the DNA of the infecting phage—and not that of the host bacterium.

This new type of RNA was also very short-lived and had a high turnover—which were exactly the features that Jacob, Monod and Pardee had predicted for their hypothetical messenger molecule. Teaming up with Matt Meselson, who had used the new technique of ultracentrifugation to show how the DNA molecule makes copies of itself, Jacob used this same method to isolate the short-lived RNA species reported by Volkin and Astrachan. In 1961, Jacob, Meselson and Sydney Brenner published a paper alongside similar work done independently by a team that included James Watson showing that this RNA was indeed the intermediate which carried information from DNA to the ribosomes during protein synthesis.[26]

In the 1960s, it was discovered that this intermediate RNA is synthesized by an enzyme called RNA polymerase, which binds next to coding regions of DNA and copies (or to use the technical term 'transcribes') them into single stranded RNA molecules. Due to the same base-pairing rules (guanine pairs with cytosine; adenine pairs with thymine - or, in the case of RNA, uracil) which hold opposite strands of the DNA double helix together, the sequence of these RNA molecules is complementary to that of the DNA region from which they were transcribed. So, if for example, the original DNA contains a guanine at a particular position in its sequence, the transcribed RNA molecule will contain a cytosine at the corresponding position, while if the original DNA sequence contains thymine, the transcribed RNA will contain uracil instead (Fig. 11.1).

Jacob called this type of RNA molecule—messenger, or mRNA—and almost 60 years later, it is a name that has become well known even outside molecular biology labs thanks to its use in the production of vaccines against SARS-CoV2. And in a bizarre twist of fate, Tübingen also played an important role in the development of this technology thanks to a team of researchers who were working at the University there, just a stone's throw from where Miescher had originally isolated nucleic acids. In 2000, they found that when RNA encoding the partial sequence of a bacterial protein was delivered within a tiny bubble of lipids into experimental mice, the animals produced an immune response against the bacterium.[27] In other words, by introducing mRNA encoding the fragment of bacterial protein into the mice, the animals had become vaccinated against the bacterium. As SARS-CoV2 began to rage across the globe in 2020, mRNA quickly became part of the daily vocabulary used in the media and press conferences, thanks to its promise of an effective method of vaccination.

[26] (Brenner, Jacob and Meselson 1961; Gros et al. 1961).

[27] (Hoerr et al. 2000).

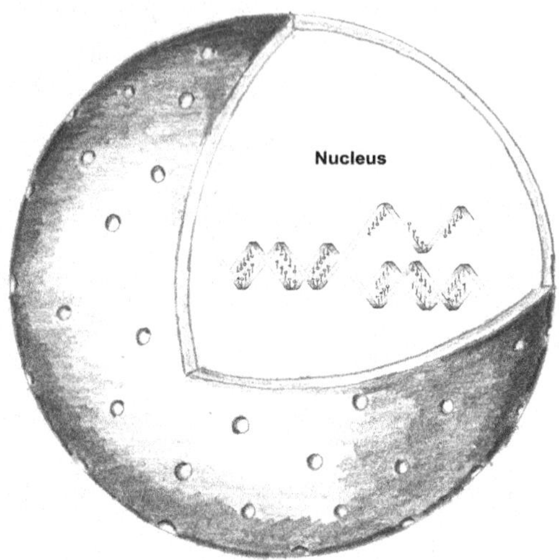

Fig. 11.1 A sequence of DNA is copied, or transcribed, into a single strand of messenger RNA (mRNA). The base sequence of this mRNA (letters shown in grey) is complementary to that of the DNA strand from which it was transcribed and contains the base uracil instead of thymine. (Diagram by K. Hall)

11.3.2 Question 2: How Many Bases?

The development of mRNA vaccines was a powerful weapon in bringing the Covid-19 pandemic under control and earned Hungarian molecular biologist Katalin Kariko (1955-) and Drew Weissmann (1959-) of the University of Pennsylvania the 2023 Nobel Prize for Physiology or Medicine for their work on the development of this technology. Nor was this the first Nobel Prize to have been awarded for work on mRNA. In the 1960s it had proven to be a powerful tool in in solving how the genetic code worked, by helping to answer the question of how many bases in a DNA sequence were needed to specify a single amino acid in a protein chain. This was a question, which had preoccupied many great minds. At the suggestion of James Watson and the British biochemist Leslie Orgel (1927–2007), Gamow had formed an elite club dedicated to addressing this question. Known as 'the RNA Tie' club, its 20 members were distinguished by each wearing a tailor-made tie pin engraved with the name of a specific amino acid. Members of this prestigious band included Erwin Chargaff, Hungarian-American physicist Edward Teller (1908–2003) and, when taking a break from playing his bongo drums, the American theoretical physicist and future Nobel laureate Richard Feynman (1918–1988). But despite this critical mass of eminent minds, the question that Gamow had first posed was not solved by the RNA Tie Club. Rather, the answer came thanks to outsiders who had not been granted access to this prestigious clique.

One of these outsiders was the American biochemist Marshall Nirenberg (1927–2010) who was described by his biographer, Franklin Portugal as being 'the most famous person you have never heard of.'[28] Certainly, few people appeared to have heard of Nirenberg, when he gave a talk on his work at the fifth International Congress of Biochemistry held in Moscow during the summer of 1961. Of the six-thousand delegates attending the meeting, only thirty-five came along to hear what Nirenberg had to say. But unlike the other 5965 delegates, they were about to enjoy the privilege of being present to witness a milestone in our understanding of how DNA directs the synthesis of proteins.

When word reached Francis Crick about Nirenberg's talk, he quickly arranged for him to give a repeat performance on the final day of the meeting. As a result, 7 years later, when Nirenberg arrived for work on the morning of 16th October 1968 he found a very pleasant surprise waiting for him. Hanging across the lab was a huge banner bearing the message 'UUU are great' (Fig. 11.2). The banner had been hung there by his research team not just as an expression of their appreciation for Nirenberg, but also to honour the news that he had just been awarded the Nobel Prize in Physiology or Medicine—an

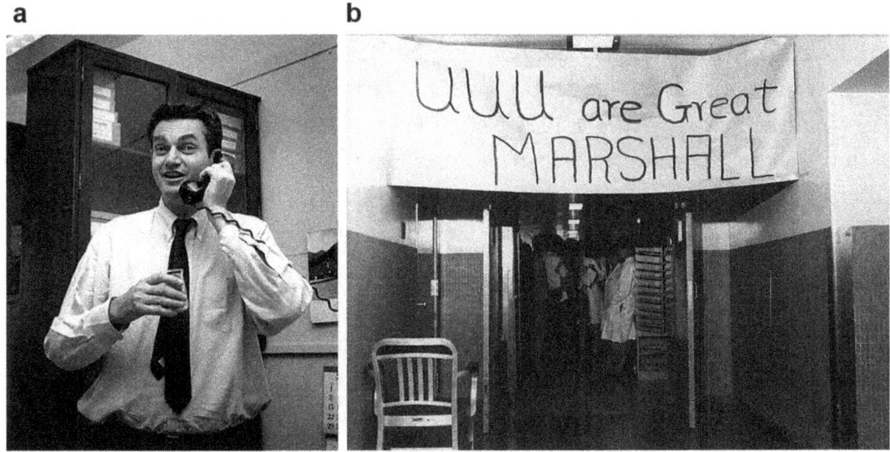

Fig. 11.2 Marshall Nirenberg (1927–2010) celebrates the news that he has been awarded a Nobel prize with a swig of champagne from a lab beaker and a banner made by his staff. US National Library of Medicine

[28] (Portugal 2015); p. xiii.

achievement which he celebrated with his staff by drinking champagne from a lab beaker.[29]

The message on the banner referred to an experiment, which had set Nirenberg on the path to receiving this accolade. In 1961, he and his post-doctoral researcher, the German biochemist Heinrich Matthaei (1929-) had synthesized strands of RNA containing only the base uracil, which is abbreviated by the single letter code U. They then added these polyuracil (polyU) RNA molecules to an in vitro experimental system that although free of cells nevertheless contained all the cellular apparatus for the synthesis of proteins. Nirenberg and Matthaei were delighted at what happened next. Addition of the polyU RNA resulted in the synthesis of a new protein chain. This itself was cause for celebration for it confirmed that RNA could indeed act as the chemical intermediate directing protein synthesis. But their experiment had also given them another invaluable piece of information. The newly synthesized protein chain resulting from the polyU RNA was made up of only one type of amino acid: phenylalanine. This suggested that somehow, the presence of the base uracil (U) in the RNA molecule directed the cellular machinery to insert only phenylalanine into the growing protein chain.

But if this was the case, why then, did Nirenberg's adoring staff not just write 'U are great' instead of 'UUU are great' on his banner? The answer lies with George Gamow's proposal that each different type of amino acid must be specified by some combination of the four different bases that make up DNA. Using a specific type of mutant bacterial virus known as T4, Francis Crick and South African biologist Sydney Brenner (1927–2019) found evidence that each amino acid in a protein chain is specified by a unit of three bases, known as a codon. Thanks to powerful methods of synthesis developed by developed by Indian-American chemist Har Gobind Khorana (1922–2011), Nirenberg and his post-doctoral fellow, the American geneticist Philip Leder (1934–2020), confirmed Crick and Brenner's discovery by synthesizing artificial molecules of RNA containing only single codons and showing that each was able to bind to only one particular type of amino acid.

When James Watson had heard Nirenberg give the repeat performance of his talk on the final day of the Moscow meeting back in 1961, he had been quick to grasp the enormity of its implications: 'From that moment on, it seemed likely that the genetic code would soon be completely cracked,' he remarked.[30]

---

[29] (Portugal 2015); p. 123.

[30] (Watson 2001); p. 265.

How the genetic code works is summarised in Fig. 11.3. In the nucleus, a particular sequence of bases in DNA are first transcribed into messenger RNA which is then transported out of the nucleus into the cytoplasm. Here it moves to the ribosomes, the sites of protein synthesis where triplet codons in the mRNA sequence are used to arrange their corresponding amino acids into position to form a growing polypeptide chain. What is particularly intriguing is that, with a few notable exceptions such as mitochondria - the organelles which generate energy within cells, the genetic code is pretty much universal, Mitochondria have their own small piece of circular DNA in which the codon that normally signals the end of protein synthesis by stopping the ribosome from translating mRNA is instead read as an instruction to insert the amino acid tryptophan into the growing protein chain. Nor is this trick confined to mitochondria. Along with several single celled organisms, certain genes in the

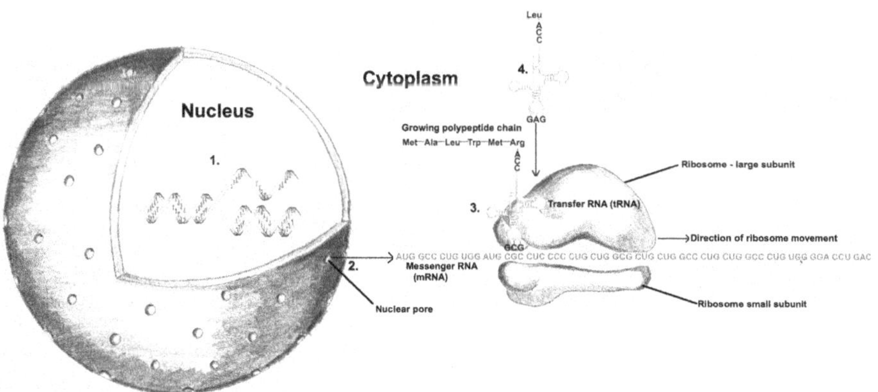

Fig. 11.3 How a sequence of bases in DNA directs the synthesis of a protein. (1) In the nucleus, a sequence of DNA that codes for a specific protein is copied (or, transcribed) into a messenger RNA (mRNA) molecule with a complementary sequence of bases. (2) The mRNA is transported through pores in the nuclear membrane into the cytoplasm and to the ribosomes, which are the sites of protein synthesis. (3) The ribosome moves along the length of the mRNA, 'reading' each triplet of bases (known as a 'codon') that represents an individual amino acid. Transfer RNA (tRNA) molecules carrying the particular amino acid corresponding to this codon dock into place. In the diagram shown, an mRNA has already been translated into the short polypeptide containing the amino acids methionine (Met), alanine (Ala), leucine (Leu), tryptophan (Trp), and methionine (Met). A tRNA carrying the amino acid arginine has recognised the mRNA codon CGC and docked into place allowing arginine to be attached to the growing polypeptide chain. As the ribosome then moves on to the next codon (CUC), this will be recognised as coding for the amino acid leucine and a tRNA carrying this amino acid will dock into place allowing it to attach to the arginine of the growing chain. (4) In this way a polypeptide chain gradually builds up to form a protein. (Diagram by K. Hall)

nuclear DNA of eukaryotes have been found to contain termination signal codons that have been reassigned to code for the insertion of amino acids modified with the element selenium.

Some pretty grand claims have been made as to why the genetic code of life on Earth should be almost universal. Readers of a certain age may well remember a late 1970s sci-fi TV series called 'Battlestar Galactica' in which each episode began with a shot of deep space while a voice proclaimed 'There are those who believe that life here began out there…far across the universe…'. It would appear that Francis Crick was one such believer. In 1973, he and biochemist Leslie Orgel (1927–2007) published a paper in which they proposed that life on earth had been seeded by microbes from space.[31] This idea, known as the 'panspermia hypothesis' was not itself novel and first been suggested by the Swedish chemist Svante Arrhenius (1859–1927) in 1908. But in their 1973 paper, Crick and Orgel were willing to stick their necks out even further:

> *Could life have started on Earth as a result of infection by microorganisms sent here deliberately by a technological society on another planet by means of a special long-range unmanned spaceship?*[32]

Regardless of the origins of the genetic code's near universality, the effort to decipher it earned Marshall Nirenberg, together with Har Gobind Khorana and the American biochemist Robert Holley (1922–1993) the 1968 Nobel Prize in Physiology or Medicine 'for their interpretation of the genetic code and its function in protein synthesis.'[33] But even by the time he won the award for this achievement, Nirenberg was already looking to the future. In an address given a year earlier, he had reflected that since the same genetic code appeared to be near universal throughout the living world it might soon become possible to reprogram organisms using chemically synthesized pieces of DNA. This would have profound consequences for all life—including ourselves:

> *When man becomes capable of instructing his own cells, he must refrain from doing so until he has sufficient wisdom to use this knowledge for the benefit of mankind. The reason I state this problem well in advance of the need to do so is because the decisions concerning the application of this knowledge ultimately must be made by society, and only an informed society can make such decisions wisely.*[34]

---

[31] (Crick and Orgel 1973).

[32] Ibid., p. 343.

[33] https://www.nobelprize.org/prizes/medicine/1968/summary/ (Accessed 21st July 2024).

[34] M. W. Nirenberg, "Man's Power to Shape His Own Biologic Destiny. Will Society Be Prepared to Use it Wisely?", Research Corporation Quarterly Bulletin (1967): p. 1, 4. http://profiles.nlm.nih.gov/ps/access/JJBCBG.pdf. Cited in Portugal 2015: p. 139.

11.4 Fears About Recombinant DNA

The need to start thinking about such decisions came about much more quickly than Nirenberg could have imagined at the time. This was thanks to the discovery made only a couple of years later of a type of bacterial enzyme that could recognise and physically cut specific sequences of DNA. Known as 'restriction enzymes' for their ability to restrict the growth of bacterial viruses, these enzymes had evolved as a primitive bacterial defence. But researchers soon realised that, far from just being a microbial curiosity, they gave scientists a powerful new tool. When used in conjunction with a suite of other enzymes, they could be used to cut DNA molecules at specific locations and physically join them together. In this way, it now became possible for the first time to take two fragments of DNA from completely unrelated organisms and physically splice them together to generate a hybrid, or 'recombinant' DNA molecule. Using such recombinant DNA molecules, researchers hoped that it might be possible to isolate DNA fragments of interest and deliberately introduce them into cells in order to study how they worked.

This was actually not a new idea. Back in 1914, the German chemist Emil Fischer (1852–1919) had proposed that artificially synthesized nucleic acids might be deliberately introduced into cells in order to learn more about how they work.[35] But the implications of recombinant DNA technology were both profound and alarming. Supposing, for example, someone was to take a sequence of DNA that was known to cause cancer in humans and, using restriction enzymes, stitch it into the genome of a virus that causes the common cold? In 1973, the American molecular biologist Paul Berg (1926–2023), whose lab had made the very first recombinant DNA molecule using genetic material from a common bacterium and the SV40 virus, published a letter in the international science journal *Nature* expressing his concerns. Signed by nine other pioneers in this field, the letter warned of 'serious concern that some of these artificial recombinant DNA molecules could prove biologically hazardous' and that 'such experiments should not be undertaken lightly.'[36]

'The Berg Letter' as it became known, quickly came to the attention of the wider news media and alarm bells began to ring. Realising that they now faced moral and ethical challenges of a similar scale to those that had confronted physicists involved with the Manhattan Project, the biologists pioneering this work held a conference at Asilomar on the Monterey Peninsula, Northern

---

[35] (Fischer 1914).

[36] (Berg et al. 1974).

California with the aim of drafting a set of guidelines by which recombinant DNA research could be regulated.

This would prove to be far from easy. Passions ran high over those few tempestuous days and tempers often flared. At one extreme were figures like Erwin Chargaff who completely opposed any such research, warning that 'You cannot recall a new form of life.'[37] At the opposite end of the spectrum were those, who were against any form of regulation on the grounds that it would hinder progress. One such voice was that of James Watson who, when presented with the argument that regulation was necessary because of the inherent risks involved, quietly exploded, saying 'We can't even measure the fucking risks.'[38] Yet after several days of argument, the moderates finally breathed a sigh of relief—they had managed to steer a course between the extremes of Chargaff and Watson to produce a draft set of guidelines. These divided recombinant DNA work into four categories based on the perceived level of risk involved and a set of containment measures deemed to be appropriate for each level of risk -ranging from wearing lab coats and abstaining from eating, drinking or smoking in the lab, all the way to working in facilities equipped with airlocks, negative pressure rooms, showers and full-body protective suits.

But their relief was to be short-lived. Thanks to having been attended by a journalist from 'Rolling Stone', a magazine which covered popular culture, the Asilomar conference and recombinant DNA were now very much in the media spotlight. And all too often the news headlines were not favourable ones. 'Those lovely people who gave us the atom bomb have another treat in store for us,' fumed the columnist Charles McCabe, 'Now they can create new forms of life, by jiggling with genes. I don't pretend to understand much about recombinant DNA, as the whole thing is called…But I do know what the act of creation is. When it is placed in the hands of men who by definition don't know what they are doing, it scares the daylights out of me…Why do we diddle ceaselessly with nature? Why will scientists persist in playing God?'[39]

The controversy that had erupted over recombinant DNA was also a gift to newspaper cartoonists. For example, a cartoon that appeared in the Boston Globe in 1977 depicted that stalwart of Hollywood B-sci-fi films, the mad scientist with wild hair and bulging eyes, racing into his lab to be greeted by a menagerie of mutant monsters which, of course, included Frankenstein's monster standing in the background.

---

[37] (Chargaff 1976); p. 938.

[38] (Rogers in Watson and Tooze, Document 2.1,36).

[39] (McCabe 1977 cited in Watson and Tooze 1981, Document 6.11, p. 165).

The relationship between science and the wider society in which it was practiced had certainly changed dramatically in the century that had passed since Miescher's discovery. When Miescher had stood shivering in the former kitchen of Tübingen castle and peering down the microscope at the leukocytes he had washed from old bandages, biology was concerned with the observation of living systems. Now it was gaining the power to control and manipulate them and their products. And the advent of this power brought with it a new demand on the scientists involved. It has been suggested that perhaps one of the reasons that Miescher's name is not better known today is because in his own time he was not a particularly good communicator and was simply far more content to live a quiet life engrossed at his lab bench. For the scientists involved in recombinant DNA research a century later, this was no longer an option. Whether they liked it or not, circumstances had forced it upon them to step out of their labs and be answerable to an increasingly fearful public about their research. Such was the situation for researchers at the Massachusetts Institute of Technology as they stood before a public hearing convened by Mayor Al Vellucci in 1977 where, in no uncertain terms, he made his feelings about what they were doing quite clear:

> *It's about time the scientists began to throw their goddamn shit right out on the table so that we can discuss it… Who the hell do the scientists think that they are that they can take federal tax dollars that are coming out of our tax returns and do research that we cannot then come in and question?*[40]

We can only speculate how Miescher would have felt at having to stand before TV cameras during proceedings that had opened with a high school choir singing 'This Land is Your Land', but it's a safe bet that he would have found it an utter nightmare. Worse still, hanging over all the researchers involved in this work was the very real possibility that the draft guidelines they had produced might become crystallised into federal law—with penalties such as heavy fines and even imprisonment for violation.

Yet by the end of the decade, the mood had changed dramatically. The much-feared federal laws never became a reality, and the alarm of the media and public that had led to circus shows such as Vellucci's hearings, subsided. But this did not mean that DNA was out of the newspapers.

Far from it, in fact. DNA was still making headlines and causing a stir—but now for a very different reason.

[40] (Lublow 1977, p. 51 cited in Watson and Tooze 1981, p. 121).

11.5 Recombinant Insulin: Biotech, Bugs, and (Very) Big Bucks

When the bell was rung on the morning of 14th October 1980 to begin the day's trading on Wall Street, dealers dived into a feeding frenzy to snap up shares in a new company called Genentech before their price rocketed even higher. Genentech's Wall Street debut was at the time, the most spectacular in stock market history and by the close of trading on that first day, its two founders, American molecular biologist Herb Boyer (1936-) and venture capitalist Bob Swanson (1947–1999) had been transformed into multimillionaires.

The secret of Genentech's success was that they had used recombinant DNA technology to produce a vital, life-saving medicine. In collaboration with researchers at City of Hope Hospital in Duarte, California, the Genentech scientists had synthesized the relatively short stretch of DNA, which encodes the amino acid sequence of insulin, the protein hormone that controls blood sugar levels and is essential for the management of type 1 diabetes. This fragment had then been inserted into *E. coli*, a common bacterium found in the human gut with the result that the modified bacterium was able to produce biologically active human insulin.

Prior to the discovery of insulin in 1922, type 1 diabetes was a fatal disease in which little could be done for patients other than to delay their inevitable death by putting them on a starvation diet. With the discovery of insulin, however, the disease was transformed from a fatal condition into a chronic one, which could be managed by daily injections with insulin recovered from pigs or cows as a side-product from the meat industry. But now, thanks to Genentech, patients could for the first time inject themselves with human insulin. This overcame problems with allergic reactions that some patients had experienced when using insulin from cows and pigs as well as removing the risk of contamination by pathogens from these sources.

By the end of the 1980s, recombinant human insulin had largely replaced insulin from cows and pigs as the main method of treatment for type 1 diabetes. Following in the wake of this success, the early 1980s witnessed a boom in biotechnology start-up companies as investors stampeded to squeeze dollars from the double-helix in the hope of founding the next Genentech. Senior staff at Cetus, another San Francisco-based technology company, may well have been kicking themselves when they saw the newspaper headlines reporting Genentech's stellar debut on Wall Street. A few years earlier when Swanson had first learned of recombinant DNA and sensed its commercial potential, he had approached Cetus with the idea of setting up a company based on the new technology only to be told that, although the field showed promise, 'it's not going to happen for a long time.'[41] Cetus executives may well have been eating their words on the morning of 14th October 1980, but if they had any fears that they had missed the boat when it came to DNA-based technology, they would soon be proven spectacularly wrong.

11.6 PCR: A Molecular Magician's Apprentice

A crucial part of Genentech's success had been their ability to link individual nucleotides together to form short strands of DNA known as oligonucleotides. These could then in turn be linked together to form the complete fragment of DNA that coded for the human insulin protein. The potential for oligonucleotide synthesis to become a powerful tool in research was quickly realised, and for one young American researcher at Cetus by the name of Kary Mullis (1944–2019), it was starting to become something of an obsession.

Mullis recalled how, one Friday evening in 1984 whilst driving up to spend the weekend in his cabin in the woods of Mendocino north of San Francisco, he suddenly pulled off the highway and began rummaging through his glove compartment in search of some paper and a pen. His girlfriend and fellow chemist Jennifer Barnett who until now had spent the journey asleep in the passenger seat, awoke to find Mullis hastily scribbling some calculations, and urged him to get back on the road. But when they eventually arrived at the cabin, Mullis's mind was still clearly elsewhere:

> *I had thought of incredible things before that somehow lost some of their sheen in the light of day. This one could wait till morning. But I didn't sleep that night. We got to my cabin and I starting drawing little diagrams on every horizontal surface that would take pen, pencil or crayon until dawn, when with the aid of a last bottle of good Mendocino county cabernet, I settled into a perplexed semiconsciousness.*[42]

The diagrams that Mullis drew all over his cabin that night would eventually earn him a Nobel Prize. Yet the principle behind the method he came up with is so disarmingly simple and straightforward it can be used at the lab bench by any undergraduate student.

---

[41] (Swanson and Smith-Hughes 2001); p. 13.

[42] https://www.nobelprizeorg/prizes/chemistry/1993/mullis/lecture/

All it takes is a couple of short synthetic pieces of DNA known as oligonucleotides that match the sequence at the extreme ends of the target to be amplified, along with a ready supply of the four nucleotides, a DNA synthesizing enzyme known as a polymerase, and a heating block. When the block is heated, the attractive forces between complementary bases on opposite strands of the DNA molecule are broken allowing the two strands to peel apart. As the block is then cooled, these same attractive forces allow the oligonucleotides to bind, or anneal, to complementary sequences on the separated single strands of the target DNA molecule. Once the oligonucleotides are bound, the block is heated to the optimal temperature for the polymerase enzyme which begins to grow new copies of the target strand from the bound oligonucleotides. On repeating this cycle, each of these new daughter strands now acts as a target for binding of the oligonucleotides and the initiation of another round replication. With each successive cycle, the number of daughter molecules copied from the original template increases exponentially such that after about 30 cycles, the original DNA fragment may well have been amplified a million-fold. Although the principle of this method is somewhat reminiscent of the magical broomsticks in the fable of 'The Magician's Apprentice', Mullis chose the more technical, but somewhat less poetic name for it of the Polymerase Chain Reaction, or PCR.

11.7 Catching Criminals with DNA

PCR has since become an essential research tool in molecular biology but in the wake of the Covid-19 pandemic, it became a household name thanks to its use in diagnostic tests for the presence of the SARS-CoV 2 virus. It is also used in forensic investigations, often in conjunction with another DNA-based innovation that was also made during the 1980s. Around the same time that Mullis was rummaging through his glove compartment, the British geneticist Alec Jeffreys (1950-) was in his laboratory at the University of Leicester, UK holding up to the light a photograph that he later described as 'a horrible, smudgy, blurry mess.'[43] The photograph showed several ladders of black bands—each of which corresponded to a fragment of human DNA of a specific size, obtained by taking DNA from three related individuals and cutting it with restriction enzymes. And although the image may have looked a mess, Jeffreys could see straightaway that it was important.

---

[43] (Zagorski 2006); p. 8919.

What Jeffreys had found was a means of using differences between the sizes of fragments produced when human DNA is cut, to establish biological identity—or, as it is more commonly known, DNA fingerprinting. Like PCR, this method has also become a household name thanks to its prominence as a plot device in TV crime dramas but had its origins in a rather curious gift to Jeffreys from the British Antarctic Survey.

This was a lump of seal meat, rich in the protein myoglobin. Like haemoglobin, myoglobin binds and releases oxygen but does so much more efficiently, allowing the seals to remain underwater for long periods. Jeffreys' initial research had been concerned with identifying specific genes by detecting the messenger RNA (mRNA) transcripts that are copied from them. Thanks to the vast amounts of myoglobin mRNA present in the muscle tissue of the seal meat, he was able to identify the corresponding gene from which they had been transcribed. This in turn enabled him to identify the human myoglobin gene, where he found something surprising. The gene contained a short section of tandem repeats of DNA made up of only 10–15 bases. Known as a 'minisatellite', these had already been reported in the research literature, but Jeffreys wondered whether they could be used to distinguish individuals. He was right in his hunch, and after having first been used to confirm the parentage of a Ghanaian boy in a disputed immigration case, Jeffreys' method of DNA fingerprinting was used 2 years later by police to successfully track down and convict a serial rapist and murderer.

With the advent of PCR, DNA fingerprinting, and drugs produced using recombinant DNA technology, the 1980s was a time of crucial innovations in the field of DNA. It was also a time when the culture of academic science and its relationship with the commercial world underwent a seismic shift. In the wake of Genentech's spectacular stock market success at the start of the decade, hopes were riding high that fortunes might be made thanks to DNA.

But before investors poured millions of dollars into biotechnology, one question needed to be answered. Was the commercial venture on which they were risking their capital built on foundations of solid intellectual property? Or, put another way, could a biological system such as a bacterium containing a recombinant DNA molecule, like the one Genentech had made to synthesise human insulin, be patented? In late 1980, much to the relief of Genentech's founders and eager investors, the US Supreme Court ruled that it could.

But what about actual DNA sequences such as the one that the scientists at Genentech had used to synthesise human insulin? Could these too be patented? Thanks to another crucial technical development of the '80 s, this was to become a very important question—and one over which passions ran high.

11.8 Life™- Patenting Genes

Thanks to having determined the amino acid sequence of insulin, the British biochemist Fred Sanger (1918–2013) already had one Nobel Prize to his name. But in 1980, he became one of only five scientists ever to win this accolade for a second time. The prize was shared between molecular biologist Paul Berg who had made the very first recombinant DNA molecule, and Walter Gilbert with whom Sanger had developed a method for determining the precise linear sequence of bases in DNA.

Sanger later described the method for DNA sequencing as having been 'the best idea I have ever had.'[44] This was, to put it mildly, a classic piece of understatement. The method enabled researchers to compare sites of interest in different organisms, to verify that products made by PCR did not contain unwanted mutations, and to identify regions involved in the regulation and control of genes—to name just a few of its many applications.

But as those same countless legions of PhD students and post-docs in molecular genetics[45] will also testify the method was laborious. It required the preparation of a rectangular slab of gel by pouring a solution of a chemical called acrylamide into a narrow gap between two glass plates. On the addition of a polymerising agent, the acrylamide solidified into a slab which, when placed in an electric field, acted as a molecular sieve to sort fragments of DNA according to their size. But such gels were notoriously fragile and had an annoying tendency to rip apart when the glass plates were separated at the end of the procedure. The number of PhD students who were spared this infuriating experience must be few and far between. As must those who, after pouring a gel and leaving it to set, have returned to find that instead of solidifying, the acrylamide has leaked out from the glass plates and formed a puddle all over the lab bench.

And a rather nasty puddle at that. In its liquid form acrylamide is a neurotoxin—and this was not the only risk to health that workers faced when using this method. In order to visualise DNA fragments as dark bands on a photographic film, the method required the use of radioactively-labelled nucleotides. And after all this, having faced down the challenges of torn gels, neurotoxins and radioactivity, the longest possible sequence that a researcher could hope to read using this method was about 300 base-pairs[46]—a mere

[44] (Sanger 1988); p. 22.

[45] We include ourselves amongst this number.

[46] We both agree that, in our days at the lab bench, neither of us ever got anywhere near this impressive figure ourselves.

drop in the ocean when compared with the approximately three billion base-pairs that are now known to make up the complete human genome.

By the end of the decade, however, automated methods of sequencing were emerging that promised to transform DNA research. Not only did these methods have the advantage of not using radioactive materials and fragile neurotoxic gels, but they also allowed researchers to sequence thousands of bases at a time.

As DNA sequencing technology improved, researchers were presented with a tantalising possibility. Might it now be possible not just to sequence a tiny part of an organism's DNA, but its entire genome? In 1995, they found that the answer was a resounding yes, when a research team published the complete genome sequence of the bacterium *Haemophilus influenzae*. Three years later came another milestone when the nematode worm, *Caenorhabditis elegans*, became the first multicellular organism to have its complete genome sequenced, followed shortly afterwards by the fruit fly, *Drosophila melanogaster*. But even before any of these projects had begun, work was already underway on the ultimate prize.

On 27th June 2000, 'The Times' newspaper of London announced the completion of the first draft of the human genome with the headline, 'Opening the Book of Life.'[47] First begun in 1990, the first draft unveiled 10 years later had been due to the efforts of two independent research projects—one a publicly funded consortium led by UK scientist and future Nobel laureate Sir John Sulston (1942–2018) together with the American physician-geneticist Francis Collins (1950-) in the USA, and the other a private venture led by American biotech-entrepreneur Craig Venter (1946-) and his company Celera.

Around the world, news media hailed the achievement as a milestone. Comparing it with the music of Bach, the sonnets of Shakespeare and the Apollo moon landing, the British evolutionary biologist Richard Dawkins (1941-) described this moment as 'one of those achievements of the human spirit that makes me proud to be human'.[48] While in an address from the White House US President Bill Clinton (1946-) proclaimed that *"Today, we are learning the language in which God created life."*[49] But as British geneticist and science broadcaster Adam Rutherford (1975-) wryly observed, 'If this was the language of God, then a holy editor would've been a great asset.'[50] For one thing, the task of sequencing the entire human genome was not even

[47] 'Opening the Book of Life', The Times (London), 27th June 2000.

[48] 'Benefits 'some distance away', Nigel Hawkes, 'The Times' (London), 27th June 2000.

[49] https://clintonwhitehouse3.archives.gov/WH/Work/062600.html

[50] (Rutherford 2016); p. 291.

complete—what was presented at the time was a draft, and one that is still undergoing revisions and updates.

Also, the claim that it had given us 'the language in which God had written the book of life' was a little optimistic. Because by the time that Clinton gave his address, it had become quite apparent that there was much more to genes and how they work than simply the synthesis of proteins according to specific sequences of DNA.

11.9 One Gene: One Enzyme?

In part, this was thanks to what Francis Crick had described as a 'mini-revolution' that had taken place just over two decades earlier.[51] Until this time, it had appeared that a gene was simply a sequence of DNA that corresponded to a particular amino acid sequence in a protein. This idea had its roots in the work of the American geneticists George Beadle (1903–1989) and Edward Tatum (1909–1975) who, in 1941, had published work showing that specific mutations induced by irradiating the bread mould *Neurospora* with X-rays affected different enzyme-controlled steps in certain metabolic pathways. This led them to propose what became known as 'the one gene, one enzyme' hypotheses according to which, each different enzyme in a given metabolic pathway was determined by a single corresponding gene. Francis Crick's proposal that the amino acid sequence of a protein was determined by the base sequence of a particular gene, together with Vernon Ingram's demonstration that changes in the amino acid sequence of haemoglobin corresponded to specific genetic mutations gave a neat explanation for this hypothesis in molecular terms. But in 1977, this simple picture was shattered.

That year, a team of researchers made a discovery that left them so stunned that, in a dramatic departure from the normally sober tone in which a scientific paper is written, they described their finding in the title of their paper as 'amazing.'[52] What they, and a number of other groups, had found was that, in a certain type of human virus, the messenger RNA encoding a particular protein is much shorter in the length than the original DNA sequence from which it had been copied. The reason for this turned out to be that, in eukaryotic cells (that is—those which, unlike prokaryotes such as bacteria, contain a membrane-bound nucleus) and their viruses, the sections of a

[51] (Crick 1979); p. 264.

[52] (Chow et al. 1977).

gene sequence that encode a particular protein (known as 'exons') are not continuous, but split up by intervening non-coding sequences that became known as 'introns.' In an astonishing feat of molecular precision surgery, the machinery of the cell copies the entire sequence of a gene—including both introns and exons—into a precursor RNA before then removing the introns and splicing together the exons to make a mature messenger RNA that is exported out of the nucleus into the cytoplasm to the sites of protein synthesis.

With the discovery of introns and exons, the simple model proposed by Beadle and Tatum that each gene specified a single protein began to fall apart very quickly—for eukaryotic cells at least. The excision of introns from an RNA molecule and the different order in which its exons could be spliced together meant that a multiplicity of different mRNAs, each encoding a different protein product, could be generated by a single gene. To date, the record is a gene encoding a protein involved in neuronal development in the fruit fly *Drosophila* which has the potential to generate approximately 38,000 possible alternative mRNA transcripts.[53]

There were more surprises to come. Twenty-three years after the discovery of introns and exons, as the first draft of the Human Genome Project neared its completion, researchers took part in a sweepstake in which they placed bets on the number of genes they expected to find. Adopting the definition of a gene to be a region of DNA that coded for a protein, their estimates ranged from 26,000 to 140,000.[54] But to date, the actual number is a modest mere 19,000—far less even than the genome of the humble maize plant which has an estimated 59,000 protein-coding genes, or the banana which has somewhere in the order of 36,000.[55] It is humbling too to consider that, despite being a much more simple organism than ourselves, the nematode worm *Caenorhabditis elegans* has roughly the same number of genes as a human being.

Even more surprising was the finding that these estimated 19,000 protein coding genes account for only about 5% of the three billion bases that make up the human genome, which raises the immediate question—what, if anything, is the other 95% doing?

---

[53] (Schmucker et al. 2000; Zipursky et al. 2006).

[54] 'Scientists cast bets on Human Genes: A Winner Will Be Picked in 2003' Nicolas Wade, New York Times, 23rd May 2000.

[55] (Messing et al. 2004; D'Hont et al. 2012).

11.10 Junk DNA: Trash or Treasure?

Nor was this phenomenon found to be confined to the human genome. Across the living world, most genomes contain a vast amount of DNA that does not encode protein products. The proportion of non-protein coding DNA is higher in eukaryotes than prokaryotes such as bacteria and came to be known as 'junk DNA'—a term whose historical origins have been the subject of debate, though not nearly as much as has its exact meaning.[56] Drawing on the analogy with junk that is often found piled up in domestic garages, several commentators have pointed out that the term has several possible meanings.[57] It could, for example, refer to broken household appliances that are beyond repair, have no further use and are destined for the local municipal waste collection centre. But it could equally refer to objects that were once useful—and may yet prove to be so again, perhaps in some new capacity.

In 2012, the debate took a new twist when several broadsheet newspapers triumphantly proclaimed the death of 'junk DNA'.[58] This was thanks to a major international study called ENCODE which, having trawled through sequence data from the human genome project, had predicted that 80% of its DNA was 'biochemically active.' But, just like 'junk DNA', the meaning of this term too was debatable.[59] The definition of a region of DNA being biochemically active was based on such criteria as whether it contained sequences predicted to bind proteins involved in transcription or generate mRNAs. But did the mere presence of such sequences in a region of DNA guarantee that they had an active functional role in the organism?[60]

Most of what has become known as 'junk DNA' contains the remnants of sequences that might once have been active but, over the course of evolutionary time, have become silent and inactive due to the accumulation of

[56] The exact origin of this term is somewhat obscure and has been the subject of much debate (see for example, https://judgestarling.tumblr.com/post/64504735261/the-origin-of-the-term-junk-dna-a-historical, It is often credited to the Japanese-American biologist Susumu Ohno (1928–2000) who in 1972 used it to describe regions of DNA that could accumulate mutations without any deleterious effect upon the organism. But 9 years before Ohno popularised the term, it had already appeared in paper on the single-celled organism Paramecium, by Charles Ehret (1916–2007) of the Division of Biological and Medical Research, Argonne National Laboratory, Illinois and Gerard de Haller of the University of Geneva who observed that 'While current evidence makes plausible the idea that all genetic material is DNA (with the possible exception of RNA viruses), it does not follow that all DNA is competent genetic material (viz. 'junk' DNA).' (Ehret and de Haller 1963; p. 39).

[57] (Graur et al. 2013); p. 586.

[58] See for example, 'Breakthrough Study Overturns Theory of 'Junk DNA' in Genome', The Guardian 5th Sept 2012; https://www.theguardian.com/science/2012/sep/05/genes-genome-junk-dna-encode

[59] (The ENCODE Consortium 2012)

[60] (Eddy 2012).

mutations. One example are sequences known as pseudogenes which, like a genetic version of the appendix in the human gut, or the coccyx, are remnants of genes that might once have been functional deep in the evolutionary past but are no longer needed, such as those encoding olfactory receptors in cetaceans.[61]

Others include vast tracts of repetitive sequences that are found throughout the genomes of most organisms. Research suggests that these may make up anywhere between 40–60% of the human genome while in domestic maize this figure is as high as 85%.[62,63] Many of these are believed to have arisen through the incorporation of DNA from a type of virus known as a retrovirus. Like tobacco mosaic virus, these viruses, which include HIV, also use RNA as their genetic material but they have evolved a cunning evolutionary trick for replication. On infecting a host cell, the virus synthesizes an enzyme called reverse transcriptase which copies the RNA genome into a double-helical DNA intermediate. This DNA then inserts itself into the host genome where it can be transcribed and translated by the cellular machinery of the host, allowing reproduction of the virus.

Over the millions of years of evolution, many such viral sequences that have found their way into the human genome have become inactive through mutation. Many, but not all. Whilst studying the genome of domestic maize the 1940 and '50s American scientist Barbara McClintock (1902-1992), discovered that some of these sequences had an intriguing property: they could copy themselves from one part of the plant genome to another. Known as transposable elements (TEs) or 'transposons', these are also found in the human genome and many different types have now been identified. Some of these can have deleterious effects and are believed to be the cause of certain cancers that arise when regulation of particular genes is disrupted as a result of the transposon inserting itself into them.[64] But by generating new combinations of genetic material, transposon activity may also have played a crucial role in the development of the placenta and the vertebrate adaptive immune system, making the steady acquisition of these mobile DNA elements over the millions of years of evolutionary history a powerful force in shaping the development of genomes.[65,66,67]

---

[61] (Cobb 2015b); p. 245.

[62] (de Koning et al. 2011).

[63] (Schnable et al. 2009).

[64] (Burns 2017).

[65] (Cornelis et al. 2014).

[66] (Koonin and Krupovic 2015).

[67] (Fedoroff 2012).

In 1983, McClintock received the Nobel Prize in Physiology or Medicine for her discovery, which may well have helped to shed light on a long-standing mystery in biology. Originally known as the C-value paradox, or more recently, the C-value Enigma, this refers to the observation that there appears to be no clear correlation between the complexity of an organism and the size of its genome as measure in base-pairs. With three billion base-pairs, the human genome is 40 times smaller than that of a lungfish and just over 5 times smaller than that of the domestic onion which has a genome of 16 billion base-pairs.[68] Within the class of amphibians, genomes can show as much as a 120-fold variation in size from frogs to salamanders, while among angiosperms this figure can be as high as a thousandfold.[69,70] Even within a genus, variation can be significant, with salamanders of the species Plethodon showing a four-fold variation in the size of their genomes.[71]

If much of an organism's genome is stuffed with what science writer Professor Matthew Cobb has memorably described as genetic 'fossils' in the form of pseudogenes and inactive transposons, then the staggering variations in size between different species such as humans and onions no longer seem quite so puzzling.[72] Meanwhile the debate continues as to whether we should refer to these regions of DNA as 'junk.' It all rather depends on how we define 'junk' but in assessing whether there might be some function to much of this genomic material, Canadian evolutionary and genome biologist T. Ryan Gregory has suggested that we apply what he has called 'The Onion Test:'

> *The onion test is a simple reality check for anyone who thinks they have come up with a universal function for non-coding DNA. Whatever your proposed function, ask yourself this question: Can I explain why an onion needs about five times more non-coding DNA for this function than a human?*[73]

The debate over just what proportion of a genome is functional—and necessary—looks set to continue and as we learn more about them, genomes will no doubt turn up yet more surprises. But one thing has become very clear—a DNA sequence does not have to code for a protein to be functional.

---

68 (Palazzo and Gregory 2014).

69 (Gregory 2003).

70 (Gregory 2005).

71 (Mizuno and Macgregor 1974).

72 (Cobb 2015a); p. 244.

73 https://www.genomicron.evolverzone.com/2007/04/onion-test.html

11.11 RNA: Revolution? Revelation? Or Both?

Some non-protein coding sequences act as switches to control the transcription of genes into mRNA. One type, known as 'promoters' are found immediately before the protein-coding part of a gene. These contain sequences that act as docking sites for a suite of proteins that physically bind to the DNA molecule to facilitate transcription of the coding DNA sequence into mRNA. Other types of regulatory sequences can be found several thousands of base pairs away from the genes that they control and are called enhancers or silencers, depending on whether they increase or repress the activity of their target gene. Through an impressive feat of chromosomal origami, these regulatory regions are brought into proximity to the coding sequences which they control. In this way, proteins that bind to specific sequences within the regulatory region can activate or repress the transcription of a specific gene.

Until the early 1990s, this was the prevalent model of gene regulation: regulatory proteins bind to specific regions of DNA and control the transcription of coding regions. But in 1993, a team of researchers studying the nematode worm *Caenorhabditis elegans* found that this was only part of the story. They were studying a gene called lin-4 that was essential for controlling the development of the worm, but what surprised them was that the lin-4 gene did not code for a protein product. Instead, it produced a short RNA that controlled the activity of a number of other genes.[74] These microRNAs (miRNA), as they became known, were at first thought to be unique to the nematode worm, but further research soon showed that they can be found across much of the animal kingdom, including humans (who to date are thought to have over 600 types of miRNA).[75]

Five years after this discovery in *C. elegans*, RNA in the nematode worm was once again causing excitement. Researchers had identified a second type of short regulatory RNA that could silence the activity of specific genes in *C. elegans* by binding to complementary regions of sequence within their messenger RNA transcripts.[76] But, just as with miRNA, these small interfering RNAs (siRNAs) as they became known, were not confined to *C. elegans* and it became quickly apparent that they were responsible for a well-known gene silencing effect in both plants and fungi.[77]

[74] (Lee et al. 1993; Wightman et al. 1993).

[75] (Pasquinelli et al. 2000; Callaway and Sanderson 2024).

[76] (Fire et al. 1998).

[77] (Romano and Macino 1992; Napoli et al. 1990; Hamilton and Baulcombe 1999).

Without all these diverse mechanisms of gene regulation, multicellular life would be impossible. To develop into a complex organism whether it be a giraffe or an oak tree, a single fertilised egg must undergo countless divisions which give rise to daughter cells. Over the course of successive divisions, these daughter cells become specialised into the components of the various tissues that make up the different organs of the developing organism—a process known as differentiation. Some cells, for example, will become skin cells, while others will become heart muscle cells (or, in the case of an oak tree, some will become leaf cells, while others become root cells). This requires that, over the course of the development of an organism from a single cell, cells differentiate at just the right time and in the right place. All of which requires a feat of choreography and timing at the level of gene control that is truly impressive. Genes involved in differentiation to become skin cells, for example, may well be silenced in cells that have differentiated to become the retina of the eye.

The discovery of the central role played by small RNA molecules in these processes came as a big surprise to a world that had, in many ways, assumed DNA to be at the heart of biology. In reflection of the importance of these discoveries, scientists Andrew Fire and Craig Mello were awarded the 2006 Nobel Prize in Physiology or Medicine for their work on RNA interference (RNAi) and siRNA, while in 2024 this award went to Victor Ambros and Gary Ruvkun for their work on miRNAs.[78]

In response to commentators who have hailed these landmarks as an RNA revolution, Craig Mello said 'it is perhaps more apt to call it an RNA 'revelation'. RNA is not taking over the cell—it has been in control all along. We just didn't realize it until now.'[79] Almost a decade and a half earlier this same view appears to have been anticipated by examiners at the University of Oxford who, in the summer of 1991, asked undergraduate students sitting their final exams in biochemistry to assess and discuss the statement 'DNA is, whilst RNA does.' For students who had been paying attention to developments in their field, this question was a gift.[80] Only 2 years earlier, Thomas R. Cech of the University of Colorado (1989-) and Sidney Altman (1939–2022) of Yale University had been awarded the Nobel Prize in Chemistry for their discovery that certain RNA molecules had catalytic properties, allowing them to act as enzymes.

---

[78] (Callaway and Sanderson 2024; Hopkin 2006).

[79] (Hopkin 2006).

[80] I (KH) was one of those who attempted this question, but I'm not sure that I was one of these diligent students.I'm hoping that this book may in some way compensate for my half-baked answer at the time.

A discussion of these 'ribozymes' as they had become known could easily have filled an outstanding exam essay at the time but an undergraduate finalist faced with this same question today, would have even more reason to smile.[81]

An entire essay could easily be written about the intriguing discovery that small regulatory RNA molecules have been found in membrane-bound vesicles called exosomes that are secreted by cells and transported throughout an organism. It is well established that cells can communicate with each other via electrical signals conducted along nerves or by chemical signalling through the action of hormones but these extracellular RNAs, or exRNA, represent a previously unknown and exciting new layer of communication in biology. Even more intriguing is that this method of communication operates not only within organisms, but between them. Human breast milk, for example, has been found to contain miRNA molecules that pass from the mother to the infant.[82]

Nor is the transfer of these short RNAs between different organisms confined to those of the same species. Several examples have now been found in which the transfer of short regulatory RNAs occurs from plants to pathogenic and symbiotic micro-organisms, nematode worms, and insects, as well as from funghi to plants. It has even been reported that human regulatory RNAs are transferred to plasmodium, the parasitic micro-organism that causes malaria.[83]

The discovery of small regulatory RNAs and that they can be secreted as exRNA has shaken up the long-held belief that genes were defined solely as stretches of DNA that code for proteins. Moreover, they also have exciting practical implications. Researchers hope that the profile of exRNAs secreted in the human body may act as new diagnostic biomarkers alerting clinicians to the early onset of diseases such as cancer and Alzheimer's.[84]

There are also hopes that the exosome vesicles in which these exRNAs are carried could prove to be a powerful new vehicle for the delivery of therapeutic agents in the targeted treatment of disease.[85] One such agent with therapeutic potential are siRNA molecules themselves. With the discovery in 2001 that siRNA molecules could be introduced into a range of mammalian cells to silence specific genes, researchers were quick to point out that RNA

[81] (https://www.nobelprize.org/prizes/chemistry/1989/summary/); (Kruger et al. 1982).

[82] (Nguyen 2020).

[83] See for example (Knip et al. 2014; Asgari 2017; Liu and Chen 2018; Chen and Rechavi 2022).

[84] (Dolgin 2020).

[85] (Keener 2020).

interference might offer a powerful new form of medical therapy.[86] This optimism that our rapidly increasing understanding of life at the genetic level would offer revolutionary new medical treatments was also shared by politicians at the time. In his address from the White House to announce the completion of the first draft of the Human Genome Project, President Clinton predicted that this new knowledge would 'revolutionise the diagnosis, prevention and treatment of most, if not all, human diseases… Doctors will increasingly be able to treat Alzheimer's, Parkinson's, diabetes and cancer by attacking their genetic roots.' Speaking via a live satellite link, UK Prime Minister Tony Blair, concurred with this rosy view, and remarked of his newborn son Leo - 'I think his life expectancy has just gone up by about 25 years.'[87]

Quite possibly so. But as we were about to discover, some of the applications of this new knowledge would also raise some very awkward social and political questions too.

[86] (Elbashir et al., 2001).

[87] The Times (London) 27th June 2000.

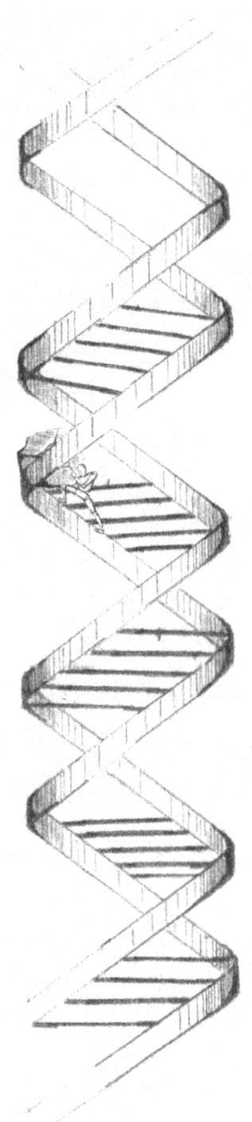

Drawing by Kersten Hall

References

Asgari, S. 2017. RNA as a means of inter-species communication and manipulation: Progresses and shortfalls. *RNA Biology* 14 (4): 389–390. https://doi.org/10.1080/15476286.2017.1306172.

Berg, P., D. Baltimore, H. Boyer, S. Cohen, R. Davis, D. Hogness, D. Nathans, et al. 1974. Potential biohazards of recombinant DNA molecules. *Science* 185:303.

Brachet, J. 1942. La Localisation Des Acides Pentose Nucleiques Dans Les Tissus Animaux et Les Oeufs d'amphibiens En Voie de Developpement. *Archives de Biologie* 53:207–257.

Brenner, S., F. Jacob, and M. Meselson. 1961. An unstable intermediate carrying information from genes to ribosomes for protein synthesis. *Nature* 190:576–581.

Burns, K. 2017. Transposable elements in human cancer. *Nature Reviews Cancer* 17:415–424.

Callaway, E., and K. Sanderson. 2024. Medicine Nobel awarded for gene-regulating "microRNAs". *Nature* 634:524–525. https://doi.org/10.1038/d41586-024-03212-9.

Caspersson, T., and J. Schultz. 1939. Pentose nucleotides in the cytoplasm of growing tissues. *Nature* 143 (3623): 602–603. https://doi.org/10.1038/143602c0.

Chargaff, E. 1976. The path to the double helix by Robert Olby (Review). *Perspectives in Biology and Medicine* 19:289–290.

Chen, X., and O. Rechavi. 2022. Plant and animal small RNA communications between cells and organisms. *Nature Reviews Molecular Cell Biology* 23 (3): 185–203. https://doi.org/10.1038/s41580-021-00425-y.

Chow, L., R. Gelinas, T. R. Broker, and R. J. Roberts. 1977. An amazing sequence arrangement at the 5′ ends of adenovirus 2 messenger RNA. *Cell* 12:1–8.

Claude, A. 1940. Particulate components of normal and tumor cells. *Science* 91 (2351): 77–78. https://doi.org/10.1126/science.91.2351.77.

Cobb, M. 2015a. *Life's Greatest Secret: The Race to Crack the Genetic Code*. London: Profile Books.

Cobb, M. 2015b. Who discovered messenger RNA? *Current Biology* 25:R523–R548.

Cornelis, G., C. Vernochet, and S. Malicorne. 2014. Retroviral envelope Syncytin capture in an ancestrally diverged mammalian clade for placentation in the primitive Afrotherian tenrecs. *Proceedings of the National Academy of Sciences of the United States of America* 111:E4332–E4341.

Crick, F. H. C. 1958. On protein synthesis. *Symposia of the Society for Experimental Biology* 12:138–163.

Crick, F. H. C. 1979. Split genes and RNA splicing. *Science* 204:264–271.

Crick, F. H. C., and L. E. Orgel. 1973. Directed panspermia. *Icarus* 19:341–346.

de Koning, A. P. J., W. Gu, T. A. Castoe, B. A. Batzer, and D. D. Pollock. 2011. Repetitive elements may comprise over two-thirds of the human genome. *PLoS Genetics* 7:e1002384. https://doi.org/10.1371/journal.pgen.1002384.

Delbrück, M., and G. Stent. 1956. On the mechanism of DNA replication. In *McCollum-Pratt symposium on the chemical basis of heredity*, ed. William D. McElroy and Bentley Glass, 699–736. Baltimore: Johns Hopkins University Press.

Dolgin, E. 2020. Could tracking RNA in body fluids reveal disease? *Nature* 582 (7812): S2–S4. https://doi.org/10.1038/d41586-020-01763-1.

Eddy, S. 2012. The C-value paradox, junk DNA and ENCODE. *Current Biology* 22:R898–R899.

Ehret, C. F., and G. De Haller. 1963. Origin, development, and maturation of organelles and organelle Systems of the Cell Surface in paramecium. *Journal of Ultrastructure Studies* 9:3–42.

Elbashir, S. M., J. Harborth, W. Lendeckel, A. Yalcin, K. Weber, and T. Tuschl. 2001. Duplexes of 21-nucleotide RNAs mediate RNA interference in cultured mammalian cells. *Nature* 411 (6836): 494–498. https://doi.org/10.1038/35078107.

Fedoroff, N. 2012. Transposable elements, epigenetics, and genome evolution. *Science* 338:758–767.

Fire, A., S. Xu, M. K. Montgomery, S. A. Kostas, S. E. Driver, and C. C. Mello. 1998. Potent and specific genetic interference by double-stranded RNA in Caenorhabditis elegans. *Nature* 391 (6669): 806–811. https://doi.org/10.1038/35888.

Fischer, E. 1914. Über Phosphosäureester des Methlyglucosids und Theophyllin-Glucosids. *Berichte Den Deutschen Chemisches Gesellschaft* 47:3193–3205.

Gamow, G. 1954. Possible relation between deoxyribonucleic acid and protein structures. *Nature* 173:318.

Graur, D., Y. Zheng, N. Price, R. B. Azevedo, R. A. Zufall, and E. Elhaik. 2013. On the immortality of television sets: "Function" in the human genome according to the evolution-free gospel of ENCODE. *Genome Biology and Evolution* 5:578–590. https://doi.org/10.1093/gbe/evt028.

Gregory, T. R. 2003. Variation across amphibian species in the size of the nuclear genome supports a pluralistic, hierarchical approach to the C-value enigma. *Biological Journal of the Linnean Society* 79:329–339.

Gregory, T. R. 2005. Synergy between sequence and size in large-scale genomics. *Nature Reviews Genetics* 6:699–708.

Gros, F., H. Hiatt, W. Gilbert, C. G. Kurland, R. W. Riseborough, and J. D. Watson. 1961. Unstable ribonucleic acid revealed by pulse labelling of Escherichia coli. *Nature* 190:581–585.

Hamilton, A. J., and D. C. Baulcombe. 1999. A species of small antisense RNA in posttranscriptional gene silencing in plants. *Science* 286 (5441): 950–952. https://doi.org/10.1126/science.286.5441.950.

Hoagland, M. B., P. C. Zamecnik, and M. L. Stephenson. 1957. Intermediate reactions in protein biosynthesis. *Biochimica Et Biophysica Acta* 24:215–216. https://doi.org/10.1016/0006-3002(57)90175-0.

Hoerr, I., R. Obst, H.-G. Rammensee, and G. Jung. 2000. In vivo application of RNA leads to induction of specific cytotoxic T lymphocytes and antibodies. *European Journal of Immunology* 30:1–7.

Hopkin, M. 2006. RNAi scoops medical Nobel. *Nature*news061002-2. https://doi.org/10.1038/news061002-2.

Judson, H. F. 1996. *The eighth day of creation*. Cold Spring Harbor, NY: Cold Spring Harbor Press.

Knip, M., M. E. Constantin, and H. Thordal-Christensen. 2014. Trans-kingdom cross-talk: Small RNAs on the move. *PLoS Genetics* 10 (9): e1004602. https://doi.org/10.1371/journal.pgen.1004602.

Koonin, E. V., and M. Krupovic. 2015. Evolution of adaptive immunity from transposable elements combined with innate immune systems. *Nature Reviews Genetics* 16:184–195.

Kruger, K., P. J. Grabowski, A. J. Zaug, J. Sands, D. E. Gottschling, and T. R. Cech. 1982. Self-splicing RNA: Autoexcision and autocyclization of the ribosomal RNA intervening sequence of Tetrahymena. *Cell* 31 (1): 147–157. https://doi.org/10.1016/0092-8674(82)90414-7.

Lee, R. C., R. L. Feinbaum, and V. Ambros. 1993. The C. elegans heterochronic gene Lin-4 encodes small RNAs with antisense complementarity to Lin-14. *Cell* 75 (5): 843–854. https://doi.org/10.1016/0092-8674(93)90529-Y.

Liu, L., and X. Chen. 2018. Intercellular and systemic trafficking of RNAs in plants. *Nature Plants* 4 (11): 869–878. https://doi.org/10.1038/s41477-018-0288-5.

Messing, J., A. K. Bharti, W. M. Karlowski, and R. A. Wing. 2004. Sequence composition and genome organization of maize. *Proceedings of the National Academy of Sciences of the United States of America* 101:14349–14354.

Mizuno, S., and H. C. Macgregor. 1974. Chromosomes, DNA sequences, and evolution in Salamaders of the genus Plethodon. *Chromosoma* 48:239096.

Napoli, C., C. Lemieux, and R. Jorgensen. 1990. Introduction of a chimeric Chalcone synthase gene into petunia results in reversible co-suppression of homologous genes in trans. *The Plant Cell* 2 (4): 279. https://doi.org/10.2307/3869076.

Palazzo, A. . F., and T. R. Gregory. 2014. The case for junk DNA. *PLoS Genetics* 10.

Pasquinelli, A. E., B. J. Reinhart, F. Slack, M. Q. Martindale, M. I. Kuroda, B. Maller, D. C. Hayward, et al. 2000. Conservation of the sequence and temporal expression of Let-7 heterochronic regulatory RNA. *Nature* 408 (6808): 86–89. https://doi.org/10.1038/35040556.

Pollock, M. R. 1970. The discovery of DNA: An ironic tale of chance, prejudice and insight third Griffith memorial lecture. *Journal of General Microbiology* 63 (1): 1–20. https://doi.org/10.1099/00221287-63-1-1.

Portugal, F. H. 2015. *The least likely man: Marshall Nirenberg and the discovery of the genetic code*. Cambridge, MA: MIT Press.

Romano, N., and G. Macino. 1992. Quelling: Transient inactivation of gene expression in *Neurospora Crassa* by transformation with homologous sequences. *Molecular Microbiology* 6 (22): 3343–3353. https://doi.org/10.1111/j.1365-2958.1992.tb02202.x.

Rutherford, A. 2016. *A brief history of everyone who ever lived*. London: Weidenfeld & Nicolson.

Sanger, F. 1988. Sequences, sequences, sequences. *Annual Review of Biochemistry* 57:1–29.

Schmucker, D., J. C. Clemens, H. Shu, C. A. Worby, J. Xiao, M. Muda, J. E. Dixon, and S. L. Zipursky. 2000. Drosophila Dscam is an axon guidance receptor exhibiting extraordinary molecular diversity. *Cell* 101:671–684.

Schnable, P. S., et al. 2009. The B73 maize genome: Complexity, diversity, and dynamics. *Science* 326:1112–1115.

Watson, J. D. 2001. *Genes, Girls and Gamow*. Oxford: Oxford University Press.
Watson, J. D., and F. H. C. Crick. 1953a. Genetical implications of the structure of deoxyribonucleic Acid. *Nature* 171:964–967.
Watson, J. D., and F. H. C. Crick. 1953b. Molecular structure of nucleic acids: A structure for deoxyribose nucleic acid. *Nature* 171:737–738.
Watson, J. D., and J. Tooze. 1981. *The DNA story: A documentary history of gene cloning*. New York: W. H. Freeman.
Wightman, B., I. Ha, and G. Ruvkun. 1993. Posttranscriptional regulation of the heterochronic gene Lin-14 by Lin-4 mediates temporal pattern formation in C. elegans. *Cell* 75 (5): 855–862. https://doi.org/10.1016/0092-8674(93)90530-4.
Zagorski, N. 2006. Profile of Alec J. Jeffreys. *Proceedings of the National Academy of Sciences of the United States of America* 103:8918–8920.
Zipursky, S. L., W. M. Wojtowicz, and D. Hattori. 2006. Got diversity? Wiring the Fly brain with Dscam. *Trends in Biochemical Sciences* 31:581–588.

12

Breaking the Glass

Contents

Abstract The 1990s saw a proliferation of biotechnology start-up companies focussed on developing protocols for the treatment of diseases that resulted from specific and well characterised mutations in a particular gene. These protocols used a delivery vector such as a replication deficient virus to introduce a healthy copy of a gene into cells carrying the mutant version. More recently however, gene therapy has become focussed not on replacing the mutant gene but directly correcting DNA sequences known to be involved in disease. This has become possible thanks to CRISPR (Clustered Regularly Interspersed Palindromic Repeats)—a technology based on a bacterial defence system, which allows direct and specific editing of genomes. This powerful technology has been used to successfully treat the blood disorder thalassaemia but raises a very difficult question: would it be ethical to modify the DNA of germ cells using CRISPR and so permanently alter the human germline? One researcher in the field, Dr. He Jiankui of Southern University of Science and Technology, Shenzhen, certainly thought so. But when he announced at a conference in 2018 that he had modified the DNA of two human embryos which had then gone to term as two baby girls, his peers in the field were horrified. This chapter uses also this recent example from the history of science to explore what exactly is meant by 'a gene' and how stories which frequently

K. Hall, R. Dahm, *The Dawn Fisherman*, Copernicus Books,
https://doi.org/10.1007/978-3-032-14219-1_12

appear in the popular press about the discovery of 'a gene for' a specific trait are based upon a flawed oversimplification of how genes operate.

Keywords DNA • CRISPR • Gene editing • Molecular computation • Digital storage • Palaeogenomics • Nanotechnology • DNA Origami • Biochemistry • Genetics • Biotechnology • Medicine • Sickle-cell anemia • Thalassaemia • Turing machine • Algorithm • Pattern recognition • Digital information • Storage medium • Silicon • Hard drives • Magnetic tape • Data storage • Nanorobots • Self-replication • DNA scaffolds • Biomolecules • Biochemical reactions • Drug delivery • Synthetic biology

Miescher's discovery opens a Pandora's Box of questions…

With the advent of genomic sequencing came a deluge of data and a pressing question—who owned this new information? The public and private research groups who had completed the first draft of the human genome differed starkly both in their scientific approach to the task, and their ultimate aims. While the former sought to make the information publicly available to researchers, the latter looked for regions of DNA that were transcribed into mRNA, i.e. protein-coding genes, with the aim of patenting them.

Passions ran high over this matter and often resulted in acrimonious divisions. Supporters of the publicly funded project objected that patenting DNA sequences would restrict research and innovation, whilst proponents of Venter's entrepreneurial approach claimed the opposite, arguing that filing patents would ensure the development of new diagnostic and therapeutic approaches. In the decade that followed, the arguments continued to rage until, in 2013, the US Supreme Court made a historic ruling.

The case involved Myriad Genetics, a company that had developed a diagnostic test to detect the presence of certain mutations in a gene known as BRCA1, which was associated with increased risk of breast cancer. In the process of developing the diagnostic test, Myriad Genetics had filed patents on the DNA sequence of the BRCA1 gene—a move, which caused some concern. Objections were raised that, if granted, these patents might severely limit the access of other researchers to work on the BRCA1 gene. Arguing that human genes were 'a product of nature', the judges on the US Supreme Court ruled—to the relief of many, and the fury of some—that their sequences could not be patented and the application by Myriad Genetics was duly rejected.[1]

As well as diagnosing predisposition for certain diseases, might new genetic information offer the possibility of treating them? President Clinton certainly believed so and, in his address of June 2000, prophesied that, with the unveiling of the first draft of the human genome, we would gain 'immense new power to heal.' Yet hopes for new medical treatments based on increased knowledge of human genes had been riding high well before then. In 1992 the San Francisco based company Genentech, whose initial success had been built on producing human insulin using recombinant DNA technology, employed similar methods to make the anti-cancer drug Herceptin. The drug works by targeting a protein known as HER2, which is involved in the regulation of cell growth.[2] In 20–30% of breast cancers, the gene encoding the HER2 protein is either amplified or over-expressed resulting in a poor prognosis for the patient. But the success of Herceptin in treating patients with the HER2 variant raised the possibility that new generations of drugs might be customised and tailored to treat a particular genetic profile—a field known as medical genomics.

Another approach which had held much promise since the late 1980s was that gene-based treatments might offer cures for diseases ranging from certain types of cancer to HIV and haemophilia. The concept of this 'gene therapy' as it became known had its origins in an experiment performed in 1962 by American molecular biologist Elizabeth Hunter Szybalska (1928–2015) and her husband, the Polish-American geneticist Waclaw Szybalski (1921–2020). Working with human cells that were unable to synthesise a particular enzyme due to a genetic deficiency, Szybalska and her husband found that this error could be corrected if DNA carrying a functional copy of the gene were to be introduced into the cells.[3] But could this be turned into a medical therapy? One major challenge was how to get a functional copy of a gene into the affected cells. What this required was some kind of delivery vehicle to carry therapeutic DNA into the target cells and as it happened, Nature had provided just that.

Viruses are pirates at a cellular level. In isolation, they are just strands of nucleic acid wrapped inside a protein coat and are unable to replicate. But over the course of millions of years they have evolved to enter host cells and

[1] 'US Supreme Court Rules Human Genes Cannot be Patented', The Guardian 13th June 2013; 'Human Genes Patent Ruling: Some Clarity But Real Problem Remains', The Guardian 13th June 2013.

[2] (Carter 1992).

[3] (Szybalska and Szybalski 1962); (Bigda and Koszalka 2013).

subvert the cellular machinery of the host to reproduce themselves. They are, therefore, the ideal delivery vehicle on which to 'piggyback' a therapeutic gene for transport into a target cell.

In 1990, approval was given for the first trial of this method to be used in the treatment of severely immunocompromised patients suffering from a condition known as ADA-SCID. In this condition a genetic error results in a deficiency of adenosine deaminase (ADA), an enzyme that is crucial for the development of a type of white blood cell known as T-cells. With T-cell development severely impaired, the patient's weakened immune system offers little defence against bacteria and viruses, hence the resulting condition is known as severe combined immunodeficiency, or (SCID).

The therapeutic strategy for this condition involved first inserting a functional copy of the gene encoding human ADA into the genetic material of a type of virus known as a retrovirus. The rationale for this choice was because when a retrovirus enters a host cell, it inserts its own genome—along with that of any foreign genes it may be carrying—into that of the host. The researchers hoped that when T-cells from the patient were infected with the modified virus, the healthy copy of the ADA gene would be plugged into the patient's genome and activated.

Four years later, the research team published the initial results of their trial showing that ADA activity had increased and that both patients who had received the treatment were now showing certain types of immune response. But although the researchers acknowledged that this 'demonstrated the potential efficacy of using gene-corrected autologous cells for the treatment of children with ADA$^-$ SCID', they remained cautious in their conclusions.[4] For several years, both children had been receiving a drug known to be effective in managing ADA-SCID, and continued to undergo this treatment throughout the trial making it difficult to draw clear conclusions about the extent to which the improvement in their condition had been due to the viral gene therapy. As a result, the authors of the paper kept their claims for gene therapy modest, noting that it 'can be a safe and effective *addition* to treatment for *some* patients with this severe immunodeficiency disease.' [our italics].[5]

The media however took the news that a genetic error had been corrected in a human being with much less sobriety. Two years before the start of the ADA-SCID trial, the French newspaper *Le Monde* had already hailed 1988 as being 'Year One of Gene Therapy' with the news of the first ever demonstration that a virus could carry DNA for a simple genetic marker into

[4] 'Autologous' means cells taken from the respective patients themselves.

[5] (Blaese 1995).

a human patient. Riding a crest of optimism in the wake of such announcements, gene therapy started to gather both momentum and vast sums of money throughout the 1990s.[6] By the middle of the decade, over 100 clinical trials had been given the go-ahead and millions of dollars were flowing into research from both private investors and the US National Institute of Health.[7] Until a tragedy brought it all crashing to a halt.

Jesse Gelsinger was a US teenager who suffered from deficiency of the liver enzyme ornithine transcarbamylase (OTC)—a rare genetic condition leading to dangerously elevated levels of ammonia which affect the nervous system. This can result in lack of energy and appetite, poorly controlled breathing and body temperature, and seizures or coma. Although Jesse was able to manage his condition through the combination of a low protein diet and a medication regime, in 1999 he volunteered to participate in a clinical trial of a new viral gene therapy vector which carried a functional copy of the OTC gene. Within only a few hours of having received the experimental treatment, Jesse died from a severe immune reaction that had been triggered in response to the viral vector. The tragedy of his death sent shockwaves through the research community and, quite rightly, prompted a much-needed re-evaluation of a field in which hype had all too easily overtaken reality. As one of the scientists involved in this work reflected, 'There's been an emphasis on glitz. It has produced a culture in which getting into clinical trials—getting into the club—has been more important than getting a meaningful result.'[8]

With the tragic death of Jesse Gelsinger, much of the hope—and hype—around gene therapy evaporated. But research into the use of modified viruses as delivery vehicles nevertheless continued. In 2003, China became the first country to give approval for the clinical use of Gendicine, a gene therapy based on an adenoviral vector and in 2012, the European Medicines Agency gave approval for Glybera, the first gene therapy product to be marketed in the EU.[9] This treated a condition caused by severe deficiency of an enzyme involved in lipid metabolism and used a different type of viral vector which was believed to be less likely to cause adverse immune responses. Known as the Adeno-Associated Virus, or AAV, this vector is also widely used as a vehicle to deliver the gene encoding the now infamous spike protein of SARS-Cov2 into human cells in certain types of vaccine against Covid-19.

[6] (LeMonde 1988).

[7] (Marshall 1995).

[8] (Marshall 1995); p. 1050.

[9] https://www.ema.europa.eu/en/news/european-medicines-agency-recommends-first-gene-therapy-approval

But even as this product appeared on the market, an alternative DNA-based therapeutic strategy was emerging, and it was one that would generate shock waves.

12.1 CRISPR: A DIY Kit for the Genome

In 2023 Victoria Gray, a woman from Mississippi, was able to do something that she had previously never dreamed might be possible. That year, Gray made her first ever trip outside the USA to travel to London where she addressed a conference of doctors, scientists and patients.[10] As a sufferer of sickle-cell anaemia, she had spent most of her life in constant pain, sometimes unable even to get out of bed.

But in 2019, Gray had participated in one of several clinical trials that used a powerful new technology known by the acronym of CRISPR. Just like the restriction enzymes that Genentech scientists had used to such effect in producing recombinant human insulin at the end of the 1970s, CRISPR too had its origins in what had once been considered a mere microbial quirk. These were curious clusters of small repeating sequences of DNA observed in the genomes of certain bacteria and from which the technique derives its name—Clustered Regular Interspersed Palindromic Repeats.[11] It turned out however that these intriguing sequences were not merely a curiosity of microbial genetics. Just like the restriction enzymes that had been discovered in the late 1960s and early 1970s, it was found that they functioned as a primitive bacterial immune system. And, just like restriction enzymes, they would turn out to be a powerful tool.

When bacteria are invaded by certain types of virus they incorporate some of the invaders' genetic material into the regions of their own genome that lie in between these clusters of repeats. This gives the bacterium a record, written in its own genome, of this encounter with the virus and the incorporated viral DNA will be passed on to its progeny to equip them against a subsequent attack by the virus. On encountering this virus a second time, the bacterium makes an RNA copy of the DNA stored from the previous infection. This RNA then directs a set of bacterial enzymes known as Cas, to cut the corresponding region of invading viral DNA thus neutralising the virus.

Argentinian-American microbiologist Luciano Marraffini (1974-) and Erik Sontheimer of Northwestern University were quick to spot that this might be

[10] (Kupferschmidt 2023).

[11] (Ishino et al. 1987); (Mojica et al. 1992); (Jansen et al. 2002); (Mojica et al. 2005).

put to uses other than defending bacteria against viruses. They pointed out that it might have 'considerable functional utility, especially if the system can function outside of its native bacterial or archaeal context.'[12]

This would prove to be a memorable understatement. In 2012, two independent teams of scientists made up of international researchers working in laboratories in California, Sweden, Austria and Lithuania published papers showing that, when provided with a specific RNA sequence, the CRISPR system could be directed to cut and edit the corresponding DNA sequence with precision. In other words, the CRISPR system was a programmable means of editing and re-writing the sequence of any given stretch of DNA.[13]

But given that CRISPR had evolved to work in bacteria, could it work in more complex eukaryotic cells? Feng Zhang, a young researcher at the Broad Institute, Cambridge, Massachusetts, who claimed the 1993 blockbuster, 'Jurassic Park', as his inspiration to become a molecular biologist, and George Church (1954-) at Harvard University showed that, in mouse and human cells, it could do just that.[14]

Suddenly researchers had at their disposal a relatively simple toolkit that could be used by any final year undergraduate student to introduce targeted changes into a DNA sequence of interest. Others were quick to spot the commercial and therapeutic potential of CRISPR and the new technology quickly brought millions of dollars of investment flooding in. it also brought headaches (as well, no doubt, as lucrative fees) for patent lawyers.

For American biochemist Jennifer Doudna (1964-) and French microbiologist Emmanuelle Charpentier (1968-), CRISPR brought the most distinguished accolade in science. In 2020, they were awarded the Nobel Prize in Chemistry for their role in having pioneered this method. But Doudna recalled that CRISPR had also brought her a nightmare. In an interview, she described having had a dream in which she found herself explaining to a pig-faced Adolf Hitler just how the new technology worked.[15]

It was not the use of CRISPR to treat patients such as Victoria Gray that was causing Doudna to lose sleep. The treatment of Gray's sickle-cell anaemia had involved removing cells from her bone marrow and making specific alterations in their DNA to reactivate the production of foetal haemoglobin.[16]

[12] (Marraffini and Sontheimer 2008).

[13] (Jinek et al. 2012); (Gasiunas et al. 2012).

[14] 'Feng Zhang : « Faire de l'ingénierie des systèmes.
biologiques pour résoudre des problèmes réels » ' Le Monde', 8th August 2016.

[15] (Doudna and Sternberg 2017); pp. 198–199.

[16] In 2023, the UK medicines regulator became the first in the world to approve Casgevy, a therapy that uses the CRISPR gene-editing tool for the treatment of sickle-cell anaemia and thalassaemia. (Wong 2023).

Although active in the foetus, production of this version of haemoglobin is normally switched off in adults. But when production of foetal haemoglobin was reactivated in Gray's bone marrow cells using CRISPR, it compensated for the defective form of the adult protein that had arisen from the sickle-cell mutation. Crucially however, because this genetic change was confined only to Gray's bone marrow cells and not her egg cells, it would never be passed on to her children.

But what if CRISPR were used to make alterations in the DNA not of somatic cells such as bone marrow, but those involved in reproduction—the eggs and the sperm? Or an embryo with the potential to develop into a human being? Any genetic changes introduced into these reproductive cells would potentially be not only inherited by all the cells of the resulting offspring but would also be carried by their descendants too. The human genetic code would have been permanently altered with a change that would be passed on down the generations. But who would have the power to decide this? At the heart of Doudna's nightmare was the fear that, in the wrong hands, CRISPR might not only herald a return to the eugenic ideologies that had proved so catastrophic in the twentieth century but also give their proponents a powerful new tool with which to bring about their vision.

Recognising the power of this new technology, bioethics commissions were founded to ponder the fallout from the new method and to issue recommendations as to what should and should not be done with it. And it did not take long before the hypothetical scenarios they had considered began to become a reality. In 2015 a paper appeared by a group of Chinese researchers claiming that they had used CRISPR to alter the beta-globin gene in a single-cell human embryo.[17] As the embryo was non-viable, it could never have come to term, which was probably just as well because the CRISPR-induced alteration had not gone quite as had been intended. Firstly, the embryos were mosaic, meaning that that they were a patchwork of edited and unedited cells. But secondly, and potentially far more seriously, the CRISPR procedure had not been as precise as had been hoped. The genomes of the edited embryos contained 'off-target' alterations other than the intended ones. Had these embryos been viable and grown to term, there would be no telling what the effect of this would have been on the developing child.

A year later, another team, led by Yong Fan of Guangzhou Medical University, reported that they too had used CRISPR to alter the genome of a non-viable human embryo, by introducing a mutation into a gene known as

---

[17] (Liang et al. 2015, p. 363).

CCR5.[18] This encodes a protein that protrudes from the surface of cell membranes and is believed to have several functions, including playing a role in inflammation, inhibiting tumour growth and limiting the damage caused by strokes. It is also believed to be a molecular gateway by which HIV can enter human cells. There are however people who lack a specific run of 32 base-pairs in the CCR5 gene and as a result, appear to have lower susceptibility to HIV infection.

But again, when Fan's group used CRISPR to introduce this mutation into the genome of human embryos, the results did not go according to plan. Not only were the embryos mosaic, but along with off-target effects, the CRISPR method had introduced new mutations into the gene. This problem had also been encountered by the team who had modified the beta-globin gene, leading them to conclude that there was a 'pressing need to further improve the fidelity and the specificity of the CRISPR/Cas9 platform, a prerequisite for any clinical applications of CRISPR/Cas9 mediated editing.'[19] Commenting on this finding, Xiaoxue Zhang, another researcher working in this area, called for caution:

> *it is time for the scientific community, public, funding bodies and governments to come together to re-define and reinforce the boundaries of gene-editing research…it is in the best interest of all parties that the research field should voluntarily avoid any study that may pose potential safety and/or ethical risks. Only by holding themselves to the highest standards will scientists retain the public's trust in biomedical research, and at the same time, provide the best service for the well-being of our society.*[20]

Unfortunately, however, some people had other ideas. The misuse of CRISPR by murderous pig-faced dictators may have caused Jennifer Doudna to lose sleep, but it is a pity that the new technology did not give Dr. He Jiankui of Southern University of Science and Technology, Shenzhen a few more sleepless nights. Had it done so, he might have been spared a spectacular fall from grace.

When He Jiankui took to the stage at a scientific meeting held in Hong Kong in 2018 and announced that two baby girls (called Nana and Lulu) had been born from embryos that he had genetically modified using CRISPR, few in the audience were surprised.[21] Rumours amongst colleagues in the field and

[18] (Kang et al. 2016).

[19] (Liang et al. 2015, p. 363).

[20] (Zhang 2015, p. 313).

[21] It was later revealed that a third child, Amy, had been born from an embryo modified by He Jiankui using CRISPR (Marx, 2021).

online postings about He Jiankui's work had been circulating for some time. But if He Jiankui was expecting this announcement to be greeted with adulation by the audience, he was in for a shock. Instead, it evoked only condemnation and alarm.

Little is still known about Nana and Lulu, but from the limited data available it appears that once again, the CRISPR alteration did not go as hoped. Both girls are a mosaic of unedited and edited cells. And in those cells which did receive the edit, the alteration made was not the one intended. Instead of the desired 32 base-pair deletion, 15 base-pairs had been deleted from Lulu's CCR5 gene, whilst the corresponding gene in Nana both gained and lost some DNA sequences. Time alone will tell what effect these changes will have on the girls but it is to be hoped that both they and their descendants who will inherit these alterations—will remain healthy.

At the end of the conference presentation, a member of the audience asked He Jiankui to explain the medical justification for altering the girls' genomic DNA in this way. It was an awkward but perceptive question for many present suspected that this experiment had served no need other than the ambition and vanity of He Jiankui.

In his presentation to the conference, Dr. He explained that his intention had been to make the girls less susceptible to infection with HIV- although as several commentators have since pointed out when discussing this incident there are far more straightforward and simpler ways of avoiding infection with HIV than editing the genome of embryos.

One of the couples who were involved in this work said that they had done so in order to spare their children the stigma and prejudice that they themselves had experienced having a family member who was HIV positive.[22] And therein lies at least part of a very big problem - which is to confuse a social and cultural problem for a scientific one. He Jiankui believed that prejudice and fear could be cured with a quick-fix glamorous technical solution and that, if science were allowed free rein, social attitudes would be forced to catch up:

> *If we are waiting for society to reach a consensus . . . it's never going to happen. . . . But once one or a couple of scientists make first kid, it's safe, healthy, then the entire society including science, ethics, law, will be accelerated. Speed up and make new rules . . . . So, I break the glass.*[23]

---

[22] Cited in (Hurlbutt 2020); p.191; (Cyranoski 2018); (Cyranoski and Ledford 2018).

[23] (Hurlbutt 2020); p. 185.

But in his rush to 'break the glass', He Jiankui had been seduced by an erroneous and vastly over-simplified concept of what genes are, and how they function.

12.2 What Exactly Is a Gene, Anyway?

This misconception has a long history. It can be traced back to the binary traits such as flower colour and seed shape that were selected by Gregor Mendel for his experiments with peas—or more accurately as historians Greg Radick and Kostas Kampourakis have shown, the way in which Mendel's work was used by British biologist William Bateson in the early twentieth century to support and promote his own theory of heredity.[24] Bateson's model of inheritance was based on the transmission of discrete 'unit-characters' and has left a legacy that is still with us today. The discrete binary traits that Mendel selected in his pea plants have become the paradigm for our understanding of genes. This has led to a simplistic concept of heredity in which every trait—whether physiological or behavioural—is determined by a 'gene for.'

And as both Radick and Kampourakis have shown, this is not only fallacious, but also dangerous. It is fallacious because it ignores the fact that genes can have very different effects depending on the details of their cellular, physiological and developmental context. This had already been recognised by William Bateson's rival in the battle over the concept of heredity—W. Frank Raphael Weldon (1860–1906) who championed an alternate view of heredity which argued that the manifestation of biological traits was not simply determined by genes acting in isolation– but crucially by the way in which these genes interact with their physiological, cellular and environmental context. Or, put another way: heredity = genes + environment.

But Bateson won the day. Not so much because he had better arguments or science, but rather because Weldon died at the tragically young age of 46 before he could finish his major body of work. And, as Radick has argued in his book 'Disputed Inheritance', Bateson's victory has left us with a powerful and problematic legacy today. The idea that our destiny is somehow dictated purely by our DNA is now so common and deeply embedded in popular culture, giving rise to the notion that any trait, no matter how complex—such as sexual orientation, propensity to become vegetarian, or criminal behaviour, can all be distilled down to a simple genetic cause. As an example, look no

[24] (Radick 2023); (Kampourakis 2024).

further than the following selection of headlines taken from British tabloid newspaper, 'The Daily Mail'—'Violent Video Game Fan? It's All in the Genes'; 'Not Eating Greens is in Your Genes'; 'Why Couch Potatoes Can Blame it on Their Genes.'[25]

But while it may be common, it is also deeply misleading. Conditions such as sickle-cell anaemia, which arise from a single change in just one gene, appear to be the exception rather than the rule. Even eye-colour, which is has long been taught in schools as a classic binary Mendelian trait, has been found to be the effect of at least 60 different regions of DNA acting in concert.[26]

With our current best estimate of the number of protein-coding genes present in the human genome standing at around only 20,000 and the subsequent discovery of genes that encode not proteins, but siRNA and miRNA, it is clear that the subtle, sophisticated, and highly variable physiological and behavioural traits we human beings exhibit are the result not of single 'genes for', but instead of a vastly complex and ever-changing interactions of multiple genes. Genes are dynamic entities—they work in different ways depending on their specific cellular, environmental, and developmental context. A gene that performs a particular function at one stage of development, or in one cell type, may play a very different role in other cells, or at different stages of development.

The picture has been further complicated by the discovery that the very same sequence of bases (known as the genotype) can give rise to radically differing traits (known as the phenotype). One example is the agouti mouse in which, despite having the same DNA sequence for a gene controlling coat colour, the animals show distinct variation in their coat colour ranging from being completely yellow to completely black. This marked variation arises not from differences in the DNA sequence, but rather on the degree to which cytosine bases in a region of DNA controlling their coat colour has been chemically modified by methyl groups.

A more tragic example of the power of DNA methylation are the two conditions known as Angelman's and Prader-Willi syndrome, both of which present with very different symptoms. Yet in both these conditions the underlying error at the genetic level is the same: one of the two parental copies of chromosome 15 carries a short deletion of bases. What determines which of the two conditions the child will develop is whether the chromosome carrying the deletion was inherited from the mother, or the father. This is because the

[25] ('The Daily Mail', 21st Feb; 14th Feb; 15th Feb 2014 respectively).

[26] (Simcoe 2021).

paternal and maternal chromosomes are chemically tagged, or 'imprinted' by the addition of methyl groups in differing patterns.

Methylation is just one example of the way in which the same base sequence of DNA (known as the genotype) can give rise to very different observable traits (known as the phenotype) due to chemical modification. Another is the attachment or removal of acetyl groups from proteins known as histones that act as a molecular scaffold to which DNA binds in the cell. In recent years, it has become ever more clear that differences in the methylation of DNA and the acetylation of the histones that package it, play a fundamental role in regulating the correct expression of genes involved in development of an organism. Under the umbrella term of 'epigenetics', such modifications ensure for example, that genes which drive the formation of an eye, are not switched on in the development of lungs, bones or kidneys.[27]

A given phenotype emerges, therefore, as a result of the complex interaction between genetics and environmental factors. All of which makes the idea that introducing a deliberate alteration into the sequence of CCR5 to confer resistance to HIV, seem incredibly naïve. If, as appears to be the case, CCR5 is involved in several different roles including mediating cell-cell contact—how could He Jiankui ever have been be sure that the changes he introduced might not have some other unanticipated—and unwanted—effects? And if a trait as apparently simple as eye colour can turn out to be so complex at the genetic level, the possibility of 'designer babies', deliberately engineered to have enhanced features such as intelligence starts to look even more naïve, misguided, and undesirable.

The idea that our destiny is somehow written in our DNA has become popular. But it's rooted in a flawed and over-simplified understanding of the gene. Not only is this idea flawed, but it also has the potential to be dangerous.[28] This is not just because it might offer a dubious rationale to carry out questionable experiments as He Jiankui did, but rather it is because the simplistic idea that complex human behavioural traits such as intelligence or sexual orientation can be determined by a single gene easily creates fertile ground and misguided justification for bigotry.[29] In his book 'DNA is not Destiny', psychologist Stephen J. Heine argues that misunderstanding the gene has the potential to cause much social harm in a way that other concepts in modern science, such as an erroneous model of the atom, do not. In school, for example, when we first encounter atoms we are often taught to think of

[27] For more on epigenetics, see (Carey 2012).

[28] Again, see (Radick 2023) and (Kampourakis 2024) for a more detailed discussion of why this is so.

[29] (Dougherty 2009; Donovan 2022; Radick 2023).

them as a miniature solar system in which electrons orbit a nucleus like planets going around a sun. Quantum physics has since shown this to be a vast over-simplification, but for most people it is one that is unlikely to have any deleterious effects on their daily lives and their interactions with others.

But, says, Heine, this is not the case when it comes to learning about genes. Misunderstanding the gene has the potential to cause much social harm in a way that misunderstanding atomic structure does not. While incorrectly thinking of atoms as miniature solar systems may be of little consequence in our daily lives, Heine warns that 'incorrectly believing that genes are like switches leads us to become fatalistic, and can result in increased sexism, racism, and irrational fears about our future disease risks.'[30] Rather than continuing to peddle the idea of there being 'a gene for', Heine proposes that 'it's far better to highlight the difficulty in understanding the complex machinery of genetics than it is to give people a false sense of understanding that leads them to make potentially costly decisions in their lives.'[31]

After Miescher's death, his uncle Wilhelm His had made a prediction. 'The appreciation of Miescher and his works will not diminish with time,' he wrote with confidence, 'instead it will grow, and the facts he has found and the ideas he has postulated are seeds which will bear fruit in the future.'[32] Over a century after Miescher's death, the innovations born from the seeds he had sown, such as recombinant human insulin, DNA sequencing, PCR, DNA fingerprinting, the Human Genome Project and CRISPR had proven His to be correct beyond his wildest imaginings.[33]

Nor were these all confined to the field of biomedicine. In 1957, Francis Crick gave a lecture to the Society for Experimental Biology at University College, London in which he predicted the emergence of a powerful new field in biology:

---

[30] (Heine 2017); p. 264.

[31] Ibid.

[32] His, Die Histochemischen Und Physiologischen Arbeiten von Friedrich Miescher, [The Histochemical and Physiological Works of Friedrich Miescher]; p. 4.

[33] One of the big challenges with writing a chapter about the latest advances in DNA-based technologies is that by the time you've finished, what you've written is already out of date! This is because this field, involving multiple disciplines such as biochemistry, physics, information technology and materials sciences, is developing so rapidly. As an example, just as we were about to submit our manuscript, we learned of a paper in the journal *Science Robotics* ('Toward three-dimensional DNA industrial nanorobots', (2023) by Feng et al., Science Robotics vol. 85) which describes the creation of self-replicating nanobots made from cuboidal DNA structures that can be programmed to assemble simple chemical structures. In their discussion, the authors of the paper describe their work as 'a step toward a manufacturing/ assembly robot and can be readily extended to more complex tasks…' Looking to the future, they speculate that 'These industrial nanorobots may find biomedical applications as artificial enzymes or organelles.' One to watch!

> *Biologists should realize that before long we shall have a subject which might be called 'protein taxonomy' - the study of the amino acid sequences of proteins of an organism and the comparison of them between species. It can be argued that these sequences are the most delicate expression possible of the phenotype of an organism and that vast amounts of evolutionary information may be hidden away within them.*[17]

The comparison of anatomical structures such as wings was a common method used by biologists when they sought to arrange organisms into taxonomic groups, or establishment of evolutionary lineages. But Crick was proposing that it might now be possible to trace evolutionary and taxonomic relationships between organisms at the molecular level by comparing similarities in the amino acid sequences of their proteins.

Today, Crick's molecular approach to taxonomy encompasses not just the comparison of protein sequences as he had originally envisaged, but also of DNA sequences. Examples include the recent discovery that the ancestors of the domestic horse first arose on the steppes of Eurasia about 4200 years ago and using comparisons of genomes to explore the evolutionary relationship between birds and reptiles.[34,35] Along with this field, which has become known as palaeogenomics, DNA analysis is now also used as a tool in archaeology where it can offer new insights into familial relationships and migration patterns—such as in studies of victims at Pompeii.[36]

These are just a couple of examples of palaeogenomics -a field which has benefited enormously from the massive increase in computing power over recent decades—a trend which looks set to continue thanks to the exciting possibility that DNA itself could be used both as a computer and a store of binary information.

12.3 DNA Computers?

A computer can be any physical system upon which algorithms can be performed. This could be a mechanical system such as the calculating machine designed in the nineteenth century by the English mathematicians Charles Babbage (1791–1871) and Ada Lovelace (1815–1852), or the electronic ones with which we are today all familiar that use a series of logical operations to

---

[34] (Thompson 2021); (Librado 2021).

[35] (O'Connor 2018). The idea of an evolutionary link between birds and reptiles was first proposed by British scientist Thomas Henry Huxley (1825–1895).

[36] (Lenharo 2024); (Pilli et al. 2024); (Callaway 2018); (Di Bernardo et al. 2009) (Cipollaro et al. 1998).

manipulate information represented as strings of ones and zeros in a binary code. But DNA presents a third possibility. In 1994, a group of researchers showed that DNA molecules could themselves be used to compute the solution to a classic problem in mathematics known as 'the travelling salesman problem.'[37] Their success led them to make a confident prediction, one which has been born out thanks to the recent announcement that tiny tiles made from DNA molecules have been used to perform the even more complex and impressive algorithm of pattern recognition[38];,[39]

> *...it is conceivable that molecular computation might compete with electronic computation in the near term...For the long term, one can only speculate about the prospects for molecular computation. It seems likely that a single molecule of DNA can encode the "instantaneous description" of a Turing machine...One can imagine the eventual emergence of a general purpose computer consisting of nothing more than a single macromolecule conjugated to a ribosome-like collection of enzymes that act on it.*[40]

And not only can DNA function as a computer, but its potential to store digital information far outstrips that of our current electronic devices. It has been estimated that by 2020, thanks to the ease with which we can now photograph our meal in a restaurant and post it onto social media, the global digital archive had reached a stunning 44 trillion gigabytes of digital information.[41] This was a tenfold increase from 2013 and if this upward trend were to continue, we will hit a big problem: there simply won't be enough silicon available to make any more chips on which all this information can be stored.

DNA may well offer a solution. It is able to store digital information at much higher density than either hard drives or magnetic tape, with one calculation estimating that if the digital information that we have so far generated could be stored at the same density as the genes in the bacterium *E. coli*, a mere 1 kg of DNA would be sufficient to house our entire digital output.[42] In addition, DNA has a much longer lifespan as a digital storage medium, one often measured in hundreds of millennia rather than years or, at best, decades for hard drives and magnetic tapes.

---

[37] (Adleman 1994).

[38] (Adleman 1994).

[39] (Phillips 2024).

[40] (Adleman 1994); p. 1024.

[41] (Extance 2016).

[42] Ibid.

The first demonstration that media such as text and images could be digitised and stored in DNA sequences came in 1988 with a collaboration between artist Joe Davis and a team of researchers at Harvard University who encoded a simple image into 35 bits. This was then synthesized as a short DNA sequence and introduced into the genome of the common gut bacterium *E. coli*. In the years that followed, there was speculation about whether DNA could be a viable alternative to electronic media for the storage of digital information.[43] One major challenge was whether the method could be scaled up. If only a handful of bits could be stored, then DNA was going to be a very unlikely challenger to hard disks and magnetic tapes.

But in 2011, a team at the University of Harvard overcame this challenge. They were not only able to convert the draft of a book containing 53,426 words from html computer code into a DNA sequence, but also able to read the book back from the DNA.[44] A year later, a team of researchers at the European Bioinformatics Institute in Hinxton, UK and Agilent Technologies in Santa Clara, California converted 739 KB of files on a computer hard-drive which included all of William Shakespeare's 154 sonnets represented in ASCII format and Dr. Martin Luther King's 1963 'I have a dream' speech from an mp3 file into DNA code. This was then synthesized as a DNA molecule, sequenced, and the original information recovered. Looking to the future, they predicted that:

> *...our DNA-based storage scheme could be scaled far beyond current global information volumes and offers a realistic technology for large-scale, long-term and infrequently accessed digital archiving.*[45]

Another obstacle to using DNA as a digital storage medium was financial. Even though costs have come down, to construct a repository for the global digital archive by artificially synthesized sequences of A, C, G, and T nucleotides would still prove to be eye-wateringly expensive. But even this may change thanks to a recent development made by researchers at Peking University in Beijing. Computational synthetic biologist Long Qian and her colleagues have shown that there is an alternative method of storing digital information in a DNA sequence. This does not involve storing the information in the sequence of A, C, G and Ts, but instead by using the nucleotides to represent 1 s and 0 s of binary code, depending on whether they have a

---

[43] See for example, (Cox 2001).

[44] (Church et al. 2012).

[45] (Goldman et al. 2013).

methyl group attached to them or not. Using DNA to represent binary information in this way potentially offers a much cheaper way by which it could be used to store digital information. And as proof of this principle, Qian and her colleagues have encoded an ancient rubbing of a tiger from the Han Dynasty, as well as a picture of panda, as a series of almost 270,000 1 s and 0 s.[46]

Alongside its immense capacity to carry out computations and information storage, DNA has also proved itself to be an incredibly versatile construction material at the molecular level. The idea of what has become known as 'DNA origami' began back in 1982 when researcher Nadrian Seeman proposed that DNA might be capable of forming branched conformations that would allow the construction of 2D and 3D structures from smaller self-assembling DNA molecules.[47] In the decades that followed, Seeman proved his point by using this method to construct 2D layers and 3D cubes from DNA.[48] These achievements, combined with the discovery that strands of DNA could be folded into arbitrary 2D and 3D structures by a process known as 'DNA origami' opened up vast possibilities for DNA-based nanotechnology.

Applications of this technology include the use of DNA scaffolds to explore interactions between biomolecules in real time, thus shedding important new light on biochemical reactions; imaging of biomolecules; providing a scaffold upon which proteins can be easily crystallised for structural analysis or used for drug delivery. Researchers have found that an impressive array of mechanical structures can be built using DNA molecules such as rotary and gearing assemblies, all of which may allow the development of DNA-based nano-machinery.

But of all the developments at the interface of DNA and computing, the biggest looks set to be the impact of artificial intelligence to analyse vast arrays of genomic data and even design novel sequences with application in fields such as drug design.[49] Exploration of this rapidly growing area is unfortunately beyond the scope of this book, but we can say with some confidence that when William His predicted Miescher's discovery would bear fruit, he was spectacularly correct. But what of his prediction that the appreciation of

[46] (Ledford 2024).

[47] (Seeman 1982).

[48] (Seeman and Chen 1991); (Winfree et al. 1998); (Paukstelis et al. 2004).

[49] For a gripping account of how AI and synthetic genomics may shape our future see 'On the Future of Species: Authoring Life by Means of Artificial Biological Intelligence' by Adrian Woolfson which came out just as we were nearing the end of the production process for this book.

Miescher himself would only grow with time? Sadly, it seems that on this matter, his powers of prophesy were not quite so accurate.

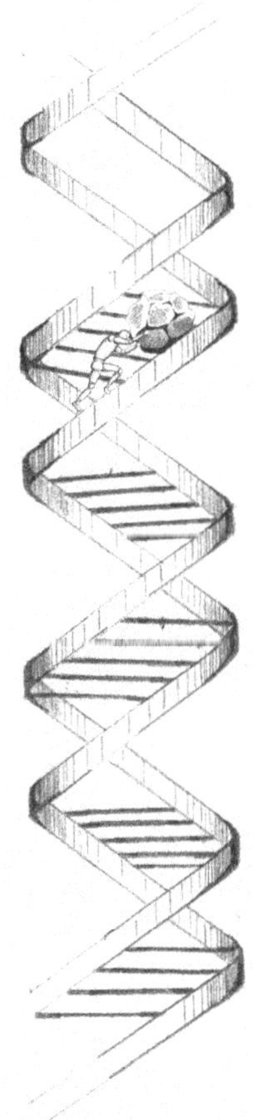

Drawing by Kersten Hall

13

The Struggle of Sisyphus

Contents

Abstract As the case of the 'CRISPR babies' shows, even scientists can be misled by oversimplifications of what genes are and how they work. But this was not just based on flawed science—it was also based on a flawed grasp of the history of science. In the popular imagination, the history of science is a story of mavericks, lone geniuses and Eureka moments. It was in this kind of hero narrative that He immersed himself, hoping perhaps to add his name to the canon. In this chapter we argue that the story of Friedrich Miescher offers an alternative—as well as a much needed corrective to such hero narratives with all their allure. Miescher's comparison of himself as Sisyphus is a far more honest and historically authentic account of how scientific knowledge evolves than that of the maverick hero. To this we can only add that the boulder is rolled steadily up the mountain slope not by one, but a multitude of Sisyphean figures. And as our power to transform and manipulate the material that Miescher discovered increases, this is surely worth remembering.

Keywords Science • History • Friedrich Miescher • DNA • Discovery • He Jiankui • Gene-editing • Controversy • Sisyphus • Mythology • Metaphor • Scientific progress • Incremental advances • Hard work • Disappointment • Persistence • Resilience • Courage • Albert Camus • Existentialism •

K. Hall, R. Dahm, *The Dawn Fisherman*, Copernicus Books,
https://doi.org/10.1007/978-3-032-14219-1_13

Philosophy • Meaning • Purpose • CRISPR • Insulin • Rosalind Franklin • James Watson • Francis Crick • Alexander Fleming • Biotechnology

Remembering Miescher – and why his story matters

Like most men in their middle-years, Nobel laureate Sydney Brenner found himself lamenting the state of the younger generation. In Brenner's case, his chief complaint was that PhD students and post-doctoral researchers had no real appreciation of the history behind their subject:

> *For most young molecular biologists, the history of their subject is divided into two epochs: the last two years and everything else before that. The present and the very recent past are perceived in sharp detail but the rest is swathed in a legendary mist where Crick, Watson, Mendel, Darwin—perhaps even Aristotle—coexist as uneasy contemporaries.*[1]

Miescher did not even make it onto Brenner's list, so the likelihood of a young molecular biologist who spends their days preparing and pipetting samples of nucleic acids (in between fretting over where their next round of research funding and job will come from), having ever heard of the man who first discovered these substances is even more remote.

But it was not always like this. In a paper given at Cold Spring Harbor in 1941, biochemist Jack Schultz credited Miescher not only with the discovery of what he referred to as 'thymonucleic acid' but for having also offered an invaluable insight into the material basis of heredity with his suggestion that isomerism might allow a molecule to represent biological traits.[2] Four years later, the British chemist John Mason Gulland acknowledged the importance of Miescher's insight that nucleic acids must have a large molecular weight, while at a Cold Spring Harbor meeting in 1947 British biochemists Edgar and Ellen Stedman presented a chemical analysis of cell nuclei in which they praised Miescher for having developed 'excellent methods' with which to isolate DNA from salmon sperm.[3] At the same meeting, Erwin Chargaff paid tribute to what he called 'a flash of insight rarely equalled in the history of the biological sciences' by which Miescher and Hoppe-Seyler had 'recognized the

[1] (Brenner 1985); p. 209.

[2] (Schultz 1941).

[3] (Gulland et al. 1945); p. 185.

importance of this group of substances almost immediately after their discovery in animal and plant cells.'[4]

But this praise for Miescher was by no means unanimous. Miescher had once described himself as being 'to the utmost, an adherent of a chemical model of inheritance' and in the eyes of some, this was a serious flaw.[5] For Alfred Mirsky (1900–1974) of the Rockefeller Institute, Miescher's insistence that only a chemical approach could solve the mysteries of the cell had been a crippling intellectual myopia that led him to ignore, and even dismiss, crucial developments in cell biology.[6] 'Miescher's desire to find explanations in physical science,' wrote Mirsky in 1968, 'did not lead him to an understanding of the biological role of nuclein.'[7] The charge of myopia was also levelled at Miescher by zoologist Ernst Mayr (1904–2005) who blamed him for having 'completely missed the significance of DNA, which he had discovered, by adopting a purely mechanical interpretation of the process of fertilization.'[8] Not that Mayr put the blame for this entirely on Miescher. 'It was one of the tragedies of biology and biochemistry,' he wrote, 'that [Wilhelm] His greatly influenced the thinking of his nephew, F. Miescher. This was in part responsible for the fact that Miescher completely missed the significance of his own discovery of nucleic acid.'[9] Mayr's assessment of Miescher reads rather like a school report written by a frustrated teacher whose student, despite clearly possessing academic talent, is consistently failing to use it:

> *It is rather sad to have to say that after his first brilliant success, Miescher's subsequent research career was an anticlimax. ... Even though it soon became obvious that the nuclein was nothing other than the chromatin of the cytologists, Miescher never looked at it as a carrier of genetic information. Instead of asking genetic questions, he asked physiological or purely chemical questions, such as: "Where does the body get all the phosphorus to synthesize the large quantities of nuciein during sperm formation?"*[10]

Mayr's charge that Miescher 'completely missed the significance of DNA' is a little unfair. It is all too easy to pass such a sweeping judgement from the vantage point of the present day and condemn Miescher for not having asked

---

[4] (Stedman and Stedman 1947); p. 225; (Chargaff 1947) p. 28.

[5] Letter LXXVIII 13th Oct 1893 His 1897 p. 122.

[6] (Mirsky 1968); p. 86.

[7] (Mirsky 1968); p. 84.

[8] (Mayr 1973); p. 132.

[9] (Mayr 1982); p. 662.

[10] (Mayr 1982); p. 810.

questions that to us may seem obvious. But the criticism of being unwilling to engage with the evidence offered by cell biology for the role of nuclein and chromatin in heredity seems to carry more weight. For Joseph Fruton (1912–2007) and Robert Olby (1933–2020), both of whom were scientists-turned-historians, Miescher's fixation on physical chemistry at the expense of cell biology left a legacy of confusion which severely delayed our understanding of how DNA functions in heredity.[11]

But how is Miescher remembered today? In tracing the historiography of Miescher, historian Sophie Veigl and her colleagues Oren Harman and Ehud Lamm argue that he has undergone a metamorphosis from being perceived as 'Miescher the contaminator' due to accusations by the likes of Kingzett and Hake in his own time, to becoming 'Miescher the confuser' thanks to the criticisms of those such as Fruton and Olby in the second half of the twentieth century. Veigl and her colleagues have also revealed how this transformation in the perception of Miescher has been shaped by tensions and seismic shifts in authority between disciplines such as cell biology and biochemistry, as well as by the academic interests of the historians who were writing about him.[12]

Whilst Sydney Brenner complained of the ignorance amongst younger scientists about the history of their subject, fellow middle-aged Nobel laureate, Walter Gilbert lamented another kind of generational forgetfulness:

> *As a successful science, molecular biology has become a set of cookbook techniques. Its very success is producing an odd sort of reaction: all of those wonderful techniques can simply be looked up in a handbook, and biologists seem to be spending their time reading techniques and then sequencing DNA. Where is the biology?'*[13]

Gilbert's complaint was that, although students and junior researchers might use a particular technique day in, day out, at the lab bench, they often did so without ever truly grasping the underlying scientific principles on which it is based. All that matters to them is that the method delivers a result. In 1980 Gilbert had himself shared the Nobel Prize in Chemistry with Fred Sanger for the invention of one such technique—a new and powerful method of DNA sequencing, but perhaps this gave his pronouncement even more authority.

Yet along with his despairing cry, 'Where is the biology?', Gilbert might equally have asked, 'Where is the history behind the biology?' In the same way that a research student might carry out the daily preparation of DNA from

---

[11] (Fruton 1999); p. 384; (Olby 2008); pp. 380–381.

[12] For a detailed and thorough discussion of the historiography of Miescher see (Veigl et al. 2020).

[13] (Gilbert 1992); p. 93.

bacterial cultures without feeling the need to grasp any of the underlying physics and chemistry behind each stage (What is the purpose of adding sodium hydroxide? What is the principle behind ultracentrifugation through a caesium chloride gradient? What substance is removed by the addition of phenol?), they probably also feel that there is equally no need to know anything about the history behind what they are doing. All that matters is that the protocol delivers clean DNA, from which data can be gleaned, research papers written, and grants won.

Historian Suzie Fisher has described this phenomenon by which a scientific technique or discovery becomes divorced, or uncoupled from its history, as 'black-boxing.' The term was first coined by the philosopher Bruno Latour (1947–2022) who borrowed it from the field of cybernetics where it describes a system about which no details other than the input and output need to be known. To show how a laboratory technique can become 'black-boxed', Fisher used the example of DNA-RNA hybridisation which was a method of using the formation of DNA-RNA hybrids through base-pairing, to detect the presence of specific DNA or RNA sequences in a sample. But although the technique was quickly adopted and widely used at the lab bench, the names of its inventors Ben Hall and Sol Spiegelman were soon forgotten.[14] As Fisher explained, once methods such as DNA-RNA hybridisation are adopted and their success established, they quickly become black-boxed—at which point 'only their input and output counts; their history or details of operation are no longer of interest to the scientists using them.'[15]

Something very similar seems to have happened to Miescher—but on a much grander scale. Eclipsed by the power and promise of the many developments and innovations that have come from his original discovery of nucleic acids, Friedrich Miescher has himself been 'black-boxed.'

Not every scientific discovery, however, is doomed by necessity to suffer this fate. Two exceptions in particular stand out, both of which have lessons to offer in understanding the fate of Miescher. The first is the discovery of the double-helical structure of DNA, which will forever be entwined with the names of James Watson and Francis Crick. The second is the discovery of insulin which, according to many textbooks (and popular UK TV quiz show 'University Challenge'), was made by Canadian scientists Frederick Banting and Charles Best. What then is the difference between Miescher and these two examples? What did Watson and Crick, or Banting and Best have, that Miescher did not?

---

[14] (Fisher 2015).

[15] Ibid. p. 49.

13.1 Discovery

One answer might be that the title of Miescher's 1869 paper 'On the Chemical Composition of Pus Cells' was hardly one to set readers pulses racing and have them on the edge of their seats.[16] Although it is certainly a factually correct summary of Miescher's research aims, it gives no indication that this work had resulted in what he considered to be a major discovery. Readers might therefore be forgiven for flicking past it. In contrast, consider the title of Watson and Crick's first paper, published in the journal *Nature* in April 1953—'Molecular structure of nucleic acids: A Structure for deoxyribose nucleic acid.'[17] Or two of the first papers published by Fred Banting and Charles Best - 'The Internal Secretion of the Pancreas', and later that same year, 'The Effect Produced on Diabetes by Extracts of Pancreas.' These titles leave readers in no doubt that they contain something worth their attention.[18]

Another reason might be that, as was pointed out by American physicist-turned philosopher Thomas Kuhn (1922–1996), the concept of discovery is not quite as simple as it might first appear. Kuhn described the popular view of scientific discovery—the sort that features in science textbooks or TV quiz show answers—as being 'a unitary event, one which, like seeing something, happens to an individual at a specifiable time and place.'[19] Understood in this way, discovery is simply a matter of mapping the name of a particular scientist onto a particular event. But as Kuhn pointed out, this is often a gross over-simplification:

> *Many scientific discoveries, particularly the most interesting and important, are not the sort of event about which the questions "Where?" and, more particularly, "When?" can appropriately be asked…Furthermore, within the rather vaguely delimited interval of internal history, there is no single moment or day which the historian, however complete his data, can identify as the point at which the discovery was made. Often, when several individuals are involved, it is even impossible unequivocally to identify any one of them as the discoverer.*[20]

It's a point that was made very eloquently by the Canadian biochemist James Collip who, with Banting, Best, and John Macleod was one of the team of

[16] (Dahm and Banerjee 2019).
[17] (Watson and Crick 1953a).
[18] (Banting and Best 1922); (Banting et al. 1922b).
[19] (Kuhn 1962); p. 760.
[20] Ibid, pp. 761–763.

Toronto scientists to develop the first medically useable insulin in 1922. Reflecting on his work many years later, Collip said:

> *Although I have been guilty of using the word "discovery" on occasion myself, generally speaking I do not like it. Most so called discoveries represent simply the last but important step in a long series of previous steps, representing contributions of many others through the years in the scientists [sic] search for truth.*[21]

The discovery of insulin happens to serve as another very good illustration of Kuhn's point—and one which offers some invaluable insights for reflecting on Miescher's place in history.[22] Until insulin was first isolated, the form of diabetes that we now recognise as type 1 was a death sentence. With a patient's body unable to produce to insulin, doctors could do little other than put them on a starvation diet to delay their inevitable death from the production of toxic ketones.

But in January 1922, this changed. Working in the laboratory of Professor John Macleod at the University of Toronto, Canadian researchers Fred Banting (1891–1941) and Charles Best (1899–1978) had spent the previous summer testing extracts made from the pancreatic tissue of cows for their anti-diabetic effects on dogs that had been rendered diabetic through surgical removal of their pancreas. The work was fraught with practical difficulties, not the least of which was Banting and Best's own inexperience, but by the end of that summer they had obtained some promising results in dogs. The next step was to test the anti-diabetic extract they had prepared in a human patient.

The opportunity to do this came quickly enough. When 13-year old Leonard Thompson had first been brought by his father to Toronto General Hospital, the boy had been at death's door due to diabetes.[23] His weight was down to 65 pounds, his hair was falling out, he was lethargic and stank of ketones. With these symptoms, progression of the disease was in one direction only.

---

[21] J.B. Collip to Mrs. M.D.Muttart, 13th August 1959. Western University, Records Collection, Collip correspondence, file 3. We would like to express our thanks to Dr. Alison Li for bringing this source to our attention.

[22] The discovery and development of CRISPR also offers another example of Kuhn's point. The more this is scrutinised, it appears not so much as a single event, but rather an unfolding process. It emerged from the observation of intriguing DNA sequences in the microbial genome; the recognition that these functioned as a primitive immune system; the discovery that this primitive immune system could be directed at specific genes of interest in bacteria using target RNA; and the discovery that this could be done not just in micro-organisms, but in eukaryotic cells.

[23] (Banting et al. 1922a); p.144.

On 11th January 1922, Leonard received an injection of the extract prepared by Banting and Best. A hundred years later, on this same date, much media coverage was devoted to how it was the centenary of Leonard Thompson having become the first diabetic patient to receive insulin. What was overlooked, however, was that this first injection given on 11th January had not been a success. Even Banting and Best admitted in one of their first scientific publications on the subject that 'no clinical benefit was evidenced.'[24] Although Leonard's blood sugar levels did come down, his body had continued to produce toxic ketones as well as suffering an adverse immune response due to impurities in the injected extract.

Two weeks later, on 23rd January, Leonard was injected again and this time, it was a success. His blood sugar levels fell, there was no production of ketones and, crucially he suffered no toxic side-effects due to impurities. So, what had changed in those 14 days since his first injection? The answer was that this second injection had been prepared not by Banting and Best, but by their colleague, James Collip (1892–1965), a biochemist with expertise in purification who had finessed a method of using alcohol to remove impurities in the pancreatic extract and isolate insulin as the active agent.

As a result in 2023, campaigners launched 'World Insulin Day' to be marked on 23rd January to raise awareness that the insulin which saved Leonard Thompson had been developed not just by Banting and Best, but also their Toronto colleagues James Collip and John Macleod. All of which raises a very important question—what exactly constitutes a discovery? In the case of insulin, at what point can we say that it was discovered? Was it in the summer of 1921 when Banting and Best found that pancreatic extracts could lower blood sugar levels in a diabetic dog? Or was it when James Collip found a means of isolating the active agent which enabled these extracts to successfully treat a human patient?

Or was it in 1908 when the German clinician Georg Zuelzer (1870–1949) claimed to have successfully treated patients in various Berlin hospitals with an extract made from pancreatic tissue on which he later successfully filed a patent? When news that the 1923 Nobel Prize for Physiology or Medicine had been awarded to Banting and his boss John J.R. MacLeod, Zuelzer protested to the Nobel Committee that it was he who should be recognised as the discoverer of insulin. Nor was he alone in his claim. Other contenders for the throne included Ernest Lyman Scott (1877–1966) at the University of Chicago, Israel Kleiner (1885–1966) of the Rockefeller Institute, John Murlin (1874–1960) of the University of Rochester, and Romanian scientist Nicolai

[24] (Banting et al. 1922a); p. 144.

Paulescu (1869–1931) (whose name would later be deservedly forever blighted for his anti-Semitic politics and role in the Holocaust in Romania) all of whom claimed to have demonstrated the antidiabetic properties of pancreatic extracts.

But if this is the case, why should it then be that when contestants on the popular weekly UK academic TV quiz show 'University Challenge' were asked to name the discoverers of insulin, the correct answer was returned as 'Fred Banting and Charles Best'? The other two members of the Toronto team who developed the first medically useful insulin, John JR Macleod and James Collip got no mention. And to include the names of Scott, Kleiner, Murlin, Paulescu, and Zuelzer in the answer would have required contestants to deliver a response more suited to an academic viva voce examination than a rapid-fire TV quiz show.[25]

The answer lies with the power of certain historical narratives not only to shape the way in which scientific research is understood and accepted, but also to exercise a grip on the public imagination that can be hard to dislodge. The protagonists of these stories have usually been inflated to larger-than-life proportions and perform heroic acts of intellectual defiance or courage. In the case of the discovery of the structure of DNA, James Watson's memoir 'The Double Helix' in which he portrayed himself and Crick as bold young mavericks who were opposed at every turn by the stubborn, dour anti-helical 'Rosy' (as he called Rosalind Franklin) became the established narrative for a long time.

The commonly accepted story of the discovery of insulin is another example of such a narrative and goes something like this: in the summer of 1921, Fred Banting and Charles Best, two young and inexperienced researchers not only triumphed against the odds to make a miraculous life-saving discovery but did so while their boss John Macleod was away on holiday in Europe. It ticks all the boxes for a Hollywood screenplay, but in the early 1980s, after studying Banting and Best's original lab notes however, the late Canadian historian Michael Bliss came to a very different conclusion. After exhaustive research with primary sources, Bliss found that the experiments performed by Banting and Best in the summer of 1921 were often inconsistent, unreliable, and poorly controlled. He showed that the orthodox account of the discovery

[25] Miescher would probably have been spinning in his grave had he known that, in another recent episode of 'University Challenge' (Southampton vs Trinity College, Cambridge), Rosalind Franklin was described as the scientist 'whose X-ray diffraction images provided key evidence for the discovery of the DNA molecule.' There is no denying that Franklin's X-ray data was crucial to the unravelling of the structure of DNA, but had Miescher not spent hours washing pus from bandages almost a century earlier, she would have had nothing to work with.

of insulin was a romanticised reconstruction—and, moreover, one which had been shaped largely by the efforts of Charles Best to diminish the vital contribution of his colleague James Collip.[26]

Until the publication of Michael Bliss's 1982 book 'The Discovery of Insulin', the popular romanticised account of the discovery of insulin held such a grip on the public imagination that the names of Banting and Best eclipsed those of the other two members of the Toronto team, John Macleod and James Collip.[27] But what of the other researchers such as Scott, Kleiner, Murlin, Paulescu and Zuelzer who all felt they too had a claim to the discovery of insulin? Why have they been forgotten?[28] In an interview given a few years before his death in 1978, Best offered an answer that was blunt, but revealing: 'None of them convinced the world of what they had. .. This is the most important thing in any discovery. You've got to convince the scientific world. And we did.'[29]

All these other researchers lacked two crucial factors, both of which were vital in helping the Toronto team convince not only the scientific world—as Best maintained was necessary—but also the wider world of the importance of their work. The first of these was industrial support (in the case of the Toronton team, from pharmaceutical company Eli Lilly, without which it is doubtful whether their promising results at the lab bench could ever have been translated into a bedside treatment for hospital patients. The other was newspaper headlines. When insulin was first used to treat patients in Toronto, headline writers were quick to proclaim it as a miracle cure for diabetes.

Clinicians at the time, however, knew that it was nothing of the sort. Insulin transformed what was an otherwise fatal condition into a long-term illness that could be managed, albeit with the potential for complications.[30] But this did not make for eye-catching triumphal headlines in the newspapers of the time.

The promise of miracle cures is the kind of heroic narrative in science that continues to exercise a powerful grip on the imagination of the public. Another is that of a frenzied race against cut-throat competition to win a glittering prize. The discovery of the double-helical structure of DNA is often

---

[26] (Bliss 1982); (Bliss 1989); (Bliss 1993).

[27] This is, thankfully, starting to change. On Sept 6th 2024 in John Macleod's home city of Aberdeen, a new memorial was unveiled which is the first in the world to bear the names of all four members of the Toronto team who developed the first medically useable insulin.

[28] It should be added that in the case of Paulescu, his role in the story of insulin has been overshadowed by his extremist anti-Semitic political writings and activities which played a part in inciting the Holocaust in Romania.

[29] Best, interviewed in (Stalvey 1971); p. 21.

[30] (Feudtner 2003).

portrayed in this way, with Watson and Crick sprinting towards the finish line as US chemist and Nobel laureate Linus Pauling from the California Institute of Technology comes breathing down their necks behind them.

Watson and Crick might well have genuinely believed themselves to be in a race with Pauling, but according to his biographer Thomas Hager they need not have worried. For Pauling faced three challenges which meant that for him at least, the so-called 'race for the double-helix' resembled not so much an Olympic 100 m sprint final, but rather a Sunday morning jog in the park with tight hamstrings.

The first of these problems was that his attention was not fully focussed on DNA, but on solving the structure of proteins; the second was that he lacked good quality X-ray pictures of DNA, and the third was simply his own pride. As Hager wrote, 'He [Pauling] simply did not feel that he needed to pursue DNA full tilt.'[31] When Pauling's wife asked him why, if the structure of DNA was so important, had he not worked harder to solve it, his reply was that he had always considered the problem to be his alone to solve and had therefore assumed that it would almost fall into his lap.[32]

Despite this, the perception that the discovery of the DNA structure was a breakneck race nevertheless still exerts a powerful grip on the popular consciousness. Perhaps the most powerful example of this is the scene in the BBC drama 'Life Story' in which Watson and Crick sprint through the streets of Cambridge to burst into 'The Eagle' pub and announce that they have found the secret of life.

These two examples of powerful historical narratives in science—one built around the discovery of what was hailed as a miracle cure, and the other of a frenzied race against cut-throat competition are instructive in considering how Miescher's work on nuclein has been remembered. In stark contrast to the heroic popular narratives around the discovery of insulin, and the double-helix, Miescher's discovery of DNA has no neat and dramatic narrative culminating in either the climactic discovery of a miracle cure or a breakneck last minute sprint to cross the finish line ahead of a competitor.[33] Instead, we have him shivering for long hours whilst diligently washing pus from discarded

[31] (Hager 1995); p. 415.

[32] (John L. Greenberg oral history interview with Linus Pauling, 10th May 1984, 23, Archives of the California Institute of Technology, Pasadena, CA, http://oralhistories.library.caltech.edu/18/1OH_Pauling.pdf); (Lake 2001); p. 558.

[33] Nor, for that matter, was the discovery of the double-helix quite the breakneck race that it is sometimes portrayed to be. With lack of access to good quality X-ray images of DNA such as 'Photo 51', and his attention divided between protein structure and that of DNA, Linus Pauling was never quite as serious a threat as Watson and Crick perceived him to be at the time.

bandages in the former kitchens of Tübingen castle before dying tragically young and crushed by a sense of failure.

As well as a gripping (albeit not entirely historically accurate) narrative of a 'winner-takes-all' race to solve the double-helical structure of DNA, Watson and Crick had another vital advantage that Miescher did not. So too, for that matter, did Banting and Best, and it was one which Charles Best had already identified in his answer to the question of why other researchers such as Zuelzer, were not remembered for their role in the discovery of insulin.

Best had recognised that, in an age of growing mass media influence, it was no longer enough for scientists to remained confined to the lab bench in the hope that their experimental results would speak for themselves. They needed to step out of the lab and persuade the world that what they were doing was important. In a world in which the mass media was becoming both ever more dominant and increasingly visual, a single image could have tremendous power.

For Banting and Best, the image which emblazoned itself into the public consciousness was a photograph of the two of them standing with one of their experimental dogs, on the roof of the Medical Building at the University of Toronto.[34] For Watson and Crick, meanwhile, the now iconic photograph of them in their laboratory gazing up reverentially at their cardboard and wire model of the double-helix is an example of such an image. The look of wonder on the young Watson's face seems to capture the spontaneous 'eureka' moment so beloved in the popular imagination, when a switch is flicked, and a light comes on.

It may come as no surprise to learn that, far from spontaneously recording the moment at which Watson and Crick glimpsed 'the secret of life', the photograph was taken several weeks later by the renowned British photographer Anthony Barrington Brown, to accompany an article that was going to appear in 'Time' magazine. But although his now iconic photograph is a retrospective reconstruction of a historical event—rather than an immediate record of it at the time it occurred, its power is nevertheless undeniable. In the public consciousness, the names of Watson and Crick would forever be interwoven with the strands of the double-helix thanks to this image.[35]

Again, Miescher has nothing to match the visual power of these photographs.[36] The few visual images to have any connection with him are hardly likely to grip an audience, as these amount only to a photograph of his rather

[34] This photo is held in the Insulin Papers, Thomas Fisher Rare Book Library, University of Toronto and can be see at: https://insulin.library.utoronto.ca/islandora/object/insulin%3AP10077

[35] (Dahm and Banerjee 2019).

[36] Ibid.

spartan kitchen laboratory in Tübingen castle, a test tube containing some salmon sperm DNA that is on display in the museum now housed in that former kitchen, and a portrait photograph of Miescher looking intense and brooding.

13.2 Cowboy Science

At the climax of the classic 1962 Western 'The Man Who Shot Liberty Valance' directed by John Ford and starring John Wayne and James Stewart, a newspaper editor is forced to choose between two competing accounts concerning the death of a ruthless outlaw several years earlier. According to one version, the outlaw was shot dead by James Stewart's character in an act of bold heroism that has since become the stuff of legend. The other version of events however is rather less heroic, more complex and very likely to tarnish the career of Stewart's character as a prominent politician. For the newspaper editor, the decision is a simple one. When faced with a choice between facts and legend he opts (spoiler alert!) to 'print the legend.'[37]

For a long time, there was a very similar attitude towards the portrayal of both the discovery of the double-helix by Watson and Crick, as well as that of insulin by Banting and Best. As we have seen however, scrutiny by historians has revealed both these episodes to have been far more complex than is popularly believed—and they turn out to be all the more interesting for it.

At which point the objection might be raised as to whether any of this really matters? Surely, it might be argued, the newspaper editor in 'The Man Who Shot Liberty Valance' had the right idea with his decision to 'print the legend.' What's so wrong after all with giving the public simple stories of scientific heroes who triumph against all the odds, and in so doing, offer inspiration?

By way of an answer, it may well be instructive to return to He Jiankui and his CRISPR-modified embryos discussed in the previous chapter. In an interview given in 2020, Dr. He recalled having experienced a moment of epiphany whilst attending a meeting of the Innovative Genomics Institute (IGI) at the University of California-Berkeley three years earlier. As one of the most junior scientists present, he had sat and listened quietly during the sessions. What had impressed him most of all was not so much the science being discussed, but an insight offered by one of the most senior scientists into the

[37] 'The Man Who Shot Liberty Valance' directed by John Ford (1962). John Ford Productions, Paramount Pictures.

nature of discovery. "[a very senior scientist] said 'many major breakthroughs are driven by one or a couple of scientists…cowboy science' recalled Dr. He, 'That strongly influenced me…you need a person to break the glass.'"[38]

Convinced that he was just such a person, He Jiankui turned to the history of science for inspiration. Or rather he turned to the popular version of the history of science, the one in which bold advances in knowledge are made thanks to lone geniuses and their Eureka moments. He immersed himself in stories of 'cowboy' scientists: figures who have since become elevated to the status of scientific legends on the pages of textbooks; mavericks and rugged individuals who were, according to popular accounts at least, not afraid to 'move fast and break things' long before Mark Zuckerberg espoused this philosophy.[39]

One such tale of scientific heroism in which He Jiankui found his inspiration was that of English clinician Edward Jenner (1749–1823). On 14th May 1796, Jenner made medical history after having noticed that milkmaids who had previously contracted cowpox, did not suffer the disfiguring scarring caused by smallpox infections. Reasoning that the cowpox had rendered them immune to smallpox, Jenner took some material from pustules on the hand of local milkmaid, Sarah Nelmes. He then introduced this material into eight-year old James Phipps by making incisions in the boy's arm. Six weeks later, when James was inoculated with a tiny dose of smallpox, he was found to be immune to the disease. Jenner had performed the very first successful vaccination—thanks to which the World Health Organisation was able to declare in 1980 that smallpox had been officially eradicated. Today the skin of the dead cow Blossom, from whom Sarah Nelmes is said to have contracted cowpox, hangs on the wall of the library at St. George's Medical School, London as a memorial to Jenner's achievement.

That at least, is the popular account widely found in textbooks. It is one in which Jenner is elevated to the status of legend and from which He Jiankui drew inspiration. But it is also one which, when subjected to scrutiny by historians, becomes somewhat more complex—and much more interesting. It has been claimed, for example, that the milkmaid element to the story was fictitious—conjured up by Jenner's friend and first biographer, John Baron to counteract criticism directed against Jenner by envious competitors and doctors who did not trust his method.[40] But perhaps more importantly, Jenner's

[38] (Hurlbutt 2020); p. 183.

[39] In 2014 Zuckerberg apparently adopted the more sobre philosophy of 'Move fast with a stable infrastructure.' (https://www.cnet.com/tech/mobile/zuckerberg-move-fast-and-break-things-isnt-how-we-operate-anymore/) Whether the same approach will be applied to CRISPR research remains to be seen.

[40] (Boylston 2013).

idea of inoculating James Phipps did not spring out of a vacuum as a 'eureka' moment, or a flash of insight by a lone genius.

The idea of inoculation had been given widespread publicity in Britain by Lady Mary Wortley Montagu (1689–1762), wife of the British consul in Constantinople, where she had observed how groups of Turkish women held 'smallpox' parties. Writing to a friend she described how, at these gatherings, 'the old woman comes with a nutshell full of the matter of the best sort of smallpox, and asks what veins you please to have open'd. She immediately rips open…and puts into the vein as much [smallpox] matter as can lie upon the head of her needle.'[41]

This form of inoculation was known as 'variolation' and its aim was to confer lasting protection against smallpox by introducing a very mild dose of the disease. On her return to Britain, Lady Mary had the physician Charles Maitland (1677–1748) perform a variolation procedure on her daughter, as did the Prince of Wales (later King George II (1683–1760) on his own two daughters. Jenner's use of cowpox was a marked improvement on this method because it used a pathogen that was benign in humans, but there has been debate as to whether he was the first person to try this. In 1774, 22 years before Jenner carried out his experiment, Benjamin Jesty (1736–1816), a dairy farmer who was familiar with local folklore that cowpox could provide immunity against smallpox had put this to the test on his own family.[42] It has also been argued that John Fewster (1738–1824), a Gloucestershire surgeon had also recognised the link between cowpox and smallpox immunity and, may well have planted the seeds of this idea in the young Jenner's mind thanks to a presentation at a local medical club of which they were both members.[43] Another source for Jenner's idea may also have been the French protestant pastor Jaques-Antoine Rabaut Pommier (1744–1820). Around 1780, Rabaut-Pommier is said to have observed that peasants in the Montepellier area where he lived, who had milked cows infected with cowpox appeared to be protected from smallpox. When Rabaut-Pommier discussed these observations with Dr. Pugh, a visiting English physician, the latter thought they might well be of interest to a friend back home—who happened to be, none other than Edward Jenner.[44]

There appear then to be at least two Edward Jenners. One is the lone, bold pioneer who defied the medical establishment and from whom He Jiankui

[41] Cited in (Porter 1997); p. 275.

[42] (Hammarsten et al. 1979).

[43] (Thurston and Williams 2015).

[44] (Theodorides 1979).

took his inspiration. The other is a more complex figure who emerges from careful historical analysis. The stark contrast between the two underlines both Thomas Kuhn's point that discovery is rarely a single event mapping neatly to one individual and the observation made by James Collip that 'most so called discoveries represent simply the last but important step in a long series of previous steps.'[45] Discoveries tend to be less a matter of a lone genius suddenly shouting 'Eureka' and more an unfolding process which emerges from a complex, tangled web woven by many different individuals and factors shaped by the intellectual, social, cultural and political landscape of the day.

In such grossly oversimplified narratives of scientific discovery the protagonists are inflated to such heroic proportions that they assume the status of demi-gods. It is tempting to dismiss this kind of narrative as mere myth, but this does a disservice to the word 'myth'. In its contemporary use, the word 'myth' is often synonymous either with an untruth or a story that is believed to be fact but has no literal or historical truth. We often hear a claim being dismissed with the words 'It's a myth that…', while the epithet of being 'legendary' meanwhile tends now to describe larger-than-life figures such as sporting heroes, reality TV stars, or pop musicians who can pack out a stadium. Science is no exception, for as we have seen, it too has its legendary figures who, on the pages of textbooks and in the popular consciousness, are elevated to the status of demi-gods who stride above the heads of mere mortals.

In its contemporary use therefore, the word 'myth' has become severely degraded. But in the pre-modern world, it had a very different meaning. A myth didn't need to be true in a literal or historic sense to have power. Rather, it was recognised as a shared story, a communal metaphor that expresses some universal and timeless aspect of what it means to be human. When Miescher compared himself with Sisyphus, he was echoing this older sense of the word, 'myth.' The fact that there never was any historical or literal truth to Sisyphus rolling the boulder up a hill in no way diminishes the power of the story. That's the whole point of myths. They don't have to be literally, or historically true to have meaning, or to tell us something important about what it is to be human and our relationship to the world.

Miescher's story may well lack the heroic grandeur found in popular accounts of the discovery of insulin or the double-helix, but it nevertheless contains elements of myth in the older, truest sense of this word. Many of the stories told by the ancient Greeks, for example, recognised that triumph and tragedy are often found in close association. Theseus is victorious in defeating the minotaur in the Cretan labyrinth but then through his own hubris on the

---

[45] (Collip to Muttart 1959).

voyage home brings about the suicide of his father, King Aegeus. In some versions of the Jason and the Argonauts myth, Jason returns in triumph from his voyage to find the golden fleece but his glory does not last. Abandoned by his wife Medea for his infidelity, Jason spends his last days alone, sitting beneath the rotting timbers of his once proud ship, the Argo, before they eventually collapse and he is crushed to death.

Miescher's story, although it is perhaps not quite so dramatic as that of Theseus or Jason, is nevertheless one in which triumph and tragedy were closely intertwined. It is the tale of a man who died burdened by a sense of failure that history would eventually prove to be thoroughly undeserved. Of a man whose success in discovering a novel cellular substance came with a hidden cost. Miescher's insight that nuclein must play some crucial role came hand in hand with an intellectual myopia that caused him to dismiss the cell biologists and their evidence that it played a crucial role in heredity. Equally tragic was that he had articulated one of the key principles by which a genetic molecule must work—that of structural variation—without ever realising that this principle applied to the very molecule he had found. And when this idea finally did become widely known, it was thanks to the work of Erwin Schrödinger, not Friedrich Miescher.

One of Miescher's few students expressed this sense of tragedy with his own memorable metaphor worthy of mythic status. He compared his former mentor to a ship carrying a priceless treasure which, just as it is approaching its home port, sinks with its precious cargo to the seabed.[46] In the century that followed Miescher's death, this sunken treasure has long since been salvaged and put on display for the world to see. But the story of Miescher himself, and how he found that treasure, has remained largely hidden—like the wreckage of the ship abandoned on the seabed long after its precious cargo has been recovered.

Telling the story of this vessel and the man who sailed it is as important to us as the more familiar story of the cargo it carried. Writing this book has been our attempt to raise that long abandoned wreck to the surface.[47] As for why

[46] (His 1897); p. 2.

[47] Just as we neared the end of writing this manuscript, we were delighted to discover that we had received a big helping hand in raising the profile of Miescher thanks to the TV dramatization of Bonnie Garmus' recent best-selling novel 'Lessons in Chemistry'. It tells the story of chemist Elizabeth Zott (played by Brie Larsen) who faces a culture of chauvinism and misogyny from her male colleagues in her university department, with the notable exception of newcomer Calvin Evans. When Evans joins forces with Elizabeth to develop her ideas about nucleic acid biosynthesis, he calls her one evening to discuss their work and she starts to tell him that she had just been jotting something down about Miescher and phosphorus. Unfortunately, she never gets to finish her sentence as, having been so distracted by Miescher and phosphorus, she has forgotten that she left the dinner in the oven—from which clouds of smoke are now billowing.

this should matter, Marshall Nirenberg offered an answer back in 1967, a year before he shared the 1968 Nobel Prize in Physiology or Medicine for his role in cracking the genetic code.

Looking to the future, Nirenberg foresaw that our increasing understanding of DNA would place a great moral responsibility on our shoulders. And he also recognised that if such power was to be exercised wisely, there was an urgent need for society to become better informed about how science works. The story of He Jiankui suggests however, that it is not just wider society, but also scientists themselves who sometimes need a more honest understanding of how science works.

It probably comes as no surprise that, as historians, we feel that study of the history of science has a crucial role to play here by challenging the simplistic narratives to which He Jiankui fell prey. It may well be a step too far to suggest that, had He Jiankui immersed himself in the story of Miescher instead of the popular account of Jenner, he might have behaved differently but Miescher's story nevertheless offers an important corrective to challenge the oversimplified and distorted view of science and its history that was to lead He Jiankui so badly astray with dreams of glory thanks to 'cowboy science'.

In undertaking the task of writing and researching this book we have, on occasion, felt some sympathy with Miescher when he compared himself to Sisyphus. It's a feeling that will no doubt also be painfully familiar to the inheritors of Miescher's legacy—all those countless graduate students and post-doctoral fellows who have found themselves sitting late at night in an otherwise deserted molecular biology lab, staring in despair at yet another experiment which did not give the result for which they had been hoping.

But as the British writer and actor Stephen Fry points out in 'Mythos', his own retelling of the Greek myths, the figure of Sisyphus need not necessarily be an image of fruitless toil and despair. It is true, says Fry, that artists and philosophers have interpreted the image of Sisyphus and his boulder as representing the absurdity of our existence in a universe that cares nothing for our struggles. But he goes on to say that Sisyphus has also been interpreted as an expression of our courage and resilience in undertaking those struggles in the first place.[48]

Another who saw hope in the image of Sisyphus was the French-Algerian writer, Albert Camus (1913–1960), whose talents and passions ranged from philosophy to goalkeeping. In his 1942 essay, 'The Myth of Sisyphus', Camus proposed that each and every one of us finds ourselves in a similar situation to Sisyphus. According to the existential philosophy of Camus, the universe and

[48] (Fry 2018); p. 282.

our place in it is devoid of any inherent meaning, purpose and value. In such a meaningless universe, it might seem at first sight as if we are all, like Sisyphus, endlessly pushing rocks up hills only to have them roll back down. And if this was so, then what on earth was the point of continuing to live?

But Camus used the myth of Sisyphus to show that there was an alternative, more hopeful response to our condition. By choosing to live, by choosing to carry on in the face of futility, by accepting the absurdity of endlessly rolling the rock up the slope, we imbue our lives with meaning and purpose. It is the very act of pushing against the boulder—not reaching the unattainable summit—that confers meaning upon our lives.

The myth of Sisyphus rolling his boulder up the mountain slope also offers a much better metaphor for how science progresses than stories of cowboy scientists who 'break the glass'. Much of science is hard toil; much of it is disappointment—as any graduate student or post-doctoral researcher will readily testify. Often, the rock slips backwards and rolls down the slope again. But unlike the boulder of Sisyphus, it never quite returns to where it started. In this way, the boulder makes its slow, steady ascent—not thanks to lone demigods and their glass-breaking genius, but instead the combined efforts of countless mere mortals—men and women many of whom remain anonymous (sadly, the latter often more so than the former) and all of whom are wracked with their own particular flaws and foibles. And although the summit will in all likelihood remain forever out of reach, this in no way diminishes the act of seeking to reach it, as Camus himself recognised:

> *'I leave Sisyphus at the foot of the mountain! One always finds one's burden again. But Sisyphus teaches the higher fidelity that negates the gods and raises rocks. He, too, concludes that all is well. The universe henceforth without a master seems to him neither sterile nor futile. Each atom of that stone, each mineral flake of that night-filled mountain, in itself forms a world. The struggle itself towards the heights is enough to fill a man's heart. One must imagine Sisyphus happy.'*[49]

There were many aspects of tragedy to Miescher's life but perhaps the greatest is that he was never able to read these words. Had he done so, they might well have brought him some comfort as his life ebbed away. He might have realised that the struggle of Sisyphus need not be one of despair but of a quiet, noble dignity. And had he only known just how many others would take on the task of slowly rolling the boulder ever upwards after his death, we might even be

[49] Camus, A. (1942); Penguin edition (1975); p. 111.

able to imagine him happy as he gazed up at the alpine mountains during his final days.

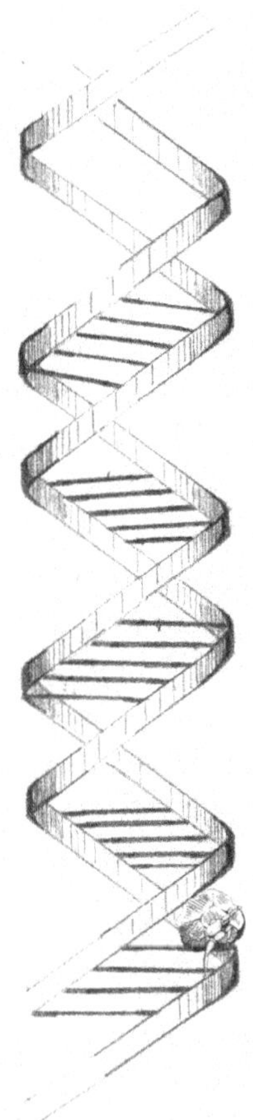

Drawing by Kersten Hall

Epilog

Having reached the end of this book, you might be forgiven for coming away with the impression that science is done almost exclusively by men—the majority of whom are white and western. Rather than being the result of a bias on our part in emphasizing their achievements over others, this is because the story we have told here reflects the makeup of the scientific community during much of the past 150 years (and to some extent still today). It may also reflect an inherent bias by the authors of some of the historical sources we consulted who might—even if only unconsciously—have ignored or downplayed the role of minorities and / or women in their time.

The good news however is that, although we cannot change the past, we can learn from it for the future. Science may well have been the preserve of mostly (white) men in the past but thankfully that is changing and science is becoming much more accessible. And when future historians write about our times, we hope that they will be able to paint a more colourful picture than the near-monochrome one that we could offer here.

K. Hall, R. Dahm, *The Dawn Fisherman*, Copernicus Books,
https://doi.org/10.1007/978-3-032-14219-1

References

Adleman, L. M. 1994. Molecular computation of solutions to combinatorial problems. *Science* 266:1021–1024.

Alexander, H. E., and G. Leidy. 1951a. Determination of inherited traits of H. influenzae by desoxyribonucleic acid fractions isolated from type-specific cells. *Journal of Experimental Medicine* 93:345–359.

Alexander, H. E., and G. Leidy. 1951b. Induction of heritable new type in type-specific strains of Hemophilus influenzae (19161). *Proceedings of the Society for Experimental Biology and Medicine* 78:625–626.

Alexander, H. E., and G. Leidy. 1953. Induction of streptomycin resistance in sensitive Hemophilus influenzae by extracts containing desoxyribosenucleic acid from resistant Hemophilus influenzae. *Journal of Experimental Medicine* 97:17–31.

Allbright, J. H. 1896. Protonuclein. *Journal of Nervous and Mental Disease* 21:431.

Altmann, R. 1889. Ueber Nucleinsäuren. *Archiv für Physiologie* 1:524–536.

Ames, D. ', and A. A. Huntley. 'The nature of the leucocytosis produced by nucleinic acid'. Journal of the American Medical Association 29 (1897): 472–478.

Asgari, S. 2017. RNA as a means of inter-species communication and manipulation: Progresses and shortfalls. *RNA Biology* 14 (4): 389–390. https://doi.org/10.1080/15476286.2017.1306172.

Astbury, W. T. 1955. In praise of wool. In *Proceedings of the International Wool Textiles Research Conference, B*, 220–243.

Astbury, W. T., and F. O. Bell. 1938a. Some recent developments in the X-ray study of proteins and related structures. *Cold-Spring Harbor Symposia on Quantitative Biology* 6:109–121.

Astbury, W. T., and F. O. Bell. 1938b. X-ray studies of thymonucleic acid. *Nature* 141:747–748.

K. Hall, R. Dahm, *The Dawn Fisherman*, Copernicus Books,
https://doi.org/10.1007/978-3-032-14219-1

Avery, O. T., C. MacLeod, and M. McCarty. 1944. Studies on the chemical nature of the substance inducing transformation of pneumococcal types: Induction of transformation by a desoxyribonucleic acid fraction isolated from pneumococcus type III. *Journal of Experimental Medicine* 79:137–158.

Baker, J. 1953. The cell theory: A restatement, history and critique: Part IV. The multiplication of cells. *Quarterly Journal of Microscopical Science* 94:407–440.

Bangham, D. H., and R. E. Franklin. 1946. Thermal expansion of coals and carbonised coals. *Transactions of the Faraday Society* 42:B289. https://doi.org/10.1039/tf946420b289.

Banting, F. G., and C. H. Best. 1922. The internal secretion of the pancreas. *Journal of Laboratory and Clinical Medicine* 7:251–266.

Banting, F. G., C. H. Best, J. B. Collip, W. R. Campbell, and A. A. Fletcher. 1922a. Pancreatic extracts in the treatment of diabetes mellitus. *Canadian Medical Association Journal* 2:141–146.

Banting, F. G., C. H. Best, J. B. Collip, W. R. Campbell, A. A. Fletcher, J. J. R. Macleod, and E. C. Noble. 1922b. The effect produced on diabetes by extracts of pancreas. *Transactions of the Association of American Physicians*1–11.

Bass, L. W. 1940. Phoebus Aaron Theodor Levene 1869-1940. *Science* 92:392–395.

Berg, P., D. Baltimore, H. Boyer, S. Cohen, R. Davis, D. Hogness, D. Nathans, et al. 1974. Potential biohazards of recombinant DNA molecules. *Science* 185:303.

Bernal, P. J. 2006. From "greasy chemistry" to "macromolecule": Thoughts on the historical development of the concept of a macromolecule. *The Journal of Chemical Education* 83:870–879.

Bigda, J. J., and P. Koszalka. 2013. Wacław Szybalski's contribution to immunotherapy: HGPRT mutation & HAT selection as first steps to gene therapy and hybrid techniques in mammalian cells. *Gene* 525:158–161.

Blaese, R. M., and K. W. Culver. 1995. T lymphocyte-directed gene therapy for ADA-SCID: Initial trial results after 4 years. *Science* 270:475–480.

Bliss, M. 1982. Texts and documents: Banting, Best's and Collip's accounts of the discovery of insulin' with an introduction by Michael Bliss. *Bulletin of the History of Medicine* 56:554–568.

Bliss, M. 1989. J.J.R. Macleod and the discovery of insulin. *Quarterly Journal of Experimental Physiology* 74:87–96.

Bliss, M. 1993. Rewriting medical history: Charles Best and the Banting and Best myth. *Journal of the History of Medicine and Allied Sciences* 48:253–274.

Boivin, A., R. Vendrely, and C. Vendrely. 1948. L'acide Désoxyribonucléique Du Noyau Cellulaire, Dépositaire Des Caractères Héréditaires; Arguments d'ordre Analytique. *Comptes Rendus De L'Académie Des Sciences* 226:1061.

Boylston, A. 2013. The origins of vaccination: Myths and reality. *Journal of the Royal Society of Medicine* 106:351–354.

Brachet, J. 1942. La Localisation Des Acides Pentose Nucleiques Dans Les Tissus Animaux et Les Oeufs d'amphibiens En Voie de Developpement. *Archives de Biologie* 53:207–257.

Breathnach, C. S. 2001. Johann Ludwig Wilhelm Thudichum 1829-1901, bane of the Protagonisers. *History of Psychiatry* 7:283–296.

Brenner, S. 1985. The rough and the smooth. *Nature* 317:209–210.

Brenner, S., F. Jacob, and M. Meselson. 1961. An unstable intermediate carrying information from genes to ribosomes for protein synthesis. *Nature* 190:576–581.

Burns, K. 2017. Transposable elements in human cancer. *Nature Reviews Cancer* 17:415–424.

Cabezas, J. A. 1994. The origins of glycobiology. *Biochemical Education* 22:3–7.

Caetano-Anollés, G., and M. J. Seufferheld. 2013. The coevolutionary roots of biochemistry and cellular organization challenge the RNA world paradigm. *Microbial Physiology* 23 (1–2): 152–177. https://doi.org/10.1159/000346551.

Callahan, M. P., K. E. Smith, H. J. Cleaves, J. Ruzicka, J. C. Stern, D. P. Glavin, C. H. House, and J. P. Dworkin. 2011. Carbonaceous meteorites contain a wide range of extraterrestrial nucleobases. *Proceedings of the National Academy of Sciences of the United States of America* 108 (34): 13995–13998. https://doi.org/10.1073/pnas.1106493108.

Callaway, E. 2018. Divided by DNA: The uneasy relationship between archaeology and ancient genomics. *Nature* 555 (7698): 573–576. https://doi.org/10.1038/d41586-018-03773-6.

Callaway, E., and K. Sanderson. 2024. Medicine Nobel awarded for gene-regulating "microRNAs". *Nature* 634:524–525. https://doi.org/10.1038/d41586-024-03212-9.

Camus, A. 1975. *The Myth of Sispyhus*. London: Penguin Books.

Carter, P. 1992. Humanization of an anti-P185her2 antibody for human cancer therapy. *Proceedings of the National Academy of Sciences of the United States of America* 89:4285–4289.

Caspersson, T., and J. Schultz. 1939. Pentose nucleotides in the cytoplasm of growing tissues. *Nature* 143 (3623): 602–603. https://doi.org/10.1038/143602c0.

Chargaff, E. 1947. On the nucleoproteins and nucleic acids of microorganisms. *Cold Spring Harbor Symposia on Quantitative Biology* 12:28–34.

Chargaff, E. 1950. Chemical specificity of nucleic acids and mechanism of their enzymatic degradation. *Experientia* 6:201–240.

Chargaff, E. 1963. *Essays on Nucleic Acids*. Amsterdam: Elsevier.

Chargaff, E. 1971. Preface to a grammar of biology. *Science* 172:637–642.

Chargaff, E. 1976a. The path to the double helix by Robert Olby (Review). *Perspectives in Biology and Medicine* 19:289–290.

Chargaff, E. 1976b. On the dangers of genetic meddling. *Science* 192:939–940.

Chargaff, E. 1978. *Heraclitean Fire: Sketches from a Life Before Nature*. New York, NY: The Rockefeller University Press.

Chen, X., and O. Rechavi. 2022. Plant and animal small RNA communications between cells and organisms. *Nature Reviews Molecular Cell Biology* 23 (3): 185–203. https://doi.org/10.1038/s41580-021-00425-y.

Chen, J. H., and N. C. Seeman. 1991. Synthesis from DNA of a molecule with the connectivity of a cube. *Nature* 350:631–633.

Chow, L., R. Gelinas, T. R. Broker, and R. J. Roberts. 1977. An amazing sequence arrangement at the 5′ ends of adenovirus 2 messenger RNA. *Cell* 12:1–8.

Church, G. M., Y. Gao, and S. Kosuri. 2012. Next-generation digital information storage in DNA. *Science* 337 (6102): 1628–1628. https://doi.org/10.1126/science.1226355.

Cipollaro, M., G. Di Bernardo, G. Galano, U. Galderisi, F. Guarino, F. Angelini, and A. Cascino. 1998. Ancient DNA in human bone remains from Pompeii archaeological site. *Biochemical and Biophysical Research Communications* 247 (3): 901–904. https://doi.org/10.1006/bbrc.1998.8881.

Claude, A. 1938. Concentration and purification of chicken tumor I agent. *Science* 87 (2264): 467–468. https://doi.org/10.1126/science.87.2264.467.

Claude, A. 1939. Chemical composition of the tumor-producing fraction of chicken tumor I. *Science* 90 (2331): 213–214. https://doi.org/10.1126/science.90.2331.213.

Claude, A. 1940. Particulate components of normal and tumor cells. *Science* 91 (2351): 77–78. https://doi.org/10.1126/science.91.2351.77.

Cobb, M. 2015a. *Life's Greatest Secret: The Race to Crack the Genetic Code*. London: Profile Books.

Cobb, M. 2015b. Who discovered messenger RNA? *Current Biology* 25:R523–R548.

Cobb, M. 2022. *The Genetic Age: Our Perilous Quest to Edit Life*. London: Profile Books.

Cobb, M., and N. Comfort. 2023. What Watson and Crick really took from Franklin. *Nature* 616:657–660.

Cogoni, C., and G. Macino. 1997. Conservation of transgene-induced post-transcriptional gene silencing in plants and fungi. *Trends in Plant Science* 2 (11): 438–443. https://doi.org/10.1016/S1360-1385(97)90028-5.

Cohen, J. S., and F. H. Portugal. 1977. *A century of DNA: A history of the discovery of the structure and function of the genetic substance*. Cambridge, MA: MIT Press.

Consden, R., A. H. Gordon, and A. J. P. Martin. 1944. Qualitative analysis of proteins: A partition chromatographic method using paper. *The Biochemical Journal* 38:224–232.

Cornelis, G., C. Vernochet, and S. Malicorne. 2014. Retroviral envelope Syncytin capture in an ancestrally diverged mammalian clade for placentation in the primitive Afrotherian tenrecs. *Proceedings of the National Academy of Sciences of the United States of America* 111:E4332–E4341.

Cox, J. P. L. 2001. Long-term data storage in DNA. *Trends in Biotechnology* 19 (7): 247–250. https://doi.org/10.1016/S0167-7799(01)01671-7.

Crescitelli, F. 1977. Friedrich Wilhelm Kühne 1837-1900. *Vision Research* 17:1317–1323.

Crick, F. H. C. 1958. On protein synthesis. *Symposia of the Society for Experimental Biology* 12:138–163.

Crick, F. H. C. 1965. Recent research in molecular Biology: Introduction. *British Medical Bulletin* 21:183–186.

Crick, F. H. C. 1979. Split genes and RNA splicing. *Science* 204:264–271.

Crick, F. H. C., and L. E. Orgel. 1973. Directed panspermia. *Icarus* 19:341–346.

Crick, F. H. C., and J. D. Watson. 1954. The complementary structure of deoxyribonucleic acid. *Proceedings of the Royal Society of London* 223:80–96.

Cyranoski, D. 2018. CRISPR-baby scientist fails to satisfy his critics. *Nature* 564:13–14.

Cyranoski, D., and H. Ledford. 2018. International outcry over genome-edited baby claim. *Nature* 563:607–608.

D'Hont, A. 2012. The Banana (Musa Acuminata) genome and the evolution of monocotyledonous plants. *Nature* 488:214–217.

Dahm, R. 2005. Friedrich Miescher and the discovery of DNA. *Developmental Biology* 278:274–288. https://doi.org/10.1016/j.ydbio.2004.11.028.

Dahm, R. 2008a. Discovering DNA: Friedrich Miescher and the early years of nucleic acid research. *Human Genetics* 122:565–581.

Dahm, R. 2008b. The discovery of DNA, circa 1869. *The Scientist.*

Dahm, R. 2008c. The first discovery of DNA. *American Scientist* 96:320–327.

Dahm, R. 2010a. *Der Vergessene Entdecker Der DNA*. Heidelberg: Spektrum der Wissenschaft.

Dahm, R. 2010b. A slip in the date of DNA's discovery. *Nature* 468:897.

Dahm, R. 2010c. From discovering to understanding. *EMBO Reports* 11 (3): 153–160.

Dahm, R., and M. Banerjee. 2019. How we forgot who discovered DNA: Why it matters how you communicate your results. *BioEssays* 41:1–3. https://doi.org/10.1002/bies.201900029.

Davies, K. 2020. *Editing humanity: The CRISPR revolution and the new era of genome editing*. New York, London: Pegasus Books.

de Koning, A. P. J., W. Gu, T. A. Castoe, B. A. Batzer, and D. D. Pollock. 2011. Repetitive elements may comprise over two-thirds of the human genome. *PLoS Genetics* 7:e1002384. https://doi.org/10.1371/journal.pgen.1002384.

Delbrück, M., and G. Stent. 1956. On the mechanism of DNA replication. In *McCollum-Pratt symposium on the chemical basis of heredity*, ed. William D. McElroy and Bentley Glass, 699–736. Baltimore: Johns Hopkins University Press.

Devlin, H. 2013. My sister, her DNA breakthrough, and the Nobel prize that never. *The Times (London).*

Di Bernardo, G., S. Del Gaudio, U. Galderisi, A. Cascino, and M. Cipollaro. 2009. Ancient DNA and family relationships in a Pompeian house. *Annals of Human Genetics* 73 (4): 429–437. https://doi.org/10.1111/j.1469-1809.2009.00520.x.

Diakonow, C. 1867. Ueber Die Phosphorhaltigen Körper Der Hühner- Und Störeier. *Medicinisch-Chemische Untersuchungen Aus Dem Laboratorium Für Angewandte Chemie Zu Tübingen* 2:221–227.

Diakonow, C. 1868. Das lecithin Im Gehirn. *Centralblatt Für Medzinische Wissenschaft* 7:97–99.

Dobni, C. B. 2008. The DNA of innovation. *Journal of Business Strategy* 29 (2): 43–50. https://doi.org/10.1108/02756660810858143.

Dolgin, E. 2020. Could tracking RNA in body fluids reveal disease? *Nature* 582 (7812): S2–S4. https://doi.org/10.1038/d41586-020-01763-1.

Donovan, B. M. 2022. Ending genetic essentialism through genetic education. *Human Genetics and Genomics Advances* 3:1–13.

Doudna, J., and S. Sternberg. 2017. *A crack in creation: The new power to control evolution*. New York, NY: Vintage (Penguin Random House).

Dougherty, M. J. 2009. Closing the gap: Inverting the genetics curriculum to ensure an informed public. *The American Journal of Human Genetics* 85:6–12.

Drabkin, D. 1958. *Thudichum: Chemist of the brain*. Philadelphia, PA: University of Pennsylvania Press.

Drews, G. 1999. Ferdinand Cohn: A founder of modern Biology. *ASM News* 65:547–553.

Eddy, S. 2012. The C-value paradox, junk DNA and ENCODE. *Current Biology* 22:R898–R899.

Ehret, C. F., and G. De Haller. 1963. Origin, development, and maturation of organelles and organelle Systems of the Cell Surface in paramecium. *Journal of Ultrastructure Studies* 9:3–42.

Eichwald, E. 1865. Beiträge Zur Chemie Der Gewebbildenen Substanzen Und Ihre Abkömmlinge: Ueber Das Mucin Besonders Der Weinbergschnecke. *Justus Liebigs Annalen Der Chemie* 134:177–211.

Elbashir, S. M., J. Harborth, W. Lendeckel, A. Yalcin, K. Weber, and T. Tuschl. 2001. Duplexes of 21-nucleotide RNAs mediate RNA interference in cultured mammalian cells. *Nature* 411 (6836): 494–498. https://doi.org/10.1038/35078107.

Extance, A. 2016. How DNA could store all the world's data. *Nature* 537 (7618): 22–24. https://doi.org/10.1038/537022a.

Fedoroff, N. 2012. Transposable elements, epigenetics, and genome evolution. *Science* 338:758–767.

Feudtner, C. 2003. *Bittersweet: Diabetes, insulin and the transformation of illness*. Chapel Hill, NC: University of North Carolina Press.

Fire, A., S. Xu, M. K. Montgomery, S. A. Kostas, S. E. Driver, and C. C. Mello. 1998. Potent and specific genetic interference by double-stranded RNA in Caenorhabditis elegans. *Nature* 391 (6669): 806–811. https://doi.org/10.1038/35888.

Fischer, E. 1914. Über Phosphosäureester des Methlyglucosids und Theophyllin-Glucosids. *Berichte Den Deutschen Chemisches Gesellschaft* 47:3193–3205.

Fischer, E. 1917. Isomerie Der Polypeptide. *Zeitschrift für Physiologische Chemie* 99:54–66.

Fisher, S. 2015. Not just "a clever way to detect whether DNA really made RNA": The invention of DNA-RNA hybridization and its outcome. *Studies in History and Philosophy of Biological and Biomedical Sciences* 53:40–52.

Flemming, W. 1880. Beiträge Zur Kenntniss Der Zelle Und Ihrer Lebenserscheinungen. *Archiv für Mikroskopische Anatomie* 18:151–259.

Flemming, W. 1882. *Zellsubstanz, Kern Und Zelltheilung*. Leipzig: F. C. W. Vogel.

Fraenkel-Conrat, H., B. Singer, and R. C. Williams. 1957. Infectivity of viral nucleic acid. *Biochimica et Biophysica Acta* 25:87–96. https://doi.org/10.1016/0006-3002(57)90422-5.

Framm, E., and J. Framm. 2022. Albrecht Kossel und die Nukleinbasen: Ein Nobelpreisträger aus Mecklenburg. *Chemie in Unserer Zeit* 56 (6): 372–377. https://doi.org/10.1002/ciuz.202010003.

Framm, E., and J. Framm. Albrecht Kossel Und Die DNA: Ein Nobelpreisträger Aus Mecklenburg, 2024.

Franklin, R. E. 1950a. The interpretation of diffuse X-ray diagrams of carbon. *Acta Crystallographica* 3 (2): 107–121. https://doi.org/10.1107/S0365110X50000264.

Franklin, R. E. 1950b. Influence of the bonding electrons on the scattering of X-rays by carbon. *Nature* 165 (4185): 71–72. https://doi.org/10.1038/165071a0.

Franklin, R. E. 1951. The structure of graphitic carbons. *Acta Crystallographica* 4 (3): 253–261. https://doi.org/10.1107/S0365110X51000842.

Fruton, J. Contrasts in scientific style: Research groups in the chemical and biochemical sciences. American Philosophical Society, Philadelphia, PA 1990.

Fruton, J. 1999. *Proteins, enzymes, genes: The interplay of chemistry and biology*. New Haven, CT: Yale University Press.

Fruton, J. 2002. A history of pepsin and related enzymes. *The Quarterly Review of Biology* 77:127–147.

Fry, S. 2018. *Mythos*.

Gamgee, A. 1895. The late professor Hoppe-Seyler I. *Nature* 52:575–576.

Gamow, G. 1954. Possible relation between deoxyribonucleic acid and protein structures. *Nature* 173:318.

Gamow, G. 1955. Information transfer in the living Cell. *Scientific American* 193:70–78.

Garman, E. F. 2020. Rosalind Franklin 1920–1958. *Acta Crystallographica Section D, Structural Biology* 76 (7): 698–701. https://doi.org/10.1107/S2059798320008827.

Gasiunas, G., R. Barrangou, P. Horvath, and V. Siksnys. 2012. Cas9-crRNA ribonucleoprotein complex mediates specific DNA cleavage for adaptive immunity in bacteria. *Proceedings of the National Academy of Sciences of the United States of America* 109:15539–15540.

Gierer, A., and G. Schramm. 1956. Infectivity of ribonucleic acid from tobacco mosaic virus. *Nature* 177 (4511): 702–703. https://doi.org/10.1038/177702a0.

Gilbert, W. 1992. A vision of the grail. In *The Code of Codes*. Cambridge, MA: Harvard University Press.

Glass, B. 1965. A century of biochemical genetics. *Proceedings of the American Philosophical Society* 109:227–236.

Glynn, J. 2012. *My Sister Rosalind Franklin*. Oxford: Oxford University Press.

Goldman, N., P. Bertone, S. Chen, C. Dessimiz, E. LeProust, B. Sipos, and E. Birney. 2013. Towards practical, high-capacity, low-maintenance information storage in synthesized DNA. *Nature* 494:77–80.

Graur, D., Y. Zheng, N. Price, R. B. Azevedo, R. A. Zufall, and E. Elhaik. 2013. On the immortality of television sets: "Function" in the human genome according to the evolution-free gospel of ENCODE. *Genome Biology and Evolution* 5:578–590. https://doi.org/10.1093/gbe/evt028.

Gregory, T. R. 2003. Variation across amphibian species in the size of the nuclear genome supports a pluralistic, hierarchical approach to the C-value enigma. *Biological Journal of the Linnean Society* 79:329–339.

Gregory, T. R. 2005. Synergy between sequence and size in large-scale genomics. *Nature Reviews Genetics* 6:699–708.

Gros, F., H. Hiatt, W. Gilbert, C. G. Kurland, R. W. Riseborough, and J. D. Watson. 1961. Unstable ribonucleic acid revealed by pulse labelling of Escherichia coli. *Nature* 190:581–585.

Gulland, J. M., G. R. Barker, and D. O. Jordan. 1945. The chemistry of the nucleic acids and nucleoproteins. *Annual Review of Biochemistry* 14:175–206.

Haeckel, E. 1866. *Generelle Morphologie Der Organismen*. Berlin: Reimer.

Haeckel, E. 1868. Monographie Der Monoren. *Jenaische Zeitschrift für Medizin und Naturwissenschaft* 4:64–137.

Hager, T. 1995. *Force of Nature: The life of Linus Pauling*. New York: Simon and Schuster.

Hall, K. T., and N. Sankaran. 2021. DNA translated: Friedrich Miescher's discovery of Nuclein in it's original context. *The British Journal for the History of Science* 54:1–9. https://doi.org/10.1017/S000708742000062X.

Hamilton, A. J., and D. C. Baulcombe. 1999. A species of small antisense RNA in posttranscriptional gene silencing in plants. *Science* 286 (5441): 950–952. https://doi.org/10.1126/science.286.5441.950.

Hammarsten, J. F., W. Tattersall, and J. E. Hammarsten. 1979. Who discovered smallpox vaccination? Edward Jenner or Benjamin Jesty? *Transactions of the American Clinical and Climatological Association* 90:44–55.

Harris, H. 1999. *The birth of the cell*. New Haven, CT: Yale University Press.

Harris, P. J. F., and I. Suarez-Martinez. 2021. Rosalind Franklin, Carbon Scientist. *Carbon* 171:289–293. https://doi.org/10.1016/j.carbon.2020.09.022.

Heine, S. J. 2017. *DNA is not destiny: The remarkable, completely misunderstood relationship between you and your genes*. New York: W. W. Norton.

Hershey, A. D., and M. Chase. 1952. Independent functions of viral protein and nucleic acid in growth of bacteriophage. *Journal of General Physiology* 36:39–56.

Hertwig, O. 1885. Das Probleme Der Befruchtung Und Der Isotropie Des Eies, Eine Theorie Der Vererbung. *Jenaische Zeitschrift Für Medizin Und Naturwissenschaft* 18:276–319.

Hertwig, O., and R. Hertwig. Untersuchungen Zur Morphologie Und Physiologie Der Zelle, 1884.

His, W. 1897. *Die Histochemischen Und Physiologischen Arbeiten von Friedrich Miescher*. Leipzig: F. C. W. Vogel.

Hoagland, M. B., P. C. Zamecnik, and M. L. Stephenson. 1957. Intermediate reactions in protein biosynthesis. *Biochimica Et Biophysica Acta* 24:215–216. https://doi.org/10.1016/0006-3002(57)90175-0.

Hoagland, M. B., M. L. Stephenson, J. F. Scott, L. I. Hecht, and P. C. Zamecnik. 1958a. A soluble ribonucleic acid intermediate in protein synthesis. *Journal of Biological Chemistry* 213:241–257.

Hoagland, M. B., M. L. Stephenson, J. F. Scott, L. I. Hecht, and P. C. Zamecnik. 1958b. A soluble ribonucleic acid intermediate in protein synthesis. *Journal of Biological Chemistry* 231 (1): 241–257. https://doi.org/10.1016/S0021-9258(19)77302-5.

Hoerr, I., R. Obst, H.-G. Rammensee, and G. Jung. 2000. In vivo application of RNA leads to induction of specific cytotoxic T lymphocytes and antibodies. *European Journal of Immunology* 30:1–7.

Hopkin, M. 2006. RNAi scoops medical Nobel. *Nature*news061002-2. https://doi.org/10.1038/news061002-2.

Hoppe-Seyler, F. 1857. Ueber Die Einwirkung Des Kohlenoxydgases Auf Das Hämatoglobulin. *Archiv für Pathologische Anatomie und Physiologie und für Klinische Medicin* 11:288–289.

Hoppe-Seyler, F. 1866a *Medicinisch-Chemische Untersuchungen: Aus Dem Laboratorium Für Angewandte Chemie Zu Tübingen.*

Hoppe-Seyler, F. 1866b. Ueber Das Vitellin, Ichthin Und Ihre Beziehung Zu Den Eiweissstoffen. *Medicinisch-Chemische Untersuchungen Aus Dem Laboratorium Für Angewandte Chemie Zu Tübingen* 1:215–220.

Hoppe-Seyler, F. 1866c. Ueber Das Vorkommen von Cholesterin Und Protagon Und Ihre Betheiligung Bei Der Bildung Des Stroma Des Rothen Blutkörperchen. *Medicinisch-Chemische Untersuchungen Aus Dem Laboratorium Für Angewandte Chemie Zu Tübingen* 1:140–150.

Hoppe-Seyler, F. 1868. Physiologische Chemie. *Jahresbericht über die Leistungen und Fortschritte in der gesamten Medizin* 1:67–101.

Hoppe-Seyler, F. 1871. Ueber die Chemische Zusammensetzung des Eiters' [on the Chemical composition of pus]. *Medicinisch-Chemische Untersuchungen* 4:486–501.

Hoppe-Seyler, F. 1876. Ueber Die Processe Der Gährungen Und Ihre Beziehungen Zum Leben Der Organismen. *E Pflüger Archiv Für Physiologie* 12:1–17.

Hotchkiss, R. D. 1995. DNA in the decade before the double helix. *Annals of the New York Academy of Sciences* 758 (1): 55–73. https://doi.org/10.1111/j.1749-6632.1995.tb24809.x.

Hunter, G. K. 1999. Phoebus Levene and the Tetranucleotide structure of nucleic acids. *Ambix* 46:73–103.

Hurlbutt, B. J. 2020. Imperatives of governance: Human genome editing and the problem of progress. *Perspectives in Biology and Medicine* 63:177–194.

Ishino, Y., H. Shinagawa, K. Makino, M. Amemura, and A. Nakata. 1987. Nucleotide sequence of the Iap gene, responsible for alkaline phosphatase isozyme conversion in Escherichia coli, and identification of the gene product. *Journal of Bacteriology* 169:5429–5433.

Jackson, A. C. 1895. The Protonucleins. *Journal of the American Medical Association* 24:872.

Jansen, R., J. D. Embden, W. Gaastra, and L. W. Schouls. 2002. Identification of genes that are associated with DNA repeats in prokaryotes. *Molecular Microbiology* 43:1565–1575.

Jinek, M., K. Chylinski, I. Fonfara, M. Hauer, J. A. Doudna, and E. Charpentier. 2012. A programmable dual-DNA-guided DNA endonuclease. *Science* 337:816–821.

Jordan, C. V. n.d. 50th anniversary of the first clinical trial with ICI 46474 (tamoxifen): Then what happened? *Endocrine-Related Cancer* 28:R11–R30.

Judson, H. F. 1996. *The eighth day of creation*. Cold Spring Harbor, NY: Cold Spring Harbor Press.

Judson, H. F. 2003. No Nobel prize for whining. *The New York Times*. Accessed 20 October 2003.

Kampourakis, K. 2024. *How we get Mendel wrong, and why it matters: Challenging the narrative of Mendelian genetics*. Boca Raton, London, New York: CRC Press Taylor & Francis Group.

Kang, X., W. He, Y. Huang, Q. Yu, Y. Chen, X. Gao, X. Sun, and Y. Fan. 2016. Introducing precise genetic modifications into human 3PN embryos by CRISPR/Cas-mediated genome editing. *Journal of Assisted Reproduction and Genetics* 33:581–588.

Kay, L. E. 2000. *Who wrote the book of life?* Stanford, CA: Stanford University Press.

Kingzett, C. T. 1878. *Animal chemistry, or the relations of chemistry to physiology and pathology*. London: Longman and Green.

Kingzett, C. T., and H. W. Hake. 1877. Physiology and its chemistry at home and abroad. *Quarterly Journal of Science* 14:91–109.

Klose, A. 2007. Victor von Bruns Und Die Sterile Verbandwatte. In *Hin Und Weg: Tübingen in Aller Welt. Tübinger Katalogue 77*, 35–45. Tübingen: Tübingen Stadtmuseum.

Klug, A. 1968. Rosalind Franklin and the discovery of the structure of DNA. *Nature* 219 (5156): 808–810. https://doi.org/10.1038/219808a0.

Klug, A. 1974. Rosalind Franklin and the double helix. *Nature* 248:787–788.

Knip, M., M. E. Constantin, and H. Thordal-Christensen. 2014. Trans-kingdom cross-talk: Small RNAs on the move. *PLoS Genetics* 10 (9): e1004602. https://doi.org/10.1371/journal.pgen.1004602.

Köhler, R. E., Jr. 1973. The enzyme theory and the origin of biochemistry. *Isis* 64:181–196.

Kölliker, A. 1885. Die Bedeutung Der Zellenkerne Für Die Vorgänge Der Vererbung. *Zeitschrift Für Wissenschaftliche Zoologie* 42:1–46.

Koonin, E. V., and M. Krupovic. 2015. Evolution of adaptive immunity from transposable elements combined with innate immune systems. *Nature Reviews Genetics* 16:184–195.

Kossel, A. 1879. Ueber Das Nuclein Der Hefe. *Zeitschrift Für Physiologische Chemie* 3:284–291.

Kossel, A. 1882. Zur Chemie Des Zellkerns. *Zeitschrift Für Physiologische Chemie* 7:7–22.

Kossel, A. 1883. Ueber Guanin. *Zeitschrift Für Physiologische Chemie* 8:404–410.

Kossel, A. 1911. The chemical composition of the cell. *Harvey Lectures Series* 7:33–51.

Kossel, A., and A. Neumann. 1894. Darstellung Und Spaltungsprodukte Der Nucleïnsäure (Adenylsäure). *Berichte der Deutschen Botanischen Gesellschaft* 27:2215–2222. https://doi.org/10.1002/cber.189402702206.

Kruger, K., P. J. Grabowski, A. J. Zaug, J. Sands, D. E. Gottschling, and T. R. Cech. 1982. Self-splicing RNA: Autoexcision and autocyclization of the ribosomal RNA intervening sequence of Tetrahymena. *Cell* 31 (1): 147–157. https://doi.org/10.1016/0092-8674(82)90414-7.

Kuhn, T. S. 1962. Historical structure of scientific discovery. *Science* 136:760–764.

Kühne, W. 1868. *Lehrbuch Der Physiologischen Chemie.* https://archive.org/details/bub_gb_DxNu12EEKh4C

Kühne, W. 1877. Nachtrag Zur Geschichte Des Trypsins. *Untersuchungen aus dem Physiologischen Institut der Universität Heidelberg* 1:325–326.

Kühne, W. 1878. Erfahrungen Und Bemerkungen Über Enzyme Und Fermente. *Untersuchungen an Dem Physiologischen Institut Der Universität Heidelberg* 1:291–324.

Kühne, W. 1882. Bemerkungen Zu Herrn Hoppe-Seyler's Darstellung Der Optochemie. *Untersuchungen Aus Dem Physiologischen Institut Der Universität Heidelberg* 2:488–492.

Kupferschmidt, K. 2023. Shadowed by past, gene-editing summit looks to future. *Science* 379:1073–1074.

Lake, J. 2001. Why Pauling didn't solve the structure of DNA. *Nature* 409 (6820): 558–558. https://doi.org/10.1038/35054717.

Leathes, J. B. 1926. Function and design. *Science* 64:387–394.

Lederberg, J. 2000. The dawning of molecular genetics. *Trends in Microbiology* 8:194–195.

Ledford, H. 2024. DNA stores data in bits after epigenetic upgrade. *Nature* 634:1029–1030. https://doi.org/10.1038/d41586-024-03443-w.

Lee, R. C., R. L. Feinbaum, and V. Ambros. 1993. The C. elegans heterochronic gene Lin-4 encodes small RNAs with antisense complementarity to Lin-14. *Cell* 75 (5): 843–854. https://doi.org/10.1016/0092-8674(93)90529-Y.

Leidy, G., E. Hahn, and H. E. Alexander. 1953. In vitro production of new types of Hemophilus influenzae. *Journal of Experimental Medicine* 97:467–482.

Lenharo, M. 2024. First DNA from Pompeii body casts illuminates who victims were. *Nature* 635:534. https://doi.org/10.1038/d41586-024-03576-y.

Levene, P. A. 1899. On the nucleoprotid of the brain (cerebronucleoprotid). *Archives of Neurology and Psychopathology* 2:3–14.
Levene, P. A. 1903. On the chemistry of the chromatin substance of the nerve-cell. *Journal of Medical Research* 10:204–211.
Levene, P. A. 1909. Über Die Hefenucleinsäure. *Biochemische Zeitschrift* 17:120–131.
Levene, P. A. 1910. On the biochemistry of nucleic acids. *Journal of the American Chemical Society* 32:231–240.
Levene, P. A. 1917. The chemical individuality of tissue elements and its biological significance. *Journal of the American Chemical Society* 39:828–837.
Levene, P. A. 1919. The structure of yeast nucleic acid IV: Ammonia hydrolysis. *Journal of Biological Chemistry* 40:415–424.
Levene, P. A. 1921. Preparation and analysis of animal nucleic acid. *Journal of Biological Chemistry* 48:177–184.
Levene, P. A., and L. W. Bass. 1931. *Nucleic acids*. New York: Chemical Catalog Company.
Levene, P. A., and W. A. Jacobs. 1909. Ueber Die Hefe-Nucleinsäuren. *Berichte Der Deutschen Chemische Gesellschaft* 42:2474–2478.
Levene, P. A., and W. A. Jacobs. 1912. On the structure of thymus nucleic acid. *Journal of Biological Chemistry* 12:411–420.
Levene, P. A., and J. A. Mandel. 1908. Über Die Konstitution Der Thymonucleinsäure. *Berichte der Deutschen Chemischen Gesellschaft* 41:1905–1908.
Levene, P. A., and R. S. Tipson. 1932. The ring structure of adenosine. *Journal of Biological Chemistry* 94:809–819.
Levene, P. A., and R. S. Tipson. 1935. The ring structure of thymidine. *Journal of Biological Chemistry* 109:623–630.
Liang, P., Y. Xu, X. Zhang, C. Ding, R. Huang, Z. Zhang, J. Lv, et al. 2015. CRISPR/Cas9-mediated gene editing in human tripronuclear zygotes. *Protein and Cell* 6:363–372.
Librado, P. 2021. The origins and spread of domestic horses from the Western Eurasian steppes. *Nature* 598:632–640.
Lister, J. 1867. On the antiseptic principle in the practice of surgery. *The Lancet* 90:353–356.
Liu, L., and X. Chen. 2018. Intercellular and systemic trafficking of RNAs in plants. *Nature Plants* 4 (11): 869–878. https://doi.org/10.1038/s41477-018-0288-5.
Lorenzano, P. 2011. What would have happened if Darwin had known Mendel (or Mendel's work)? *History and Philosophy of the Life Sciences* 33:3–48.
Macleod, J. J. R. 1899. Zur Kenntnis Des Phosphor Im Muskel. *Zeitschrift Für Physiologische Chemie* 28:535–558.
Maddox, B. 2002. *Rosalind Franklin: The dark lady of DNA*. New York: Harper Collins.
Manton, I. 1945. Comments on chromosome structure. *Nature* 155:471–473.
Markel, H. 2021. *The secret of life: Rosalind Franklin, James Watson, Francis Crick and the Discovery of the Double Helix*. New York: W. W. Norton.

Marraffini, L. A., and E. J. Sontheimer. 2008. CRISPR interference limits horizontal gene transfer in staphylococci by targeting DNA. *Science* 322:1843–1845.

Marshall, E. 1995. Gene therapy's growing pains. *Science* 269:1050–1055.

Marx, V. 2021. The CRISPR children. *Nature Biotechnology* 39 (12): 1486–1490. https://doi.org/10.1038/s41587-021-01138-5.

Mathews, P. 1927. Professor Albrecht Kossel. *Science* 66.

Mayr, E. n.d. *The growth of biological thought: Diversity, evolution, and inheritance*. Cambridge, MA: Belknap Press of Harvard University Press.

Mazzarello, P. 1999. A unifying concept: The history of Cell theory. *Nature Cell Biology* 1:E13–E15.

McCaskey, G. W. 1906. The role of Nuclein in the animal economy. *Journal of the American Medical Association* 47:1800–1804.

McClintock, C. T. 1895. The disease resisting powers of the body: A review of the foundations of Nuclein and serum therapy. *Journal of the American Medical Association* 24:535–543.

Mendel, G. J. 2020. *Experiments on plant hybrids: Versuche Über Pflanzen-Hybriden: New translation by Staffan Müller-Wille & Kersten Hall with commentary*. Masaryk: Masaryk University Press.

Merke, F. 1973. Bemerkungen Zu Vergessen, Grundlegenden Physiologischen Untersuchungen Am Wandernden Lachs Durch Den Basler Physiologen Friedrich Miescher. *Gesnerus* 30:47–52.

Messing, J., A. K. Bharti, W. M. Karlowski, and R. A. Wing. 2004. Sequence composition and genome organization of maize. *Proceedings of the National Academy of Sciences of the United States of America* 101:14349–14354.

Mielewczik, M., D. P. Francis, B. Studer, M. V. Simunek, and U. Hossfelder. 2017. Die Rezeption von Gregor Mendels Hybridisierungsversuchen Im 19. Jahrhundert—Eine Bio-Bibliographische Studie. In *150 Jahre Mendelsche Regeln: Vom Erbsenzählen Zum Gen-Editieren. Nova Acta Leopoldina*, vol. 143, 83–134. Stuttgart: Wissenschaftliche Verlagsgesellschaft.

Miescher, F. 1871. Ueber Die Chemische Zusammensetzung Der Eiterzellen. *Medicinisch-Chemische Untersuchungen Aus Dem Laboratorium Für Angewandte Chemie Zu Tübingen* 4:441–461.

Miescher, F. 1874. Die Spermatozoen Einiger Wirbeltiere. Ein Beitrag Zur Histochemie. *Verhandlungen Der Naturforschenden Gesellschaft in Basel* 6:138–208.

Miescher, F. 1880. Statistische Und Biologische Beiträge Zur Kenntnis Vom Leben Des Rheinlachses Im Süsswasser. In *Abdruck Aus Dem Wissenschaftlichen Anhang Zum Katalog Der Schweizerischen Betheiligung an Der Internationalen Fischereiausstellung in Berlin*.

Miescher, F., and O. Schmiedeberg. 1896. Physiologische-Chemisch Untersuchungen Über Die Lachsmilch. *Naunyn-Schmiedeberg's Archives of Pharmacology* 37:100–155.

Mirsky, A. 1968. The discovery of DNA. *Scientific American* 218:78–90.

Mizuno, S., and H. C. Macgregor. 1974. Chromosomes, DNA sequences, and evolution in Salamaders of the genus Plethodon. *Chromosoma* 48:239096.

Mojica, M. J., G. Juez, and F. Rodriguez-Valera. 1992. Transcription at different salinities of Haloferax Mediterranei sequences adjacent to partially modified PstI sites. *Molecular Microbiology* 9:613–621.

Mojica, F. J., C. Diez-Villasenor, J. Garcia-Martinez, and E. Soria. 2005. Intervening sequences of regularly spaced prokaryotic repeats derive from foreign genetic elements. *Journal of Molecular Evolution* 60:174–182.

Napoli, C., C. Lemieux, and R. Jorgensen. 1990. Introduction of a chimeric Chalcone synthase gene into petunia results in reversible co-suppression of homologous genes in trans. *The Plant Cell* 2 (4): 279. https://doi.org/10.2307/3869076.

Nelkin, D., and M. S. Lindee. 1995. *The DNA mystique: The gene as a cultural icon.* New York: W. H. Freeman.

O'Connor, R. E. 2018. Reconstruction of the Diapsid ancestral genome permits chromosome evolution tracing in avian and non-avian dinosaurs. *Nature Communications* 9. https://doi.org/10.1038/s41467-018-04267-9.

Ohno, S. 1972. So much "junk" in our genome. *Brookhaven Symposium in Biology* 23:366–370.

Olby, R. 1979. Mendel—No Mendelian? *History of Science* 17:53–72.

Olby, R. 1994. *The path to the double helix: The discovery of DNA.* Garden City, NY: Dover Publications.

Olby, R. 2009. *Francis Crick: Hunter of life's secrets.* Cold Spring Harbor Press, NY: Cold Spring Harbor Press.

Osborne, T. B., and I. F. Harris. 1902. Die Nucleinsäure Des Weizenembryos. *Hoppe-Seylers Zeitschrift Für Physiologische Chemie* 36 (2–3): 85–133.

Palazzo, A. . F., and T. R. Gregory. 2014. The case for junk DNA. *PLoS Genetics* 10.

Parke, J. L. 1866. Ueber Die Chemische Constitution Des Eidotters. *Medicinisch-Chemische Untersuchungen Aus Dem Laboratorium Für Angewandte Chemie Zu Tübingen* 1:209–214.

Pasquinelli, A. E., B. J. Reinhart, F. Slack, M. Q. Martindale, M. I. Kuroda, B. Maller, D. C. Hayward, et al. 2000. Conservation of the sequence and temporal expression of Let-7 heterochronic regulatory RNA. *Nature* 408 (6808): 86–89. https://doi.org/10.1038/35040556.

Paukstelis, P. J., J. Nowakowski, J. J. Birktoft, and N. C. Seeman. 2004. Crystal structure of a continuous three-dimensional DNA lattice. *Chemical Biology* 11:1119–1126.

Pauling, L. 1987. Schrödinger's contribution to chemistry and Biology. In *Schrodinger: Centenary Celebration of a Polymath.* Cambridge: Cambridge University Press.

Pennisi, E. 2012. ENCODE project writes eulogy for junk DNA. *Science* 337:1159–1161.

Perutz, M. 1987. Erwin Schrödinger's what is life? And molecular biology. In *Schrodinger: Centenary Celebration of a Polymath.* Cambridge: Cambridge University Press.

Pflüger, E. 1877. Die Physiologie Und Ihre Zukunft. *Deutsche Medizinische Wochenschrift*597–598.

Pflüger, E. 1878. Die Physiologie Und Ihre Zukunft. *Archiv Für Physiologie* 15:361–365.

Phillips, A. 2024. Self-assembling DNA recognises patterns. *Nature* 625:454.

Pilli, E., S. Vai, V. C. Moses, S. Morelli, M. Lari, A. Modi, M. A. Diroma, et al. 2024. Ancient DNA challenges prevailing interpretations of the Pompeii plaster casts. *Current Biology* 34:5307. https://doi.org/10.1016/j.cub.2024.10.007.

Pollock, M. R. 1970. The discovery of DNA: An ironic tale of chance, prejudice and insight third Griffith memorial lecture. *Journal of General Microbiology* 63 (1): 1–20. https://doi.org/10.1099/00221287-63-1-1.

Porter, R. 1997. *The greatest benefit to mankind: A medical history of humanity from antiquity to the present.* New York: Harper Collins.

Portugal, F. H. 2010. Oswald T. Avery: Nobel laureate or Noble luminary? *Perspectives in Biology and Medicine* 53:558–570.

Portugal, F. H. 2015. *The least likely man: Marshall Nirenberg and the discovery of the genetic code.* Cambridge, MA: MIT Press.

Radick, G. M. 2023. *Disputed inheritance: The battle over Mendel and the future of biology.* Chicago, IL: University of Chicago Press.

Rasmussen, N. 2014. *Gene jockeys: Life science and the rise of biotech Enterprise.* Baltimore, MD: Johns Hopkins University Press.

Robertson, M. P., and G. F. Joyce. 2012. The origins of the RNA world. *Cold Spring Harbor Perspectives in Biology* 4 (5): a003608. https://doi.org/10.1101/cshperspect.a003608.

Romano, N., and G. Macino. 1992. Quelling: Transient inactivation of gene expression in *Neurospora Crassa* by transformation with homologous sequences. *Molecular Microbiology* 6 (22): 3343–3353. https://doi.org/10.1111/j.1365-2958.1992.tb02202.x.

Rutherford, A. 2016. *A brief history of everyone who ever lived.* London: Weidenfeld & Nicolson.

Saltzwedel, G. 1977. Victor von Bruns (1812-1883): Leben Und Werk. *Beiträge zur Geschichte der Eberhard-Karls Universität Tübingen* 13:1–214.

Sanger, F. 1988. Sequences, sequences, sequences. *Annual Review of Biochemistry* 57:1–29.

Schmucker, D., J. C. Clemens, H. Shu, C. A. Worby, J. Xiao, M. Muda, J. E. Dixon, and S. L. Zipursky. 2000. Drosophila Dscam is an axon guidance receptor exhibiting extraordinary molecular diversity. *Cell* 101:671–684.

Schnable, P. S., et al. 2009. The B73 maize genome: Complexity, diversity, and dynamics. *Science* 326:1112–1115.

Schrödinger, E. 1944. *What is life?* Cambridge: Cambridge University Press.

Schultz, J. 1941. The evidence for the nucleoprotein nature of the gene. *Cold Spring Harbor Symposia on Quantitative Biology* 9:55–65.

Schwann, T. 1911. Mikroskopische Untersuchungen Über Die Uebereinstimmung in Der Struktur Und Dem Wachstum Der Thiere Und Pflanzen. *Nature* 86:41.

Seeman, N. C. 1982. Nucleic acid junctions and lattices. *Journal of Theoretical Biology* 99:237–247.

Seeman, N. C., and H. F. Sleiman. 2017. DNA nanotechnology. *Nature Reviews Materials* 3. https://doi.org/10.1038/natrevmats.2017.68.

Sheth, J. N. 2011. The double helix of marketing: The complementary relationship between marketing history and marketing theory. *Marketing Theory* 11 (4): 503–505. https://doi.org/10.1177/1470593111418805.

Signer, R. 1969. Die Entwicklung Der Erforschung Der Nukleinsäuren. *Bulletin Der Schweizerischen Akademie Der Medizinischen Wissenschaften*25–31.

Silver, G. A. 1987. Virchow, the heroic model in medicine: Health policy by accolade. *American Journal of Public Health* 77:82–88.

Simcoe, M. 2021. Genome-wide association study in almost 195,000 individuals identifies 50 previously unidentified genetic loci for eye color. *Science Advances* 7.

Stalvey, R. M. 1971. A chat with Dr. Charles Best. *Nutrition Today* 6:5–7.

Stedman, E., and E. Stedman. 1947. The chemical nature and functions of the components of cell nuclei. *Cold Spring Harbor Symposia on Quantitative Biology* 12:224–236.

Stent, G. S. 1968. That was the molecular biology that was. *Science* 160:390–395.

Strasburger, E. 1909. The minute structure of cells in relation to heredity. In *Darwin and modern science*, ed. A. C. Seward, 102–111. Cambridge: Cambridge University Press.

Summers, O. T. 1895. Leucocytes and Nucleins. *Journal of the American Medical Association* 24:963–966.

Suter, F. 1944. Prof. F. Miescher. Persönlichkeit Und Lehrer. *Helvetica Physiologica et Pharmacologia Acta Supplementa* 2:5–17.

Sutton, W. 1902. On the morphology of the chromosome Group in Brachystola Magna. *Biological Bulletin* 4:24–39.

Sutton, W. 1903. The chromosomes in heredity. *Biological Bulletin* 4:231–251.

Symonds, N. 1986. What is life? Schrödinger's influence on Biology. *The Quarterly Review of Biology* 61:221–226.

Szybalska, E. H., and W. Szybalski. 1962. Genetics of human cell lines, IV. DNA-mediated heritable transformation of a biochemical trait. *Proceedings of the National Academy of Sciences* 48:2026–2034.

Theodorides, J. 1979. Rabaut-Pommier, A neglected precursor of Jenner. *Medical History* 23:479–480.

Theß, A., and R. Dahm. 2019. F . Mieschers Entdeckung der DNA—Unbekannter Meilenstein der Biochemie. *BIOspektrum*793–794. https://doi.org/10.1007/s12268-019-1311-8.

Thess, A., I. Hoerr, B. Yazdan Panah, G. Jung, and R. Dahm. 2021. Historic nucleic acids isolated by Friedrich Miescher contain RNA besides DNA. *Biological Chemistry* 402:1179–1185.

Thieffry, D., and R. M. Burian. 1996. Jean Brachet's alternative scheme for protein synthesis. *Trends in Biochemical Sciences* 21 (3): 114–117. https://doi.org/10.1016/S0968-0004(96)10015-3.

Thompson, T. 2021. Ancient DNA points to origins of modern domestic horses. *Nature* 598:550.

Thudichum, J. L. W. 1867. *Dr. Thudichum on researches intended to promote an improved treatment of disease*, 152–294. Report of Chief Medical Officer of Privy Council Appendix 7.

Thudichum, J. L. W. 1874. *Researches on the Chemical constituents of the brain*, 113–247. Reports of the Chief Medical Officer of the Privy Council.

Thudichum, J. L. W. 1879a. Note and experiments on the alleged existence in the brain of a body termed "Protagon". *Annals of Chemical Medicine—Including the Application of Chemistry to Physiology, Pathology, Therapeutics, Pharmacy, Toxicology and Hygiene* 1:254–263.

Thudichum, J. L. W. 1879b. Preface. *Annals of Chemical Medicine—Including the application of Chemistry to Physiology, Pathology, Therapeutics, Pharmacy, Toxicology and Hygiene* 1:v–x.

Thudichum, J. L. W. 1881. On modern text-books as impediments to the progress of animal chemistry. (A deduction). *Annals of Chemical Medicine* 2:183–189.

Thudichum, J. L. W. 1884. *Physiological chemistry of the brain*. London: Bailliere, Tindall and Cox.

Thurston, L., and G. Williams. 2015. An examination of John Fewster's role in the discovery of smallpox vaccination. *Journal of the Royal College of Physicians of Edinburgh* 45:173–179.

Timofeef-Ressovsky, N., K. G. Zimmer, and M. Delbruck. 1935. The nature of genetic mutations and the structure of the gene. *Nachrichten Aus Der Biologie* 1:189–245.

Veigl, S. J., O. Harman, and E. Lamm. 2020. Friedrich Miescher's discovery in the historiography of genetics: From contamination to confusion, from Nuclein to DNA. *Journal of the History of Biology* 53:451–484. https://doi.org/10.1007/s10739-020-09608-3.

Vischer, E., S. Zamenhof, and E. Chargaff. 1949. Microbial nucleic acids: The Desoxypentose nucleic acids of avian tubercle bacilli and yeast. *Journal of Biological Chemistry* 177:429–438.

von Jaksch, R. 1876. Ueber Das Vorkommen von Nuclein Im Menschengehirn. *Pflüger's Archive: European Journal of Physiology* 13:469–473.

von Liebig, J. F. 1842. *Animal chemistry: Or, organic chemistry in its applications to physiology and pathology*. Luton: Taylor & Walton.

Waldeyer, W. 1888. Ueber Karyokinese Und Ihre Beziehungen Zu Den Befruchtungsvorgängen. *Archiv für Mikroskopische Anatomie* 32:1–122.

Waller, J. 2002. *Fabulous science: Fact and fiction in the history of scientific discovery*. Oxford: Oxford University Press.

Watson, J. D. 1968. *The double-helix*. London: Weidenfeld & Nicolson.

Watson, J. D. 2001. *Genes, Girls and Gamow*. Oxford: Oxford University Press.

Watson, J. D., and F. H. C. Crick. 1953a. Genetical implications of the structure of deoxyribonucleic Acid. *Nature* 171:964–967.

Watson, J. D., and F. H. C. Crick. 1953b. Molecular structure of nucleic acids: A structure for deoxyribose nucleic acid. *Nature* 171:737–738.

Watson, J. D., and J. Tooze. 1981. *The DNA story: A documentary history of gene cloning*. New York: W. H. Freeman.

Watt, J. D., and R. E. Franklin. 1957. Changes in the structure of carbon during oxidation. *Nature* 180 (4596): 1190–1191. https://doi.org/10.1038/1801190a0.

Weiner, D. B. 1968. *Francois-Vincent Raspail, 1794–1878. Scientist and Reformer*. New York: Columbia University Press.

Weismann, A. 1883. *Über Die Vererbung*. Jena: Fischer.

Weismann, A. 1891. *Essays upon heredity and kindred biological problems*. Oxford: Clarendon Press.

Welch, G. R. 1995. T.H. Huxley and the protoplasmic theory of life: 100 years later. *Trends in Biochemical Sciences* 20:481–485.

Wightman, B., I. Ha, and G. Ruvkun. 1993. Posttranscriptional regulation of the heterochronic gene Lin-14 by Lin-4 mediates temporal pattern formation in C. elegans. *Cell* 75 (5): 855–862. https://doi.org/10.1016/0092-8674(93)90530-4.

Wilkins, M. 2003. *The third man of the double helix*. Oxford: Oxford University Press.

Williams, N. 2016. Irene Manton, Erwin Schrödinger and the puzzle of chromosome structure. *Journal of the History of Biology* 49:425–459.

Williams, G. 2019. *Unravelling the double helix: The lost heroes of DNA*. London: Weidenfeld and Nicholson.

Wilson, E. B. 1896. *The cell in development and heredity*. 1st ed. London: Macmillan.

Wilson, E. B. 1911. *The cell in development and heredity*. 2nd ed. London: Macmillan.

Wilson, E. B. 1925. *The cell in development and heredity*. 3rd ed. London: Macmillan.

Winfree, E., F. Liu, L. A. Wenzler, and N. C. Seeman. 1998. Design and self-assembly of two-dimensional DNA crystals. *Nature* 394:539–544.

Witkowski, J. A. 1986. Schrödinger's "what is life?": Entropy, order and hereditary code-scripts. *Trends in Biochemical Sciences* 11:266–268.

Witze, A. 2025. Asteroid fragments upend theory of how life on earth bloomed. *Nature*. https://doi.org/10.1038/d41586-025-00264-3.

Wolpert, L. 1996. The evolution of "the Cell theory". *Current Biology* 6:225–228.

Wong, C. 2023. UK first to approve CRISPR treatment for diseases: What you need to know. *Nature* 623 (7988): 676–677. https://doi.org/10.1038/d41586-023-03590-6.

Worm-Müller, J. 1874. Zur Kentniss Der Nuclein. *Pflügers Archiv: European Journal of Physiology* 8:190–194.

Wrinch, D. M. 1934. Chromosome behaviour in terms of Protein pattern. *Nature* 134:978–979.

Young-Simpson, J. 1867. Carbolic acid and its compounds in surgery. *The Lancet* 90 (November): 546–549.

Zacharias, E. 1881a. Ueber Die Chemische Beschaffenheit Des Zellkerns. *Botanische Zeitung* 39:170–177.

Zacharias, E. 1881b. Ueber Die Spermatozoiden. *Botanische Zeitung* 39:827–840.

Zacharias, E. 1882a. Ueber Den Zellkern. *Botanische Zeitung* 40:628–650.

Zacharias, E. 1882b. Ueber Den Zellkern. *Botanische Zeitung* 40:612–615.

Zacharias, E. 1888a. Ueber Kern- Und Zelltheilung. *Botanische Zeitung* 46:51–65.

Zacharias, E. 1888b. Ueber Kern- Und Zelltheilung. *Botanische Zeitung* 46:438–451.

Zacharias, E. 1888c. Ueber Strasburger's Schrift "Kern- Und Zelltheilung Im Pflanzenreiche" Jena 188'. *Botanische Zeitung* 46:438–451.

Zacharias, E. 1888d. Ueber Strasburger's Schrift "Kern- Und Zelltheilung Im Pflanzenreiche" Jena 188'. *Botanische Zeitung* 46:35–40.

Zagorski, N. 2006. Profile of Alec J. Jeffreys. *Proceedings of the National Academy of Sciences of the United States of America* 103:8918–8920.

Zamecnik, P. C., and E. B. Keller. 1954. Relation between phosphate energy donors and incorporation of labeled amino acids into proteins. *Journal of Biological Chemistry* 209 (1): 337–354. https://doi.org/10.1016/S0021-9258(18)65561-9.

Zhang, X. 2015. Urgency to Rein in the gene-editing technology. *Protein and Cell* 6:313.

Zipursky, S. L., W. M. Wojtowicz, and D. Hattori. 2006. Got diversity? Wiring the Fly brain with Dscam. *Trends in Biochemical Sciences* 31:581–588.

Further Reading

For reasons of brevity we have, unfortunately been able to offer only a brief introduction to several of the subjects and ideas discussed in later chapters of this book. For readers who are keen to know more, the following titles are a good place to explore these themes in more depth—but we would also strongly recommend 'The DNA Papers'—a series of podcasts presented by the Consortium for the History of Science, Technology and Medicine (CHSTM), in which a round table of guests discuss landmark historical papers and characters in the story of DNA. Miescher is the subject of the very first of these episodes, which can be found at the website of the CHSTM – https://www.chstm.org/video/144#21741

Cobb, M. 2015. *Life's greatest secret: The race to crack the genetic code*. London: Profile Books.

Cobb, M. 2022. *The genetic age: Our perilous quest to edit life*. London: Profile Books.

Cohen, J. S., and F. H. Portugal. 1977. *A century of DNA: A history of the discovery of the structure and function of the genetic substance*. Cambridge, MA: MIT Press.

Davies, K. 2020. *Editing humanity: The CRISPR revolution and the new era of genome editing*. New York, London: Pegasus Books.

Doudna, J., and S. Sternberg. 2017. *A crack in creation: The new power to control evolution*. New York, NY: Vintage (Penguin Random House).

Glynn, J. 2012. *My Sister Rosalind Franklin*. Oxford: Oxford University Press.
Hall, K. T. 2014. *The man in the Monkeynut coat: William Astbury and how wool wove a forgotten road to the double-helix*. Oxford: Oxford University Press.
Judson, H. F. 1996. *The eighth day of creation*. Cold Spring Harbor, NY: Cold Spring Harbor Press.
Kampourakis, K. 2024. *How we get Mendel wrong, and why it matters: Challenging the narrative of Mendelian genetics*. Boca Raton, London, New York: CRC Press Taylor & Francis Group.
Maddox, B. 2002. *Rosalind Franklin: The dark lady of DNA*. New York: Harper Collins.
Portugal, F. H. 2015. *The least likely man: Marshall Nirenberg and the discovery of the genetic code*. Cambridge, MA: MIT Press.
Radick, G. M. 2023. *Disputed inheritance: The Battle over Mendel and the future of biology*. Chicago, IL: University of Chicago Press.
Wilkins, M. 2003. *The third man of the double helix*. Oxford: Oxford University Press.
Williams, G. 2019. *Unravelling the double helix: The lost heroes of DNA*. London: Weidenfeld and Nicholson.

GPSR Compliance
The European Union's (EU) General Product Safety Regulation (GPSR) is a set of rules that requires consumer products to be safe and our obligations to ensure this.

If you have any concerns about our products, you can contact us on

ProductSafety@springernature.com

In case Publisher is established outside the EU, the EU authorized representative is:

Springer Nature Customer Service Center GmbH
Europaplatz 3
69115 Heidelberg, Germany

www.ingramcontent.com/pod-product-compliance
Lightning Source LLC
Chambersburg PA
CBHW061424100826
49614CB00016B/350

* 9 7 8 3 0 3 2 1 4 2 1 8 4 *